T0255947

Fuzzy Statistical Inferences Based on Fuzzy Random Variables

Fuzzy Statistical Inferences Based on Fuzzy Random Variables

Gholamreza Hesamian

CRC Press is an imprint of the
Taylor & Francis Group, an informa business
A CHAPMAN & HALL BOOK

First edition published 2022
by CRC Press
6000 Broken Sound Parkway NW, Suite 300, Boca Raton, FL 33487-2742

and by CRC Press
2 Park Square, Milton Park, Abingdon, Oxon, OX14 4RN

CRC Press is an imprint of Taylor & Francis Group, LLC

Library of Congress Cataloging-in-Publication Data

Names: Hesamian, Gholamreza, author.
Title: Fuzzy statistical inferences based on fuzzy random variables / Gholamreza Hesamian.
Description: First edition. | Boca Raton : CRC Press, 2022. | Includes bibliographical references and index.
Identifiers: LCCN 2021041840 (print) | LCCN 2021041841 (ebook) | ISBN 9781032162225 (hardback) | ISBN 9781032164533 (paperback) | ISBN 9781003248644 (ebook)
Subjects: LCSH: Fuzzy statistics. | Random variables. | Inference.
Classification: LCC QA276.5 .H42 2022 (print) | LCC QA276.5 (ebook) | DDC 519.5--dc23/eng/20211112
LC record available at https://lccn.loc.gov/2021041840
LC ebook record available at https://lccn.loc.gov/2021041841

ISBN: 978-1-032-16222-5 (hbk)
ISBN: 978-1-032-16453-3 (pbk)
ISBN: 978-1-003-24864-4 (ebk)

DOI: 10.1201/9781003248644

Typeset in CMR10
by KnowledgeWorks Global Ltd.

To my parents,
Sedigheh and Mohammad

And special thanks to my wife, Samin

Contents

Foreword

Probability and statistics are concerned with the chance to analyze events and analysis of data and stochastic models. Therefore, the quantitative description of events and data is essential for statistical inferences.

This book tries to extend some common statistical inferences adopted with new ideas and simple techniques as much as possible for the probability of fuzzy events and statistics with fuzzy data. For simplicity in theoretic and calculations, all fuzzy statistical procedures were conducted on triangular fuzzy numbers. These methods essentially rely on some fuzzy statistical methods recently published by the author. Many of these methods also modified and developed as much as possible.

Compared to similar fuzzy statistics books, we tried to cover a set of frequently used statistical techniques used in applied sciences. Unlike other existing fuzzy statistics books, this book tried to gather almost all essential statistical methods such as exact and fuzzy probabilities, probabilistic inequalities, limit theorems, hypothesis tests, quality control, reliability theory, analysis of variance, and descriptive statistics for fuzzy data. In each section, the proposed techniques were also compared with other existing ones. Each section was then ended via some theoretical and applied exercises. The required optimization procedure can be done by any mathematical software. However, in this book the Mathematica and MATLAB software were utilized in calculations and plotting the figures. The available programs can be obtained from the author.

It is my hope that this book will serve as a solid starting point for analyzing fuzzy quantities and thus enlarging the domain of applicability of the field of probability and statistics.

Preface

Probability and statistics are two fascinating topics on the relationship between mathematics and real-life applications. The probabilistic methods provide a rigorous framework for modeling uncertainty due to randomness. Statistics can be classified into two different categories namely descriptive statistics and statistical inference. Descriptive statistics describe the data, whereas statistical inference consists in the use of statistics to draw conclusions about some unknown aspect of a population based on a random sample from that population. It also helps to assess the relationship between the dependent and independent variables. In statistical inference, the data are taken from the sample and allow you to generalize the population.

However, the methods of probability theory and statistics typically involve processing imprecision of two distinct types arising from a lack of knowledge due to inherent vagueness in concepts themselves which, in the sense of classical probability and statistical inferences, may be well defined. Therefore, the conventional methods are usually inadequate for dealing with certain kinds of imprecision called fuzzy sets. There are many situations in the conventional probability theory that an event and/or a parameter are fuzzy quantities. Moreover, there are many practical problems that require dealing with observations that represent inherently imprecise or linguistic characteristics. In such cases, fuzzy sets may be more effective in encoding such quantities rather than precise ones.

It is now generally accepted that statistical theory and the theory of fuzzy sets are both used to study uncertainties where both types of uncertainties including randomness and fuzziness have occurred. Thus, to produce suitable statistical inferences dealing with imprecise information, we need to model the imprecise information and extend the usual probability theory and statistics to imprecise environments.

This book is an honest attempt to produce an unified techniques for the elementary probability theory based on fuzzy events induced by a family of density functions induced by fuzzy parameters. The main attempt of this book is to introduce some new techniques for the most commonly used statistical inferences based on fuzzy data. This book is designed as a practical guide for scientists to help them solve statistical problems involving fuzzy data and fuzzy events. Consequently, the book has a practical bent and contains lots of examples. A familiarity with probability and statistics is assumed for this book. The book is intended for a statistical scientist at the Bachelor of Science level or above for a specific course on probability and statistics. It also has applications to other applied sciences such as engineering and social sciences.

Payame Noor University, Department of Statistics, Tehran 19395-3697, Iran.
Gholamreza Hesamian
September 2021

List of Figures

List of Figures

List of Tables

Symbols

Symbol Description

Symbol	Description
A	Crisp set
$\widetilde{A}$	Fuzzy set
FN	Fuzzy number
$\widetilde{A}$	Fuzzy set
TFN	Triangular fuzzy number
STFN	Symmetric triangular fuzzy number
$\widetilde{A}[\alpha]$	α-cut of a fuzzy number $\widetilde{A}$
$\widetilde{A}_{\alpha}^{L}$	Lower bound of the α-cut of a fuzzy number $\widetilde{A}$
$\widetilde{A}_{\alpha}^{U}$	Upper bound of the α-cut of a fuzzy number $\widetilde{A}$
$\{x_1, x_2, ...\}$	Set of elements $x_1, x_2, ...$
$\mathbb{R}$	Set of real numbers
$\mathbb{X}$	Universal set
$\mathbb{Z}$	Set of all integer numbers
$\mathbb{N}$	Set of all natural numbers
$\mathbb{F}(\mathbb{X})$	Set of all fuzzy sets on $\mathbb{X}$
$\mathbb{F}(\mathbb{R})$	Set of all fuzzy numbers on $\mathbb{R}$
$\widetilde{A}_{\alpha}$	α-value of a fuzzy number $\widetilde{A}$
$\widetilde{A}(x)$	Membership degree of $\widetilde{A}$ at x
$\lvert\widetilde{A}\rvert$	Cardinal number of a fuzzy set $\widetilde{A}$
$\widetilde{A} \cup \widetilde{B}$	Union of fuzzy sets $\widetilde{A}$ and $\widetilde{B}$
$\widetilde{A} \cap \widetilde{B}$	Intersection of fuzzy sets $\widetilde{A}$ and $\widetilde{B}$
$\widetilde{A} - \widetilde{B}$	Difference between two fuzzy sets $\widetilde{A}$ and $\widetilde{B}$
$\widetilde{A} \odot \widetilde{B}$	Product of two fuzzy sets $\widetilde{A}$ and $\widetilde{B}$
$\widetilde{A} \triangle \widetilde{B}$	Symmetric difference of $\widetilde{A}$ and $\widetilde{B}$
$\widetilde{A}^c$	Complement of $\widetilde{A}$
$\widetilde{A} \oplus \widetilde{B}$	Summation of two fuzzy sets $\widetilde{A}$ and $\widetilde{B}$
$\widetilde{A} \ominus \widetilde{B}$	Difference of two fuzzy sets $\widetilde{A}$ and $\widetilde{B}$
$\widetilde{A} \otimes \widetilde{B}$	Multiply of two fuzzy sets $\widetilde{A}$ and $\widetilde{B}$
$\widetilde{A} \oslash \widetilde{B}$	Deviation of two fuzzy sets $\widetilde{A}$ and $\widetilde{B}$
$g(\widetilde{A})$	Image of $\widetilde{A}$ under the mapping $g(.)$
$\liminf \widetilde{A}_n$	Limes inferior of a countable collection of fuzzy sets $\{\widetilde{A}_n\}_{n\in\mathbb{N}}$

Symbol	Description
$\limsup \widetilde{A}_n$	Limes superior of a countable collection of fuzzy sets $\{\widetilde{A}_n\}_{n\in\mathbb{N}}$
$\widetilde{A}_1 \times \widetilde{A}_2 \times \ldots \times \widetilde{A}_n$	Cartesian product of the fuzzy sets $\widetilde{A}_1, \widetilde{A}_2, , \ldots, \widetilde{A}_n$
$S(\widetilde{A}, \widetilde{B})$	Similarity measure between two triangular fuzzy numbers of $\widetilde{A}$ and $\widetilde{B}$
$P_d(\widetilde{A} \succ \widetilde{B})$	Preference degree that '$\widetilde{A}$ is greater than $\widetilde{B}$'
$d(\widetilde{A}, \widetilde{B})$	Exact distance between two fuzzy numbers of $\widetilde{A}$ and $\widetilde{B}$
$\widetilde{D}(\widetilde{A}, \widetilde{B})$	Fuzzy distance between two fuzzy numbers of $\widetilde{A}$ and $\widetilde{B}$
$\widetilde{A} \ominus_G \widetilde{B}$	Generalized difference between two fuzzy numbers of $\widetilde{A}$ and $\widetilde{B}$
$\widetilde{\max}(\widetilde{A}, \widetilde{B})$	Maximum of the two fuzzy numbers of $\widetilde{A}$ and $\widetilde{B}$
$\widetilde{\min}(\widetilde{A}, \widetilde{B})$	Minimum of the two fuzzy numbers of $\widetilde{A}$ and $\widetilde{B}$
$\|\|\widetilde{A}\|\|$	absolute value a fuzzy number of $\widetilde{A}$
P	Probability measure
Ω	Set of outcomes of an experiment
$\mathcal{A}$	A σ-filed of Ω
$(\Omega, \mathcal{A}, P)$	Probability space
$\mathcal{B}(\mathbb{R})$	Set of all Borel subsets of $\mathbb{R}$
X	Random variable
P_X	Probability measure induced by X
$(\mathbb{R}, \mathbb{F}_m(\mathbb{R}))$	Fuzzy measurable space
$P_X(\widetilde{A})$	Exact probability of a fuzzy event of $\widetilde{A}$
$f_X^{\boldsymbol{\theta}}$	Parametric density function of X
$=^d$	identical distribution function
E_X	Expectation with respect to X
$\widetilde{\boldsymbol{\theta}}$	Fuzzy parameter
$P_X^{\widetilde{\boldsymbol{\theta}}}(\widetilde{A})$	Exact probability of $\widetilde{A}$ with fuzzy parameter
$\widetilde{P}_X^{\widetilde{\boldsymbol{\theta}}}(\widetilde{A})$	Fuzzy probability of $\widetilde{A}$ with fuzzy parameter
$P_{\boldsymbol{X}}(\widetilde{\boldsymbol{A}})$	Probability of a fuzzy product event $\widetilde{\boldsymbol{A}}$
$P_X(\widetilde{A}\|\widetilde{B})$	Conditional probability of $\widetilde{A}$ given $\widetilde{B}$
$\widetilde{\overline{x}}$	Fuzzy arithmetic mean
$\widetilde{\overline{x}}_g$	Fuzzy geometric mean
$\widetilde{\overline{x}}_h$	Fuzzy harmonic mean
$\widetilde{m}$	Fuzzy median
$\widetilde{H}_p$	Fuzzy quantile
$\widetilde{R}$	Fuzzy sample range
$\widetilde{IQR}$	Fuzzy interquartile range
S_D	Mean absolute deviation from the fuzzy mean
S^2	Standard deviation of a set of fuzzy data

$\widetilde{S}_k$	Fuzzy (Pearson) measure of skewness
m_t	Exact measure of skewness
g	Measure of kurtosis of fuzzy data
CDF	Cumulative distribution function
$\widetilde{X}$	Fuzzy random variable (**FRV**)
$\widetilde{E}(\widetilde{X})$	Fuzzy expectation of $\widetilde{X}$
FRS	Ruzzy random sample
var$(\widetilde{X})$	Exact variance of $\widetilde{X}$
Cov$(\widetilde{X}, \widetilde{Y})$	Covariance of two **FRV**s of $\widetilde{X}$ and $\widetilde{Y}$
$\boldsymbol{\rho}(\widetilde{X}, \widetilde{Y})$	Covariance of two **FRV**s of $\widetilde{X}$ and $\widetilde{Y}$
$\widetilde{P}(\widetilde{X} \in (a, b))$	Fuzzy probability that $\widetilde{X} \in (a, b)$
$\liminf_{n\to\infty} \widetilde{X}_n$	lim inf of a sequence of **FRV**s $\{\widetilde{X}_n\}_{n\geq 1}$
$\limsup_{n\to\infty} \widetilde{X}_n$	lim sup of a sequence of **FRV**s $\{\widetilde{X}_n\}_{n\geq 1}$
$\widetilde{F}_{\widetilde{X}}(x)$	Fuzzy cumulative distribution function of $\widetilde{X}$ at x (**FCDF**)
LSFRV	Location-scale fuzzy random variable
SFRV	Scale fuzzy random variable
$\widetilde{X}_n \to^c \widetilde{X}$	Converges completely
$\widetilde{X}_n \to^p \widetilde{X}$	Converges in probability
$\widetilde{X}_n \to^{a.s} \widetilde{X}$	Converges almost surely
$\widetilde{X}_n \to^d \widetilde{X}$	Converges in distribution
$\widetilde{X}_n \to^{\widetilde{L}^p} \widetilde{X}$	converges in $\widetilde{L}^p$
SLLN	Strong Low of Large Numbers
GFRV	Generalized fuzzy random variable
TFRV	Triangular fuzzy random variable
TNFRV	Triangular normal fuzzy random variable
$\widetilde{H}_0$	Fuzzy null hypothesis
$\widetilde{H}_1$	Fuzzy alternative hypothesis
$\widetilde{T}$	Fuzzy sign-test Statistics
D_R	Degree of rejection of $\widetilde{H}_0$
D_A	Degree of acceptance of $\widetilde{H}_0$
δ	Significance level
$\widetilde{M}$	Fuzzy median
$\widetilde{W}$	Fuzzy Wilcoxon test statistics
$\sqrt{n}\widetilde{D}_n$	Fuzzy one-sample Kolmogorov-Smirnov test
FECDF	Fuzzy empirical cumulative distribution function
$\sqrt{\frac{mn}{m+n}}\widetilde{D}_{mn}$	Fuzzy Two-sample Kolmogorov-Smirnov test
$\widetilde{H}$	Fuzzy Kruskal-Wallis test
Cr	Credibility measure
ANOVA	Analysis of variance
SST	Sum of squares total of fuzzy data
SSR	Sum of squares

Symbol	Description
	between variations of fuzzy data
SSE	Sum of squares within variations of fuzzy data
χ^2_r	Chi-square distribution with r degrees of freedom
t_r	t-student distribution with r degrees of freedom
F_{ν_1,ν_2}	F-fisher distribution with r degrees of freedom of ν_1 and ν_2
$\chi^2_{r,\delta}$	δ-quantile of the chi-squared distribution with r degrees of freedom
$t_{r,\delta}$	δ-quantile of the t-distribution with r degrees of freedom
$F_{\nu_1,\nu_2,\delta}$	δ-quantile of the f-distribution with ν_1 and ν_2 degrees of freedom
$\widetilde{LCL}$	Fuzzy lower central limit
$\widetilde{CL}$	Fuzzy central limit
$\widetilde{LCL}$	Fuzzy upper central limit
$\widetilde{\overline{x}}$-chart	Fuzzy $\overline{x}$-chart
s-chart	s-chart based on fuzzy random variables
D_v	Degree of violence
EWMA	Exponentially-weighted moving average
$\widetilde{C}(u,v)$	Fuzzy process capability
$\widetilde{d}_a(\widetilde{A},\widetilde{B})$	Fuzzy triangular absolute error distance
$\widetilde{\widetilde{C}}(u,v)$	Fuzzy estimated value of $\widetilde{C}(u,v)$
LTIFRV	Life time fuzzy random variable
FRF	Fuzzy reliability function
IMTTF	Fuzzy mean time to failure
FMRLF	Fuzzy mean residual life function
$\widetilde{R}_S$	Fuzzy system reliability
$\widetilde{\widetilde{\beta}}_0$	Fuzzy intercept
β_j	Regression coefficient
MSE	Mean square error
RMSE	Root mean square error
MSM	Mean similarity measure
$K(.)$	Kernel function
h	Bandwidth
GCV	Generalized cross-validation criterion
ACF	Autocorrelation function
$\widetilde{\boldsymbol{X}}_{\boldsymbol{T}}$	Fuzzy time series data
$\mathbf{cov}(\widetilde{X}_t,\widetilde{X}_{t-k})$	Autocovariance of the generating fuzzy time series
$\mathbf{cov}(\widetilde{X}_t,\widetilde{X}_{t-k})$	Autocorrelation of the generating fuzzy time series
$\widehat{\mathbf{c}}_k$	sample autocovariance function of the generating fuzzy time series

Part I

Fuzzy Statistical Inferences Based on Fuzzy Random Variables

1

Introduction

In this chapter, the basic definitions and properties from fuzzy sets and fuzzy numbers needed for the book are recalled and discussed. We utilize such definitions and properties to produce some probabilistic reasoning and statistical inferences in the fuzzy domain the next chapter.

1.1 Fuzzy Set

Let $\mathbb{X}$ be an universal in the universe of discourse and it includes all possible elements related with the given problem. The classical set theory is built on the fundamental concept of 'set'. A set is defined as collection of objects, which share certain characteristics. A set A (of $\mathbb{X}$) can be represented a set by enumerating its elements as $A = \{x_1, x_2, ..., x_n\}$. There exists a membership function, which may be also used to define a set as a mapping of $\mathbb{X}$ into $\{0, 1\}$. Such membership function for a set A can be represents by

$$A(x) = \begin{cases} 1, & x \in A, \\ 0, & x \notin A. \end{cases} \tag{1.1}$$

Therefore, each individual element in a set is a member or an element of the set. However, in real-life we come across many situations where membership and non-membership in a set are not clearly defined. For instance, consider the set of $A = \{\text{all real numbers near to 1}\}$ in the context 'positive values less than 5'. In classical set theory, one may imagine a membership function to describe A as

$$A(x) = \begin{cases} 1, & x \in [1-a, 1+a], \\ 0, & x \notin (1-a, 1+a). \end{cases} \tag{1.2}$$

where $a \in (0, 1]$. As a result, a value which is exactly a belongs to A and is considered to be 'all real numbers near to 1', but the values of $[0, a - 0.001]$ do not belong to A. In addition, all values in $[1-a, 1+a]$ may provide different memberships to describe 'near to 1'. The value of 1 is completely consistence to 'near to 1' but the other values partially inherit the imprecise concept of 'all real numbers near to 1'. These distinction are mathematically correct, but practically unreasonable. Therefore, A is not well defined in the sense

DOI: 10.1201/9781003248644-1

of classical set theory and cannot be precisely represented by 0 or 1. The ambiguity in this situation follows from that the boundaries of A is imprecise so as to characterize the A in a precise and rigorous way. For this purpose, instead of a sharp cut at the exact values of 0.9 and 1.1 (as well as all values in $[1-a, 1+a]$), different memberships of 'near to 1' can be given via a membership function on $[0,1]$ instead of $\{0,1\}$. In this regard, one may utilize the following membership function to describe 'All numbers near to 1' as

$$A(x) = \begin{cases} x, & 0 \leq x \leq 1, \\ 1-x, & 1 \leq x \leq 2, \\ 0, & \mathbb{R} - [1,2]. \end{cases} \tag{1.3}$$

For example, the value of 0.25 is considered to be 'all real numbers near to 1' with 'degree 0.25' and with 'degree 0.95' for the value of 0.25 and 0.95 according to the membership function of f. Therefore, we cannot exclude the all values in $[0,1]$ from A, nor include them completely. Thus A is a generalization of the classical characteristic function, which can be used to conclude that a value 'is' or 'is not' a member of 'all real numbers near to 1'. Thus, such membership functions can provide a suitable mathematical tool for modeling such imprecision situations. Unlike the classical sets, the fuzzy set allows flexible membership degrees to each element of sets. That means that a fuzzy set contains elements that have varying degrees of membership in the set. Therefore, it provides a potential procedure to appreciate how uncertainty originating from human thinking can affect scientific problems.

Fuzzy sets were introduced by Zadeh [219] for the first time as an extension of classical sets. During the last two decades, fuzzy set theory has been successfully employed in working with numerous practical applications (for instance, see [12, 157, 185, 187, 223, 215]).

Definition 1.1 ([41, 157, 187]) *A fuzzy set of $\mathbb{X}$ (universe of discourse) is a mapping $\widetilde{A} : \mathbb{X} \to [0,1]$, which assigns to each $x \in \mathbb{X}$ a degree of membership $0 \leq \widetilde{A}(x) \leq 1$. The set of all fuzzy sets on $\mathbb{X}$ is denoted by $\mathbb{F}(\mathbb{X})$.*

Definition 1.2 ([41, 114, 157]) *$\widetilde{A}$ is said to be a convex fuzzy set on $\mathbb{X}$ if and only if for all $x, y \in \mathbb{X}$ and $\lambda \in [0,1]$, $\widetilde{A}(\lambda x + (1-\lambda)y) \geq \min\{\widetilde{A}(x), \widetilde{A}(y)\}$. A fuzzy set $\widetilde{A}$ is called normal fuzzy set if there exist $x_0 \in \mathbb{X}$ such that $\widetilde{A}(x_0) = 1$.*

Definition 1.3 ([41, 114, 157]) *For each $\alpha \in (0,1]$, $\{x \in \mathbb{X} \mid \widetilde{A}(x) \geq \alpha\}$ of $\mathbb{X}$ is called the α-cut of $\widetilde{A}$ and is denoted by $\widetilde{A}[\alpha]$. In addition, $\widetilde{A}[0] = Cl\{x \in \mathbb{X} : \widetilde{A}(x) > 0\}$ is called the support of $\widetilde{A}$ where $Cl(A)$ is the closure of A [140].*

For every $\alpha \in (0,1]$, the lower and upper bounds of $\widetilde{A}[\alpha]$ are denoted by $\widetilde{A}^L_\alpha$ and $\widetilde{A}^U_\alpha$, respectively.

Lemma 1.1 *A fuzzy set $\widetilde{A}$ is convex if and only if $\{\widetilde{A}[\alpha]\}_{\alpha \in [0,1]}$ is a collection of non-empty nested sets on $\mathbb{X}$.*

Proof *For a proof see [114].*

Having a collection of non-empty nested sets $\{\widetilde{A}[\alpha]\}_{\alpha\in[0,1]}$ on $\mathbb{X}$, the membership function of $\widetilde{A}(x)$ can be evaluated by the following lemma.

Lemma 1.2 *Let $\{\widetilde{A}[\alpha]\}_{\alpha\in[0,1]}$ be a collection of non-empty nested sets on $\mathbb{X}$. Then:*

$$\widetilde{A}(x) = \sup\{\alpha \in [0,1] : x \in \widetilde{A}[\alpha]\}. \tag{1.4}$$

Proof *See [114, 123, 157, 187] to prove this result.*

Definition 1.4 *We say $\widetilde{A}$ is a discrete fuzzy set if it is convex, normal and $\widetilde{A}(x)$ is not zero only at a finite number of $x \in \mathbb{X} = \{x_1, x_2, \ldots, x_n\}$. For such cases, $\widetilde{A}$ is typically denoted by $\widetilde{A} = \{\frac{a_1}{x_1}, \frac{a_2}{x_2}, \ldots, \frac{a_n}{x_n}\}$ where $\widetilde{A}(x_i) = a_i \in (0,1]$ are the membership degrees of x_i.*

Definition 1.5 *A fuzzy set $\widetilde{A}$ is called positive fuzzy set ($\widetilde{A} > 0$) if $\inf \widetilde{A}[0] > 0$, and it is negative fuzzy set ($\widetilde{A} < 0$) if $\sup \widetilde{A}[0] < 0$.*

Example 1.1 *Consider the set 'small' in $\mathbb{X}$ consisting of natural numbers less than 10. Such an imprecise set can be described via the following fuzzy set:*

$$\widetilde{A} = \{\frac{1}{1}, \frac{0.9}{2}, \frac{0.8}{3}, \frac{0.5}{4}, \frac{0.3}{5}, \frac{0.1}{6}, \frac{0}{\{7,8,9\}}\}.$$

For all $x, y \in \mathbb{X}$ and $\lambda \in [0,1]$, it is easy to check that $\widetilde{A}(\lambda x + (1-\lambda)y) \geq \min\{\widetilde{A}(x), \widetilde{A}(y)\}$. $\widetilde{A}$ is also a normal fuzzy set. Accordingly, $\widetilde{A}$ is a (discrete) fuzzy set.

Since an element can partially belong to a fuzzy set, a natural generalization of the classical notion of cardinality is to weigh each element by its membership degree, which resulted in the following definition for cardinality of a fuzzy set.

Definition 1.6 *[41, 114] For any fuzzy set $\widetilde{A}$ defined on a finite universal set $\mathbb{X}$, the cardinality of $\widetilde{A}$ is defined by $|\widetilde{A}| = \sum_{x\in\mathbb{X}} \widetilde{A}(x)$.*

Example 1.2 *Let $\mathbb{X} = \{1,2,3,4,5\}$ and $\widetilde{A} = \{\frac{0.3}{1}, \frac{0.5}{2}, \frac{0.7}{3}, \frac{1}{4}, \frac{0.8}{5}\}$. Then $|\widetilde{A}| = \sum_x \widetilde{A}(x) = 0.3 + 0.5 + 0.7 + 1 + 0.8 = 3.3$.*

The inclusion measure is an important concept in the area of fuzzy sets and indicates the degree to which a given fuzzy set belongs to a non-fuzzy set [34].

Definition 1.7 *Let $\widetilde{A} \in \mathbb{F}(\mathbb{X})$ be a fuzzy set and I be a subset of $\mathbb{R}$. Then, the degree to which $\widetilde{A}$ belongs to I is defined by $d(\widetilde{A} \in I) = \frac{\sum_{x\in I} \widetilde{A}(x)}{|\widetilde{A}|}$.*

Therefore, $d(\widetilde{A} \in I)$ is a relation between fuzzy sets $\widetilde{A}$ and classical set of I that indicates the degree to which $\widetilde{A}$ belongs to I.

Lemma 1.3 *Let $\widetilde{A} \in \mathbb{F}(\mathbb{X})$ and $I \subseteq \mathbb{R}$. Then:*

1) If $\{I_j\}_{j=1}^k$ is a collection of disjoint sets on $\mathbb{X}$, then $\sum_{j=1}^k d(\widetilde{A} \in I_j) = 1$.

2) $d(\widetilde{A} \in I) = 1$ if and only if $\widetilde{A}[0] \subseteq I$.

Proof *Assuming $\{I_j\}_{j=1}^k$ is a sequence of disjoint sets on $\mathbb{X}$, implies that*

$$\begin{aligned}\sum_{j=1}^{k} d(\widetilde{A} \in I_j) &= \frac{1}{|\widetilde{A}|} \sum_{j=1}^{k} \sum_{x \in I_j} \widetilde{A}(x) \\ &= \frac{1}{|\widetilde{A}|} \sum_{x \in \cup_{j=1}^k I_j} \widetilde{A}(x) \\ &= \frac{1}{|\widetilde{A}|} |\widetilde{A}| = 1.\end{aligned}$$

Now, if $\widetilde{A}[0] \subseteq I$ then:

$$d(\widetilde{A} \in I) = \frac{\sum_{x \in \widetilde{A}[0]} \widetilde{A}(x) + \sum_{x \in (I - \widetilde{A}[0])} \widetilde{A}(x)}{|\widetilde{A}|} = \frac{|\widetilde{A}|}{|\widetilde{A}|} = 1.$$

Conversely, $d(\widetilde{A} \in I) = 1$ yields $\sum_{x \in I^c} \widetilde{A}(x) = 0$ where I^c shows the compliment set of I. This simply concludes that $\widetilde{A}[0] \subseteq I$ and verifies the proof.

Remark 1.1 *It should be pointed out that $d(\widetilde{A} \in I)$ may be interpreted as the probability that $\widetilde{A}$ belongs to I. Moreover, based on a given collection of disjoint sets $\{I_j\}_{j=1}^k$, from Lemma 1.3, it is natural to say that $\widetilde{A} \in I_{j*}$ if $d(\widetilde{A} \in I_{j*}) = \max_{j=1}^k d(\widetilde{A} \in I_j)$. In particular, assume that I and J are two sets of $\mathbb{X}$ in which $I \cup J = \mathbb{X}$ and $I \cap J = \emptyset$. Therefore, if $d(\widetilde{A} \in I) \geq 0.5$ then $d(\widetilde{A} \in J) < 0.5$. In this case, one can say $\widetilde{A}$ belongs to I.*

Example 1.3 *Let $\widetilde{A} = \{\frac{0.4}{3}, \frac{0.6}{4}, \frac{1}{5}, \frac{0.8}{6}, \frac{0.5}{7}\}$ be a fuzzy set on $\mathbb{X} = \{1, 2, 3, 4, 5, 6\ , 7, 8, 9, 10\}$. Then, assuming $I_1 = \{1, 2, 3, 4, \}$, $I_2 = \{5, 6\}$, and $I_3 = \{7, 8, 9, 10\}$ implies that $d(\widetilde{A} \in I_1) = 0.303$, $d(\widetilde{A} \in I_2) = 0.545$, and $d(\widetilde{A} \in I_1) = 0.152$. Since $d(\widetilde{A} \in I_2) \geq 0.5$, we get that $\widetilde{A} \in I_2$.*

1.1.1 Operations of fuzzy sets

Although the set-theoretic operations (union, intersection, and complement) possess some rigorous axiomatic properties, the conventional operations of sets can be extended for fuzzy sets [12, 157, 185, 187, 215, 223]. In this section, some well-established operations of fuzzy sets are recalled and discussed.

Definition 1.8 *For two fuzzy sets $\widetilde{A}$ and $\widetilde{B}$, we say:*

1. *$\widetilde{A}$ is included in $\widetilde{B}$, say $\widetilde{A} \subseteq \widetilde{B}$, if $\widetilde{A}(x) \leq \widetilde{B}(x)$, for all $x \in \mathbb{X}$.*
2. *$\emptyset$ is the empty (fuzzy) set if $\emptyset(x) = 0$ for any $x \in \mathbb{X}$.*
3. *$\widetilde{A}^c$ is the complement of fuzzy set $\widetilde{A}$ with the membership function $\widetilde{A}^c(x) = 1 - \widetilde{A}(x)$, for all $x \in \mathbb{X}$.*
4. *$\widetilde{A} \cup \widetilde{B}$ is the union of fuzzy sets $\widetilde{A}$ and $\widetilde{B}$ with the membership function $(\widetilde{A} \cup \widetilde{B})(x) = \max\{\widetilde{A}(x), \widetilde{B}(x)\}$, for all $x \in \mathbb{X}$.*
5. *$\widetilde{A} \cap \widetilde{B}$ is the intersection of fuzzy sets $\widetilde{A}$ and $\widetilde{B}$ with the membership function $(\widetilde{A} \cap \widetilde{B})(x) = \min\{\widetilde{A}(x), \widetilde{B}(x)\}$, for all $x \in \mathbb{X}$.*
6. *$\widetilde{A} - \widetilde{B}$ is the difference between two fuzzy sets $\widetilde{A}$ and $\widetilde{B}$ with the membership function of $(\widetilde{A} - \widetilde{B})(x) = \max\{0, \widetilde{A}(x) - \widetilde{B}(x)\}$.*
7. *$\widetilde{A} \odot \widetilde{B}$ is the product of two fuzzy sets $\widetilde{A}$ and $\widetilde{B}$ with the membership function of $(\widetilde{A} \odot \widetilde{B})(x) = \widetilde{A}(x)\widetilde{B}(x)$.*
8. *$\widetilde{A} \triangle \widetilde{B}$ is the symmetric difference of $\widetilde{A}$ and $\widetilde{B}$ with the membership function of $(\widetilde{A} \triangle \widetilde{B})(x) = |\widetilde{A}(x) - \widetilde{B}(x)|$.*

Example 1.4 *Let $\mathbb{X} = \{1, 2, 3, 4\}$, $\widetilde{A} = \{\frac{0.2}{1}, \frac{0.7}{2}, \frac{1}{3}, \frac{0.6}{4}\}$, and $\widetilde{B} = \{\frac{0.9}{1}, \frac{1}{2}, \frac{0.8}{3}, \frac{0.5}{4}\}$. Then, it can be easy to verify that:*

$$\widetilde{A}^c = \left\{\frac{0.8}{1}, \frac{0.3}{2}, \frac{0}{3}, \frac{0.4}{4}\right\},$$

$$\begin{aligned}\widetilde{A} \cup \widetilde{B} &= \left\{\frac{\max\{0.2, 0.9\}}{1}, \frac{\max\{0.7, 1\}}{2}, \frac{\max\{1, 0.8\}}{3}, \frac{\max\{0.6, 0.5\}}{4}\right\} \\ &= \{\frac{0.9}{1}, \frac{1}{2}, \frac{1}{3}, \frac{0.6}{4}\},\end{aligned}$$

$$\begin{aligned}\widetilde{A} \cap \widetilde{B} &= \left\{\frac{\min\{0.2, 0.9\}}{1}, \frac{\min\{0.7, 1\}}{2}, \frac{\min\{1, 0.8\}}{3}, \frac{\min\{0.6, 0.5\}}{4}\right\} \\ &= \left\{\frac{0.2}{1}, \frac{0.7}{2}, \frac{0.8}{3}, \frac{0.5}{4}\right\},\end{aligned}$$

$$\begin{aligned}\widetilde{A} - \widetilde{B} = &\left\{\frac{\max\{0, 0.2 - 0.9\}}{1}, \frac{\max\{0, 0.7 - 1\}}{2}, \frac{\max\{0, 1 - 0.8\}}{3}, \right. \\ &\left. \frac{\max\{0, 0.6 - 0.5\}}{4}\right\} \\ = &\left\{\frac{0}{1}, \frac{0}{2}, \frac{0.2}{3}, \frac{0.1}{4}\right\},\end{aligned}$$

$$\widetilde{A} \odot \widetilde{B} = \left\{\frac{0.2 \times 0.9}{1}, \frac{0.7 \times 1}{2}, \frac{1 \times 0.8}{3}, \frac{0.6 \times 0.5}{4}\right\}$$
$$= \left\{\frac{0.18}{1}, \frac{0.7}{2}, \frac{0.8}{3}, \frac{0.03}{4}\right\},$$

$$\widetilde{A} \triangle \widetilde{B} = \left\{\frac{|0.2 - 0.9|}{1}, \frac{|0.7 - 1|}{2}, \frac{|1 - 0.8|}{3}, \frac{|0.6 - 0.5|}{4}\right\}$$
$$= \left\{\frac{0.7}{1}, \frac{0.3}{2}, \frac{0.2}{3}, \frac{0.1}{4}\right\}.$$

Lemma 1.4 *For three fuzzy sets $\widetilde{A}$, $\widetilde{B}$, and $\widetilde{C}$,*

1. *Distributivity:*

$$\widetilde{A} \cap (\widetilde{B} \cup \widetilde{C}) = (\widetilde{A} \cap \widetilde{B}) \cup (\widetilde{A} \cap \widetilde{C}), \tag{1.5}$$

$$\widetilde{A} \cup (\widetilde{B} \cap \widetilde{C}) = (\widetilde{A} \cup \widetilde{B}) \cap (\widetilde{A} \cup \widetilde{C}). \tag{1.6}$$

2. *De Morgan Laws:*

$$(\widetilde{A} \cup \widetilde{B})^c = \widetilde{A}^c \cap \widetilde{B}^c, \tag{1.7}$$

$$(\widetilde{A} \cap \widetilde{B})^c = \widetilde{A}^c \cup \widetilde{B}^c. \tag{1.8}$$

Proof *To see a proof of these results see [41, 114].*

Theorem 1.1 *For three fuzzy sets $\widetilde{A}$, $\widetilde{B}$, and $\widetilde{C}$, the following hold.*

1) $\widetilde{A} \odot (\widetilde{B} \cup \widetilde{C}) = (\widetilde{A} \odot \widetilde{B}) \cup (\widetilde{A} \odot \widetilde{C})$.

2) $\widetilde{A} \odot (\widetilde{B} \cap \widetilde{C}) = (\widetilde{A} \odot \widetilde{B}) \cap (\widetilde{A} \odot \widetilde{C})$.

3) $\widetilde{A} \odot (\widetilde{B} - \widetilde{C}) = (\widetilde{A} \odot \widetilde{B}) - (\widetilde{A} \odot \widetilde{C})$.

Proof *Note that* $\widetilde{A} \odot (\widetilde{B} \cup \widetilde{C})(x) = \widetilde{A}(x) \max\{\widetilde{B}(x), \widetilde{C}(x)\} = \max\{\widetilde{A}(x)\widetilde{B}(x), \widetilde{A}(x)\widetilde{C}(x)\}$. *This verifies part (1). The second part can be shown similarly. Further, we have* $(\widetilde{A} \odot (\widetilde{B} - \widetilde{C}))(x) = \widetilde{A}(x) \max\{0, \widetilde{B}(x) - \widetilde{C}(x)\} = \max\{0, \widetilde{A}(x)\widetilde{B}(x) - \widetilde{A}(x)\widetilde{C}(x)\}$. *This concludes that* $\widetilde{A} \odot (\widetilde{B} - \widetilde{C}) = (\widetilde{A} \odot \widetilde{B}) - (\widetilde{A} \odot \widetilde{C})$.

Theorem 1.2 *For two fuzzy sets $\widetilde{A}$ and $\widetilde{B}$ and for every $\alpha \in (0, 1]$, the following results show the connections between fuzzy sets and their α-cuts:*

1. *If $\widetilde{A} \subseteq \widetilde{B}$ then $\widetilde{A}[\alpha] \subseteq \widetilde{B}[\alpha]$.*
2. $\widetilde{A}^c[\alpha] = (\widetilde{A}\overline{[1 - \alpha]})^c$ *where* $\widetilde{A}\overline{[1 - \alpha]} = \{x : \widetilde{A}(x) > 1 - \alpha\}$.

3. $(\widetilde{A} \cup \widetilde{B})[\alpha] = \widetilde{A}[\alpha] \cup \widetilde{B}[\alpha]$.

4. $(\widetilde{A} \cap \widetilde{B})[\alpha] = \widetilde{A}[\alpha] \cap \widetilde{B}[\alpha]$.

Proof *For a comprehensive proof see [41, 114].*

Lemma 1.5 *For two fuzzy sets $\widetilde{A}$ and $\widetilde{B}$, $|\widetilde{A} \cup \widetilde{B}| = |\widetilde{A}| + |\widetilde{B}| - |\widetilde{A} \cap \widetilde{B}|$.*

Proof *According to the definition of $\widetilde{A} \cup \widetilde{B}$, we easily have:*

$$\begin{aligned} |\widetilde{A} \cup \widetilde{B}| &= \sum_{x \in \mathbb{R}} (\widetilde{A} \cup \widetilde{B})(x) \\ &= \sum_{x \in \mathbb{R}} (\widetilde{A}(x) + \widetilde{B}(x) - (\widetilde{A} \cap \widetilde{B})(x)) \\ &= |\widetilde{A}| + |\widetilde{B}| - |\widetilde{A} \cap \widetilde{B}|. \end{aligned}$$

Definition 1.9 *(Extension principle [41, 114]) Suppose that $g : \mathbb{X} \to \mathbb{Y}$ is a real-valued function and $\widetilde{A}$ is a fuzzy set on $\mathbb{F}(\mathbb{X})$. The image of the fuzzy set $\widetilde{A}$ under the mapping $g(.)$ can be expressed as a fuzzy set $g(\widetilde{A}) \in \mathbb{F}(\mathbb{Y})$ with the following membership function:*

$$g(\widetilde{A})(y) = \begin{cases} \sup_{x:y=g(x)} \widetilde{A}(y), & g^{-1}(y) \neq \emptyset, \\ 0, & g^{-1}(y) = \emptyset. \end{cases} \tag{1.9}$$

Example 1.5 *Consider a fuzzy set on $\mathbb{Z} = \{..., -2, -1, 0, 1, 2, ...\}$ stands for 'about 1' with the membership function of $\widetilde{A} = \{\frac{0.4}{-2}, \frac{0.6}{-1}, \frac{0.8}{0}, \frac{1}{1}, \frac{0.8}{2}, \frac{0.6}{3}, \frac{0.4}{4}\}$. Letting $g(x) = x^4$, we have $y \in \{0, 1, 16, 81, 256\}$ and thus:*

$$\begin{aligned} g(\widetilde{A})(0) &= \widetilde{A}(0) = 0.8, \\ g(\widetilde{A})(1) &= \max\{\widetilde{A}(-1), \widetilde{A}(1)\} = \max\{0.6, 1\} = 1, \\ g(\widetilde{A})(16) &= \max\{\widetilde{A}(-2), \widetilde{A}(2)\} = \max\{0.4, 0.8\} = 0.8, \\ g(\widetilde{A})(81) &= \widetilde{A}(3) = 0.6, \\ g(\widetilde{A})(256) &= \widetilde{A}(4) = 0.4. \end{aligned}$$

From above, the membership function of $g(\widetilde{A}) = (\widetilde{A})^4$ can be obtained as:

$$(\widetilde{A})^4 = \{\frac{0.8}{0}, \frac{1}{1}, \frac{0.8}{16}, \frac{0.6}{81}, \frac{0.4}{256}\}.$$

In order to use fuzzy sets in any real-life one should be able to perform arithmetic operations including addition, subtraction, multiplication, and division in computational process, which is called fuzzy arithmetic. The usual arithmetic operations on real numbers can be extended to the ones defined on fuzzy sets based on the Extension principle.

Definition 1.10 *[41, 114] The arithmetic operations between two fuzzy sets of $\widetilde{A}$ and $\widetilde{B}$ is defined as a fuzzy set $\widetilde{A} \circ \widetilde{B}$ with the following membership function:*

$$(\widetilde{A} \circ \widetilde{B})(z) = \sup_{x,y:z=x \bullet y} \min\{\widetilde{A}(x), \widetilde{B}(y)\}. \tag{1.10}$$

The symbols $\circ = \oplus, \ominus, \otimes, \oslash$ denotes the extended arithmetic operations $\bullet = +, -, \times, /$ in a fuzzy domain.

Next, a notion of convergence a collection of fuzzy sets is given.

Definition 1.11 *Let $\{\widetilde{A}_n\}_{n\in\mathbb{N}}$ be a countable collection of fuzzy sets and $\widetilde{A}$ be a fuzzy set. Then, we say:*

1. *$\widetilde{A}_n \to \widetilde{A}$ if $\lim_{n\to\infty} \widetilde{A}_n(x) = \widetilde{A}(x)$ for all $x \in \mathbb{R}$.*
2. *$\widetilde{A}_n \uparrow \widetilde{A}$ if $\{\widetilde{A}_n\}_{n\in\mathbb{N}}$ is an increasing countable collection of fuzzy sets and $\lim_{n\to\infty} \widetilde{A}_n(x) = \widetilde{A}(x)$ for all $x \in \mathbb{R}$.*
3. *$\widetilde{A}_n \downarrow \widetilde{A}$ if $\{\widetilde{A}_n\}_{n\in\mathbb{N}}$ is a decreasing countable collection of fuzzy sets and $\lim_{n\to\infty} \widetilde{A}_n(x) = \widetilde{A}(x)$ for all $x \in \mathbb{R}$.*

Example 1.6 *Let $\mathbb{X} = \{-3, -2, -1, 0, 1, 2, 3\}$ and consider a sequence of discrete fuzzy set on $\mathbb{X}$ with $\widetilde{A}_n(x) = (1 - (1/nx^2))^n$, $x \in \mathbb{X}$. Since $\lim_{n\to\infty} \widetilde{A}_n(x) = e^{-x^2}$, we have $\lim_{n\to\infty} \widetilde{A}_n(x) = \widetilde{A}(x) = e^{-x^2}$ with the following membership function:*

$$\widetilde{A} = \{\frac{-3}{e^{-9}}, \frac{-2}{e^{-4}}, \frac{-1}{e^{-1}}, \frac{0}{1}, \frac{1}{e^{-1}}, \frac{2}{e^{-4}}, \frac{3}{e^{-9}}\}.$$

Proposition 1.1 *For a a countable collection of fuzzy sets $\{\widetilde{A}_n\}_{n\in\mathbb{N}}$, $\widetilde{A}_n \to \widetilde{A}$ if and only if $\widetilde{A}_n[\alpha] \to \widetilde{A}[\alpha]$ for any $\alpha \in (0, 1]$.*

Proof *The proof is left for the reader.*

Definition 1.12 *The limes superior and limes inferior of a countable collection of fuzzy sets $\{\widetilde{A}_n\}_{n\in\mathbb{N}}$ are defined to be fuzzy sets $\limsup \widetilde{A}_n = \cap_{n\geq 1} \cup_{m\geq n} \widetilde{A}_m$ and $\liminf \widetilde{A}_n = \cup_{n\geq 1} \cap_{m\geq n} \widetilde{A}_m$, respectively.*

Example 1.7 *Let $\mathbb{X} = \{-3/4, -2/4, -1/4, 0, 1/4, 2/4, 3/4\}$ and consider a sequence of discrete fuzzy set on $\mathbb{X}$ with $\widetilde{A}_n(x) = \max\{0, 1 - (n/(n+1))|x|\}$, $x \in \mathbb{X}$. Then, for a fixed $x \in \mathbb{X}$, we have*

$$\begin{aligned}(\cup_{m\geq n}\widetilde{A}_m)(x) &= \max_{m\geq n} \widetilde{A}_m(x) \\ &= \max\{0, \max_{m\geq n}(1 - (m/(m+1))|x|\} \\ &= \max\{0, 1 - (n/(n+1))|x|\}.\end{aligned}$$

This concludes that

$$\begin{aligned}(\limsup \widetilde{A}_n)(x) &= (\cap_{n\geq 1} \cup_{m\geq n} \widetilde{A}_m)(x) \\ &= \min_{n\geq 1}(\cup_{m\geq n}\widetilde{A}_m)(x) \\ &= \min_{n\geq 1} \max\{0, 1-(n/(n+1))|x|\} \\ &= \max\{0, 1-|x|\} = \widetilde{A}(x).\end{aligned}$$

Therefore,

$$\limsup \widetilde{A}_n = \{\frac{-3/4}{1/4}, \frac{-2/4}{2/4}, \frac{-1/4}{3/4}, \frac{0}{1}, \frac{1/4}{3/4}, \frac{2/4}{2/4}, \frac{3/4}{1/4}\}.$$

Theorem 1.3 *Let $\{\widetilde{A}_n\}_{n\in\mathbb{N}}$ be a countable collection of fuzzy sets. Then, the following are valid.*

1. $\liminf \widetilde{A}_n \subseteq \limsup \widetilde{A}_n$.
2. $(\liminf \widetilde{A}_n)^c = \limsup \widetilde{A}_n^c$.
3. $(\limsup \widetilde{A}_n)^c = \liminf \widetilde{A}_n^c$.

Proof *First note that* $(\liminf \widetilde{A}_n)[\alpha] = \cup_{n\geq 1} \cap_{m\geq n} \widetilde{A}_n[\alpha]$. *That concludes that* $(\liminf \widetilde{A}_n)[\alpha] \subseteq \cap_{n\geq 1} \cup_{m\geq n} \widetilde{A}_n[\alpha] = (\limsup \widetilde{A}_n)[\alpha]$ *for any* $\alpha \in [0,1]$ *or* $\liminf \widetilde{A}_n \subseteq \limsup \widetilde{A}_n$. *In addition, for any* $\alpha \in [0,1]$, $(\liminf \widetilde{A}_n)^c[\alpha] = (\cup_{n\geq 1} \cap_{m\geq n} \widetilde{A}[\overline{1-\alpha}])^c = \cap_{n\geq 1} \cup_{m\geq n} (\widetilde{A}_n[\overline{1-\alpha}])^c = \cap_{n\geq 1} \cup_{m\geq n} \widetilde{A}_n^c[\alpha]$. *This means* $(\liminf \widetilde{A}_n)^c = \limsup \widetilde{A}_n^c$. *Similarly, it can be shown that* $(\limsup \widetilde{A}_n)^c = \liminf \widetilde{A}_n^c$.

Theorem 1.4 *Let $\{\widetilde{A}_n\}_{n\in\mathbb{N}}$ and $\{\widetilde{B}_n\}_{n\in\mathbb{N}}$ be two countable collections of fuzzy sets. Then:*

1) $\limsup(\widetilde{A} \oplus \widetilde{B})_n \subseteq (\limsup \widetilde{A}_n \oplus \limsup \widetilde{B}_n)$.

2) $\liminf(\widetilde{A} \oplus \widetilde{B})_n \supseteq (\liminf \widetilde{A}_n \oplus \liminf \widetilde{B}_n)$.

Proof *For any* $\alpha \in (0,1]$, $\limsup(\widetilde{A}_n \oplus \widetilde{B}_n)[\alpha] = \cap_{n\geq 1} \cup_{m\geq n} (\widetilde{A}_n[\alpha] + \widetilde{B}_n[\alpha])$. *Now, assume that x is an arbitrary element of* $\limsup(\widetilde{A}_n \oplus \widetilde{B}_n)[\alpha]$. *Therefore, for any $n \geq 1$, there exist a $m \geq n$ such that $x = y + z$ where $y \in \widetilde{A}_n[\alpha]$ and $z \in \widetilde{B}_n[\alpha]$. This concludes that* $x \in (\limsup \widetilde{A}_n \oplus \limsup \widetilde{B}_n)[\alpha]$ *which providing (1). The second part can be verified similarly.*

Proposition 1.2 *For a countable collection of fuzzy sets $\{\widetilde{A}_n\}_{n\in\mathbb{N}}$:*

1) $\lim_{n\to\infty} \cup_{k\geq n}\widetilde{A}_k = \limsup \widetilde{A}_n$.

2) $\lim_{n\to\infty} \cap_{k\geq n}\widetilde{A}_k = \liminf \widetilde{A}_n$.

Proof *Let* $\widetilde{B}_k^n = \cup_{k\geq n}\widetilde{A}_k$, *for any* $\alpha \in (0,1]$. *For (1), since* $\widetilde{B}_k^n[\alpha] \downarrow \cap_{n\geq 1} \cup_{k\geq n} \widetilde{A}_k[\alpha] = \limsup \widetilde{A}_n[\alpha]$, *we easily have* $\lim_{n\to\infty} \cup_{k\geq n}\widetilde{A}_k = \limsup \widetilde{A}_n$ *by Lemma 1.1. The second part can be verified similarly.*

Proposition 1.3 *Let* $\{\widetilde{A}_n\}_{n\in\mathbb{N}}$ *be a countable collection of fuzzy sets. Then* $\widetilde{A}_n \to \widetilde{A}$ *if and only if* $\liminf \widetilde{A}_n = \limsup \widetilde{A}_n = \widetilde{A}$.

Proof *First assume that* $\widetilde{A}_n \to \widetilde{A}$. *This simply concludes that* $\widetilde{A}_n[\alpha] \to \widetilde{A}[\alpha]$ *for any* $\alpha \in (0,1]$ *which implies that* $\liminf \widetilde{A}_n[\alpha] = \limsup \widetilde{A}_n = \widetilde{A}[\alpha]$ *for any* $\alpha \in [0,1]$. *Thus* $\liminf \widetilde{A}_n = \limsup \widetilde{A}_n = \widetilde{A}$ *and thus the 'if part' is verified. The 'only if part' can be shown similarly.*

Next, a common notion of a real-valued function of a fuzzy set is given.

1.1.2 Cartesian product of fuzzy sets

The Cartesian products of some sets mean the product of some non-empty sets in a systematic way. Here, a common definition of the fuzzy product of (discrete) fuzzy is recalled, and some of its main properties are investigated. All such concepts were gathered from [100].

Definition 1.13 *Let* $\widetilde{A}_i \in \mathbb{F}(\mathbb{X}_i)$, $i = 1, 2, \ldots, n$. *A fuzzy product set on* $\mathbb{X} = \mathbb{X}_1 \times \mathbb{X}_2 \times \ldots \times \mathbb{X}_n$ *is defined as a Cartesian product of the fuzzy set* $\widetilde{\boldsymbol{A}} = \widetilde{A}_1 \times \widetilde{A}_2 \times \ldots \times \widetilde{A}_n$ *with the following membership function:*

$$\widetilde{\boldsymbol{A}}(\boldsymbol{x}) = \min_{i=1}^{n}\{\widetilde{A}_i(x_i)\}, \ x_i \in \mathbb{X}_i. \tag{1.11}$$

Example 1.8 *Let* $\mathbb{X}_1 = \mathbb{X}_2 = \{2,4,6\}$, $\widetilde{A}_1 = \{\frac{0.7}{2}, \frac{1}{4}, \frac{0.6}{6}\}$ *and* $\widetilde{A}_2 = \{\frac{1}{2}, \frac{0.8}{4}\}$. *Then, the membership function of fuzzy product of* $\widetilde{A}_1$ *and* $\widetilde{A}_2$ *can be shown by:*

$$\widetilde{A}_1 \times \widetilde{A}_2 = \{\frac{0.7}{(2,2)}, \frac{0.7}{(2,4)}, \frac{1}{(4,2)}, \frac{0.8}{(4,4)}, \frac{0.6}{(6,2)}, \frac{0.6}{(6,4)}\}.$$

Proposition 1.4 *If* $\{\widetilde{\boldsymbol{A}}_j = \widetilde{A}_{1j} \times \widetilde{A}_{2j} \times \ldots \times \widetilde{A}_{nj}\}_{j=1}^m$ *is a sequence of fuzzy product sets then* $\bigcup_{j=1}^m \widetilde{\boldsymbol{A}}_j = (\bigcup_{j=1}^m \widetilde{A}_{1j}) \times \ldots \times (\bigcup_{j=1}^m \widetilde{A}_{nj})$.

Proof *For a fixed* $\boldsymbol{x} = (x_1, x_2, \ldots, x_n) \in \mathbb{X}^n$, *we have:*

$$\begin{aligned}
(\bigcup_{j=1}^m \widetilde{\boldsymbol{A}}_j)(\boldsymbol{x}) &= \max_{j=1}^{m} \min\{\widetilde{A}_{1j}(x_1), \widetilde{A}_{2j}(x_2), \ldots, \widetilde{A}_{nj}(x_n)\} \qquad (1.12)\\
&= \min\{\max_{j=1}^{m} \widetilde{A}_{1j}(x_1), \max_{j=1}^{m} \widetilde{A}_{2j}(x_2), \ldots, \max_{j=1}^{m} \widetilde{A}_{nj}(x_n)\}\\
&= (\bigcup_{j=1}^m \widetilde{A}_{1j})(x_1) \times (\bigcup_{j=1}^m \widetilde{A}_{2j})(x_2) \times \ldots \times (\bigcup_{j=1}^m \widetilde{A}_{nj})(x_n).
\end{aligned}$$

It completes the proof.

Proposition 1.5 *If* $\{\widetilde{\boldsymbol{A}}_j = \widetilde{A}_{1j} \times \widetilde{A}_{2j} \times ... \times \widetilde{A}_{nj}\}_{j=1}^m$ *is a sequence of fuzzy product sets then* $\bigcap_{j=1}^m \widetilde{\boldsymbol{A}}_j = (\bigcap_{j=1}^m \widetilde{A}_{1j}) \times \ldots \times (\bigcap_{j=1}^m \widetilde{A}_{nj})$.

Proof *For a fixed* $\boldsymbol{x} = (x_1, x_2, ..., x_n) \in \mathbb{X}^n$, *it is easily seen that:*

$$(\bigcap_{j=1}^{m} \widetilde{\boldsymbol{A}}_j)(\boldsymbol{x}) = \min_{j=1}^{m} \min\{\widetilde{A}_{1j}(x_1), \widetilde{A}_{2j}(x_2), ..., \widetilde{A}_{nj}(x_n)\} =$$

$$\min\{\min_{j=1}^{m} \widetilde{A}_{1j}(x_1), \min_{j=1}^{m} \widetilde{A}_{2j}(x_2), ..., \min_{j=1}^{m} \widetilde{A}_{nj}(x_n)\} =$$

$$(\bigcap_{j=1}^{m} \widetilde{A}_{1j})(x_1) \times (\bigcap_{j=1}^{m} \widetilde{A}_{2j})(x_2) \times ... \times (\bigcap_{j=1}^{m} \widetilde{A}_{nj})(x_n).$$

Lemma 1.6 *Let* $\{\widetilde{A}_i\}_{i=1}^n$ *be a finite collection of fuzzy sets. For a given fuzzy set* $\widetilde{B}$, *we have the following results:*

1. $(\cup_{i=1}^n \widetilde{A}_i) \times \widetilde{B} = \cup_{i=1}^n (\widetilde{A}_i \times \widetilde{B})$.
2. $\widetilde{B} \times (\cup_{i=1}^n \widetilde{A}_i) = \cup_{i=1}^n (\widetilde{B} \times \widetilde{A}_i)$.
3. $(\cap_{i=1}^n \widetilde{A}_i) \times \widetilde{B} = \cap_{i=1}^n (\widetilde{A}_i \times \widetilde{B})$.
4. $\widetilde{B} \times (\cap_{i=1}^n \widetilde{A}_i) = \cap_{i=1}^n (\widetilde{B} \times \widetilde{A}_i)$.

Proof *The first part is proved only. The others are left for readers. For this purpose, a fixed* $\boldsymbol{x} = (x_1, x_2, ..., x_n) \in \mathbb{X}^n$ *and* $x \in \mathbb{X}$, *we have:*

$$((\cup_{i=1}^n \widetilde{A}_i) \times \widetilde{B})(\boldsymbol{x}, x) = \min\{\max\{\widetilde{A}_1(x_1), \widetilde{A}_2(x_2), ..., \widetilde{A}_n(x_n)\}, \widetilde{B}(x)\} =$$

$$\max\{\min\{\widetilde{A}_1(x_1), \widetilde{B}(x)\}, ..., \min\{\widetilde{A}_n(x_n), \widetilde{B}(x)\}\} = (\cup_{i=1}^n (\widetilde{A}_i \times \widetilde{B}))(\boldsymbol{x}, x).$$

This confirm that $(\cup_{i=1}^n \widetilde{A}_i) \times \widetilde{B} = \cup_{i=1}^n (\widetilde{A}_i \times \widetilde{B})$.

1.2 Fuzzy Numbers

Fuzzy numbers generalize classical real numbers and roughly speaking a fuzzy number is a fuzzy set of the real line that has some additional properties. That is the domain of a fuzzy number is a specified set of real numbers and its range is the span of non-negative real numbers between 0 and 1. Each numerical value in the domain is assigned a specific degree of membership where 0 represents the smallest possible grade, and 1 is the largest possible degree. A unified definition of a fuzzy number is given below.

Definition 1.14 ([41, 114, 157]) *A fuzzy set* $\widetilde{A}$ *of* $\mathbb{R}$ *is called a fuzzy number (**FN**) if it is normal and the set* $\widetilde{A}[\alpha] = \{x \in \mathbb{R} : \widetilde{A}(x) \geq \alpha\}$ *is a non-empty nested closed interval in* $\mathbb{R}$, *for every* $\alpha \in [0, 1]$.

This interval is denoted by $\widetilde{A}[\alpha] = [\widetilde{A}^L_\alpha, \widetilde{A}^U_\alpha]$, where $\widetilde{A}^L_\alpha = \inf\{x : x \in \widetilde{A}[\alpha]\}$ and $\widetilde{A}^U_\alpha = \sup\{x : x \in \widetilde{A}[\alpha]\}$. For this book, take $\mathcal{F}(\mathbb{R})$ as denoting the set of all **FN**s.

The arithmetic operations between two **FN**s $\widetilde{A}$ and $\widetilde{B}$ can also be evaluated using arithmetic operations of their α-cuts.

Theorem 1.5 *Let $\widetilde{A}$ and $\widetilde{B}$ be two* **FNs***. Then, for every $\alpha \in (0,1]$:*

1. $(\widetilde{A} \oplus \widetilde{B})[\alpha] = [\widetilde{A}^L_\alpha + \widetilde{B}^L_\alpha, \widetilde{A}^U_\alpha + \widetilde{B}^U_\alpha]$.
2. $(\widetilde{A} \ominus \widetilde{B})[\alpha] = [\widetilde{A}^L_\alpha - \widetilde{B}^U_\alpha, \widetilde{A}^U_\alpha - \widetilde{B}^L_\alpha]$.
3. $(\widetilde{A} \otimes \widetilde{B})[\alpha] = [(\widetilde{A} \otimes \widetilde{B})^L_\alpha, (\widetilde{A} \otimes \widetilde{B})^U_\alpha]$ *where*

$$(\widetilde{A} \otimes \widetilde{B})^L_\alpha = \min\{\widetilde{A}^L_\alpha \widetilde{B}^L_\alpha, \widetilde{A}^L_\alpha \widetilde{B}^U_\alpha, \widetilde{A}^U_\alpha \widetilde{B}^L_\alpha, \widetilde{A}^U_\alpha \widetilde{B}^U_\alpha\}, \tag{1.13}$$

$$(\widetilde{A} \otimes \widetilde{B})^U_\alpha = \max\{\widetilde{A}^L_\alpha \widetilde{B}^L_\alpha, \widetilde{A}^L_\alpha \widetilde{B}^U_\alpha, \widetilde{A}^U_\alpha \widetilde{B}^L_\alpha, \widetilde{A}^U_\alpha \widetilde{B}^U_\alpha\}, \tag{1.14}$$

4. $(\widetilde{A} \oslash \widetilde{B})[\alpha] = [(\widetilde{A} \oslash \widetilde{B})^L_\alpha, (\widetilde{A} \oslash \widetilde{B})^U_\alpha]$ *where*

$$(\widetilde{A} \oslash \widetilde{B})^L_\alpha = \min\{\widetilde{A}^L_\alpha / \widetilde{B}^L_\alpha, \widetilde{A}^L_\alpha / \widetilde{B}^U_\alpha, \widetilde{A}^U_\alpha / \widetilde{B}^L_\alpha, \widetilde{A}^U_\alpha / \widetilde{B}^U_\alpha\}, \tag{1.15}$$

$$(\widetilde{A} \oslash \widetilde{B})^U_\alpha = \max\{\widetilde{A}^L_\alpha / \widetilde{B}^L_\alpha, \widetilde{A}^L_\alpha / \widetilde{B}^U_\alpha, \widetilde{A}^U_\alpha / \widetilde{B}^L_\alpha, \widetilde{A}^U_\alpha / \widetilde{B}^U_\alpha\}, \tag{1.16}$$

provided that $0 \notin \widetilde{B}[0]$.

Proof *See [41, 114, 157, 187], for instance.*

By the above results and Lemma 1.2, the membership functions of $(\widetilde{A} \circ \widetilde{B})$ can be evaluated as:

$$(\widetilde{A} \circ \widetilde{B})(x) = \sup\{\alpha \in [0,1] : x \in (\widetilde{A} \circ \widetilde{B})[\alpha]\}. \tag{1.17}$$

where $\circ = \oplus, \ominus, \otimes, \oslash$ denotes the extended arithmetic operations $\bullet = +, -, \times, /$ in a fuzzy domain.

Several definitions have been proposed to capture the information contained in a (unimodal) **FN** to simplify the **FN**'s representation using a functional parametric form known as LR-fuzzy numbers (LR-**FN**) $\widetilde{A} = (a; l_a, r_a)_{LR}$ [41, 114, 123]. The membership function of a LR-**FN** $\widetilde{A}$ is defined by:

$$\widetilde{A}(x) = \begin{cases} L(\frac{a-x}{l_a}), & a - l_a \leq x \leq a, \\ R(\frac{x-a}{r_a}), & a \leq x \leq a + r_a, \\ 0, & x \in \mathbb{R} - [a - l_a, a + r_a], \end{cases} \tag{1.18}$$

where $l_a > 0$ is the left spread, $r_a > 0$ is the right spread, and L and R represent reference functions defining the left and the right shapes of the **FN**, respectively. The left and the right shapes $L, R : [0,1] \to [0,1]$ should also satisfy the following conditions:

1. $L(1) = R(1) = 0$,
2. $L(0) = R(0) = 1$, and
3. $L(x)$ and $R(x)$ are continuous and monotone decreasing functions on $[0, 1]$.

Moreover, a *LR*-**FN** becomes an *L*-**FN** when $L(x) = R(x)$. It is easy to check that the α-cut of a *LR*-**FN** $\widetilde{A}$ is:

$$\widetilde{A}[\alpha] = [\widetilde{A}_\alpha^L, \widetilde{A}_\alpha^U] = [a - L^{-1}(\alpha)l_a, a + R^{-1}(\alpha)r_a], \quad \alpha \in [0, 1].$$

Some common operations between two *LR*-**FN**s of are given as follows.

Theorem 1.6 *For two LR-**FNs** $\widetilde{A} = (a; l_a, r_a)_{LR}$ and $\widetilde{B} = (b; l_b, r_b)_{LR}$,*

1) Addition: $\widetilde{A} \oplus \widetilde{B} = (a + b; l_a + l_b, r_a + r_b)_{LR}$.

2) Difference: $\widetilde{A} \ominus \widetilde{B} = (a - b; l_a + r_b, r_a + l_b)_{RL}$.

3) The scalar multiplication:

$$\lambda \otimes \widetilde{A} = \begin{cases} (\lambda a; \lambda l_a, \lambda r_a)_{LR}, & \lambda > 0, \\ (\lambda a; -\lambda r_a, -\lambda l_a)_{RL}, & \lambda < 0. \end{cases}$$

Proof *See [41, 114, 123] to have a proof.*

In next the chapters, we employ the most commonly used *LR*-**FN**s so-called triangular fuzzy numbers (**TFNs**) to handle imprecision included in data set. Note that the membership function of a **TFN**, denoted by $\widetilde{A} = (a; l_a, r_a)_T$, is given by:

$$\widetilde{A}(x) = \begin{cases} \frac{x-(a-l_a)}{l_a}, & a - l_a \leq x \leq a, \\ \frac{a+r_a-x}{r_a}, & a \leq x \leq a + r_a, \\ 0, & x \in \mathbb{R} - [a - l_a, a + r_a]. \end{cases} \tag{1.19}$$

It is worth noting that a **TFN** $\widetilde{A} = (a; l_a, r_a)_T$ can also be rewritten as $\widetilde{A} = (a^L, a, a^U)_T$ where $a^L = a - l_a$ and $a^U = a + r_a$. In this regards, the membership function of $\widetilde{A}$ is:

$$\widetilde{A}(x) = \begin{cases} \frac{x-a^L}{a-a^L}, & a^L \leq x \leq a, \\ \frac{a^U-x}{a^U-a}, & a \leq x \leq a^U, \\ 0, & x \in \mathbb{R} - [a^L, a^U]. \end{cases} \tag{1.20}$$

For simplicity, a symmetric **TFN** is represented by **STFN** throughout this book.

Here, an example is provided to obtain the function of a *LR*-**FN**.

Example 1.9 *Let* $g(x) = x^2$ *and* $\widetilde{A} = (0; 2, 1)_{LR}$ *with* $L(x) = \max\{0, 1 - x\}$ *and* $R(x) = \sqrt{1 - x^2}$. *First note that the membership function of* $\widetilde{A}$ *is:*

$$\widetilde{A}(x) = \begin{cases} \frac{x+2}{2}, & -2 \leq x \leq 0, \\ \sqrt{1 - x^2}, & 0 \leq x \leq 1, \\ 0, & x \in \mathbb{R} - [-2, 1]. \end{cases} \tag{1.21}$$

Then, it is easy to check that:

$$\begin{aligned} g(\widetilde{A})(y) &= \sup_{x:y=x^2} \widetilde{A}(y) \\ &= \begin{cases} \max\{\widetilde{A}(\sqrt{y}), \widetilde{A}(-\sqrt{y})\}, & 0 \leq y \leq 1, \\ \widetilde{A}(-\sqrt{y}), & 1 < y \leq 4, \end{cases} \\ &= \begin{cases} \max\{\frac{2-\sqrt{y}}{2}, \sqrt{1 - y}\}, & 0 \leq y \leq 1, \\ \frac{2-\sqrt{y}}{2} & 1 < y \leq 4, \end{cases} \\ &= \begin{cases} \sqrt{1 - y}, & 0 \leq y \leq 0.64, \\ \frac{2-\sqrt{y}}{2}, & 0.64 < y \leq 4. \end{cases} \end{aligned}$$

The membership function of $g(\widetilde{A}) = \widetilde{A}^2$ *is plotted in Fig. 1.1.*

An example of investigating of limit superior of a sequence of **TFN**s is given by the following example

Example 1.10 *Consider a countable sequence of **TFN**s defined as* $\widetilde{A}_n = (0; l_{a_n}, r_{a_n})_T$ *where* $l_{a_n} = 1/n$ *and* $r_{a_n} = 2 - 1/n$. *Based on Proposition 1.3, we have*

$$\begin{aligned} \limsup \widetilde{A}_n[\alpha] &= \cap_{n \geq 1} \cup_{m \geq n} [-(1/m)(1 - \alpha), (2 - 1/m)(1 - \alpha)] \\ &= \cap_{n \geq 1} [-(1/n)(1 - \alpha), 2(1 - \alpha)] \\ &= [0, 2(1 - \alpha)] \\ &= \liminf \widetilde{A}_n[\alpha], \end{aligned}$$

which concludes that $\lim_{n \to \infty} \widetilde{A}_n = (0; 0, 2)_T$.

1.2.1 A similarity measure

Similarity measures are very important and useful tool to determine the amount of similarity between two or more fuzzy quantities. In almost every field of science and engineering, the concept of similarity measure has important significance [8]. Such measures are used to compare different kinds of objects in real-life applications [16, 20, 26, 76, 124, 171, 176, 193]. Here, a similarity measure between two **TFN**s is introduced. We will employ such criteria as a performance measure in Chapter 7.

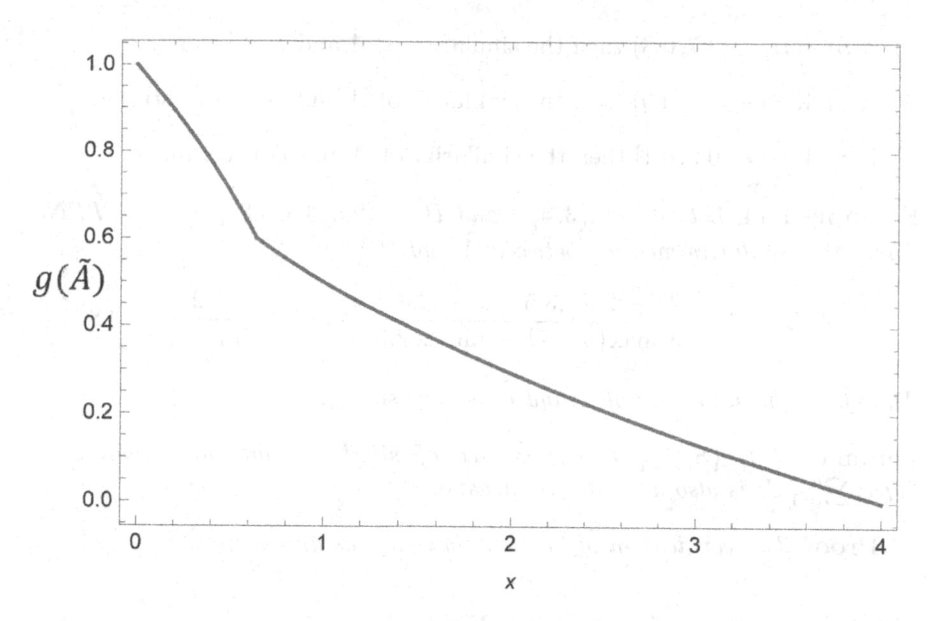

FIGURE 1.1
Plot of $g(\widetilde{A})$ in Example 1.9.

Definition 1.15 *Let $\widetilde{A} = (a^L, a, a^U)_T$ and $\widetilde{B} = (b^L, b, b^U)_T$ be two* **TFNs***. The similarity measure between $\widetilde{A}$ and $\widetilde{B}$ is defined as:*

$$S(\widetilde{A}, \widetilde{B}) = 1 - \frac{|a^L - b^L| + |a - b| + |a^U - b^U|}{3(\max\{a^U, b^U\} - \min\{a^L, b^L\})}. \tag{1.22}$$

Proposition 1.6 *Let $\widetilde{A} = (a^L, a, a^U)_T$ and $\widetilde{B} = (b^L, b, b^U)_T$ be two* **TFNs***. Then:*

1) $S(\widetilde{A}, \widetilde{B}) \in [0, 1]$.

2) $S(\widetilde{A}, \widetilde{B}) = S(\widetilde{B}, \widetilde{A})$.

3) $S(\widetilde{A}, \widetilde{B}) = 1$ *if and only if* $\widetilde{A} = \widetilde{B}$.

Proof *The proof is left for the reader.*

According to Proposition 1.6, we can say that the values of $S(\widetilde{A}, \widetilde{B}) < 0.5$ show the degree of dissimilarity while the values of $S(\widetilde{A}, \widetilde{B}) \geq 0.5$ represent the degree of similarity. In this regard, one can say that:

1) If $S(\widetilde{A}, \widetilde{B}) \in [0.5, 0.6)$ then $\widetilde{A}$ and $\widetilde{B}$ are more or less similar.

2) If $S(\widetilde{A}, \widetilde{B}) \in [0.6, 0.7)$ then the similarity of $\widetilde{A}$ and $\widetilde{B}$ is moderate.

3) If $S(\widetilde{A},\widetilde{B}) \in [0.7, 0.8)$ then the similarity of $\widetilde{A}$ and $\widetilde{B}$ is strong.

4) If $S(\widetilde{A},\widetilde{B}) \in [0.8, 0.9)$ then the similarity of $\widetilde{A}$ and $\widetilde{B}$ is very strong.

3) If $S(\widetilde{A},\widetilde{B}) \in [0.9, 0.1]$ then the similarity of $\widetilde{A}$ and $\widetilde{B}$ is complete.

Example 1.11 *Let $\widetilde{A} = (2,3,4)_T$ and $\widetilde{B} = (2.5, 3.5, 3.8)_T$ be two **TFNs**. Then, the similarity measure between $\widetilde{A}$ and $\widetilde{B}$ is:*

$$S(\widetilde{A},\widetilde{B}) = 1 - \frac{|2.5-2| + |3.5-3| + |3.8-4|}{3(\max\{4, 3.7\} - \min\{2, 2.5\})} = 1 - \frac{1.2}{3(4-2)} = 0.8.$$

Therefore, the similarity of $\widetilde{A}$ and $\widetilde{B}$ is very strong.

Lemma 1.7 *If $\{S_i\}_{i=1}^n$ is a sequence of similarity measures then $S = (1/n)\sum_{i=1}^n S_i$ is also a similarity measure.*

Proof *The verification of this lemma is left as an exercise.*

1.2.2 A criteria for ranking FNs

The problem of ordering **FNs** has received considerable attention in real-life applications. For instance, in many statistical hypothesis tests with fuzzy data, fuzzy hypotheses are based on comparing observed fuzzy test statistics and critical values . Therefore there is a need for a criterion to compare such quantities. Because of the nature of the measurement, many different strategies have been proposed for ranking of **FNs** [14, 22, 27, 42, 155, 178, 192, 203, 200, 214]. Another method of ranking **FNs** relies on preference degree. This method provides a decision-maker with a preference degree to measure how a **FN** is greater than another [216, 143, 87]. Here, a common preference criterion to rank **FNs** is recalled [143].

Definition 1.16 *For two **FNs** $\widetilde{A}$ and $\widetilde{B}$, define:*

$$\Delta_{\widetilde{A}\widetilde{B}} = \int_{\{\alpha:\widetilde{A}_\alpha^L \geq \widetilde{B}_\alpha^L\}} (\widetilde{A}_\alpha^L - \widetilde{B}_\alpha^L)d\alpha + \int_{\{\alpha:\widetilde{A}_\alpha^U \geq \widetilde{B}_\alpha^U\}} (\widetilde{A}_\alpha^U - \widetilde{B}_\alpha^U)d\alpha. \quad (1.23)$$

The preference degree that '$\widetilde{A}$ is greater than $\widetilde{B}$' is defined by:

$$P_d(\widetilde{A} \succ \widetilde{B}) = \begin{cases} 0.5, & \widetilde{A} = \widetilde{B}, \\ \frac{\Delta_{\widetilde{A}\widetilde{B}}}{\Delta_{\widetilde{A}\widetilde{B}} + \Delta_{\widetilde{B}\widetilde{A}}}, & \widetilde{A} \neq \widetilde{B}. \end{cases} \quad (1.24)$$

Definition 1.17 *For two **FNs** $\widetilde{A}$ and $\widetilde{B}$, it is said that:*

1. *$\widetilde{A}$ is greater than $\widetilde{B}$, denoting by $\widetilde{A} \succ_{P_d} \widetilde{B}$, if $P_d(\widetilde{A} \succ \widetilde{B}) > 0.5$.*

2. *$\widetilde{A}$ is equivalent to $\widetilde{B}$, denoting by $\widetilde{A} \simeq_{P_d} \widetilde{B}$, if $P_d(\widetilde{A} \succ \widetilde{B}) = P_d(\widetilde{B} \succ \widetilde{A}) = 0.5$.*

3. $\widetilde{A}$ *is greater than or equal to* $\widetilde{B}$*, denoting by* $\widetilde{A} \succeq_{P_d} \widetilde{B}$*, if* $\widetilde{A} \succ_{P_d} \widetilde{B}$ *or* $\widetilde{A} \simeq_{P_d} \widetilde{B}$.

The preference criterion P_d meets the following properties.

Proposition 1.7 *let* $\widetilde{A}$*,* $\widetilde{B}$*, and* $\widetilde{C}$ *be three* ***FNs****. Then:*

1) P_d *is reciprocal, i.e.,* $P_d(\widetilde{A} \succ \widetilde{B}) = 1 - P_d(\widetilde{B} \succ \widetilde{A})$,

2) P_d *is reflexive, i.e.,* $\widetilde{A} \succeq_{P_d} \widetilde{A}$,

3) P_d *is transitive, i.e.,* $\widetilde{A} \succeq_{P_d} \widetilde{B}$ *and* $\widetilde{B} \succeq_{P_d} \widetilde{C}$ *imply* $\widetilde{A} \succeq_{P_d} \widetilde{C}$,

4) $P_d(\widetilde{A} \succeq \widetilde{B}) = 1$ *if and only if* $\widetilde{B}^L_\alpha \leq \widetilde{A}^L_\alpha$ *and* $\widetilde{B}^U_\alpha \leq \widetilde{A}^U_\alpha$ *for all* $\alpha \in [0, 1]$.

Proof *The parts (1), (2), and (4) are immediately followed. To prove (3), note that* $P_d(\widetilde{A} \succeq \widetilde{B}) \geq 0.5$ *if and only if* $\int_0^1 (M_{\widetilde{A}}[\alpha] - M_{\widetilde{B}}[\alpha])d\alpha \geq 0$ *where* $M_{\widetilde{A}}[\alpha] = (\widetilde{A}^L_\alpha + \widetilde{A}^U_\alpha)/2$ *and* $M_{\widetilde{B}}[\alpha] = (\widetilde{B}^L_\alpha + \widetilde{B}^U_\alpha)/2$*. Therefore,* $\widetilde{A} \succeq_{P_d} \widetilde{B}$ *and* $\widetilde{B} \succeq_{P_d} \widetilde{C}$ *concludes that:*

$$\int_0^1 (M_{\widetilde{A}}[\alpha] - M_{\widetilde{C}}[\alpha])d\alpha = \int_0^1 (M_{\widetilde{A}}[\alpha] - M_{\widetilde{B}}[\alpha])d\alpha + \int_0^1 (M_{\widetilde{B}}[\alpha] - M_{\widetilde{C}}[\alpha])d\alpha \geq 0.$$

That is $\widetilde{A} \succeq_{P_d} \widetilde{C}$.

Example 1.12 *Consider two* ***TFNs*** $\widetilde{A} = (a^L, a, a^U)_T$ *and* $\widetilde{B} = (b^L, b, b^U)_T$ *for three specific cases as shown in Fig 1.2. First, assume that* $b^L < a^L < a < b < b^U < a^U$ *((A) in Fig. (1.2)). To compute* $P_d(\widetilde{B} \succeq \widetilde{A})$*,* $\Delta_{\widetilde{A}\widetilde{B}}$ *and* $\Delta_{\widetilde{B}\widetilde{A}}$ *can evaluate as follows:*

$$\begin{aligned}
\Delta_{\widetilde{A}\widetilde{B}} &= \int_0^{\frac{a^L - b^L}{b - a + a^L - b^L}} \left(a - b + ((b - b^L) - (a - a^L))(1 - \alpha)\right)d\alpha \\
&+ \int_0^{\frac{a^U - b^U}{b - a + a^U - b^U}} \left((b - a) + ((b^U - b) - (a^U - a))(1 - \alpha)\right)(1 - \alpha)d\alpha \\
&= \frac{(a^L - b^L)^2}{2(b - a + a^L - b^L)} + \frac{(a^U - b^U)^2}{2(b - a + a^U - b^U)},
\end{aligned}$$

and

$$\begin{aligned}
\Delta_{\widetilde{B}\widetilde{A}} &= \int_{\frac{a^L - b^L}{b - a + a^L - b^L}}^1 (b - a + ((a - a^L) - (b - b^L))(1 - \alpha)d\alpha \\
&+ \int_{\frac{a^U - b^U}{b - a + a^U - b^U}}^1 \left((b - a) + ((b^U - b) - (a^U - a))(1 - \alpha)\right)(1 - \alpha)d\alpha \\
&= \frac{(a - b)^2}{2(b - a + a^L - b^L)} + \frac{(a - b)^2}{2(b - a + a^U - b^U)}.
\end{aligned}$$

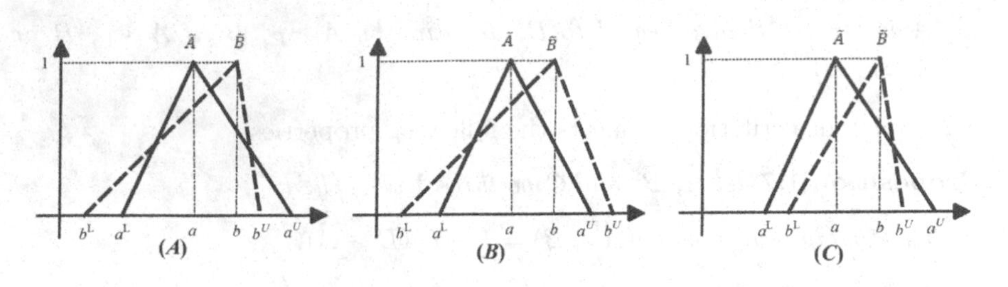

FIGURE 1.2
Membership functions of $\widetilde{A}$ and $\widetilde{B}$ in three cases in Example 1.12.

Therefore:

$$P_d(\widetilde{B} \succeq \widetilde{A}) = \frac{\Delta_{\widetilde{B}\widetilde{A}}}{\Delta_{\widetilde{A}\widetilde{B}} + \Delta_{\widetilde{B}\widetilde{A}}},$$

where

1) $\Delta_{\widetilde{B}\widetilde{A}} = (a-b)^2(2(b-a) + a^L - b^L + a^U - b^U)$.

2) $\Delta_{\widetilde{A}\widetilde{B}} = (b - a + a^U - b^U)(a^L - b^L)^2 + (b - a + a^L - b^L)(a^U - b^U)^2$.

Similarly, the preference degree to which $\widetilde{A}$ *is greater that* $\widetilde{B}$ *for cases* (1) *and* (2) *can be summarized as follows:*

(1) *If* $b^L < a^L < a < b < a^U < b^U$ *((B) in Fig.* (1.2)*), then:*

$$P_d(\widetilde{B} \succeq \widetilde{A}) = \frac{\frac{(b-a+b^L-a^L)(b-a+a^U-b^U)+(a-b)^2}{2(b-a+a^U-b^U)}}{\frac{(b-a+b^L-a^L)(b-a+a^U-b^U)+(a-b)^2}{2(b-a+a^U-b^U)} + \frac{(a^U-b^U)^2}{2(b-a+a^U-b^U)}}.$$

(2) *If* $a^L < b^L < a < b < b^U < a^U$ *((C) in Fig.* (1.2)*), then:*

$$P_d(\widetilde{B} \succeq \widetilde{A}) = \frac{\frac{(b-a+b^U-a^U)(b-a+a^L-b^L)+(a-b)^2}{2(b-a+a^L-b^L)}}{\frac{(b-a+b^U-a^U)(b-a+a^L-b^L)+(a-b)^2}{2(b-a+a^L-b^L)} + \frac{(a^L-b^L)^2}{2(b-a+a^L-b^L)}},$$

For all cases (1), (2), *and* (3), *it should be noted that* $P_d(\widetilde{B} \succeq \widetilde{A}) \in (0.5, 1)$.

Remark 1.2 *Using the preference degree* P_d *defined for each ordered pair of* ***FN****s, it should be noted that we can sort a set of* n ***FN****s* $\widetilde{A}_i$ *by sorting* $\{\widetilde{A}_1, \widetilde{A}_2, \ldots, \widetilde{A}_n\}$ *via calculating* $\binom{n}{2}$ *preference degrees into* $\{\widetilde{A}_{k_1}, \widetilde{A}_{k_2}, \ldots, \widetilde{A}_{k_n}\}$ *so that for any* $i < j$, $P_d(\widetilde{A}_j \succeq \widetilde{A}_i) \geq 0.5$. *The feasibility of the sorting is guaranteed by Proposition 1.7. Based on the sorting results, therefore,* $\widetilde{A}_{k_n}$ *is the most preferred choice,* $\widetilde{A}_{k_{(n-1)}}$ *is the second, etc.*

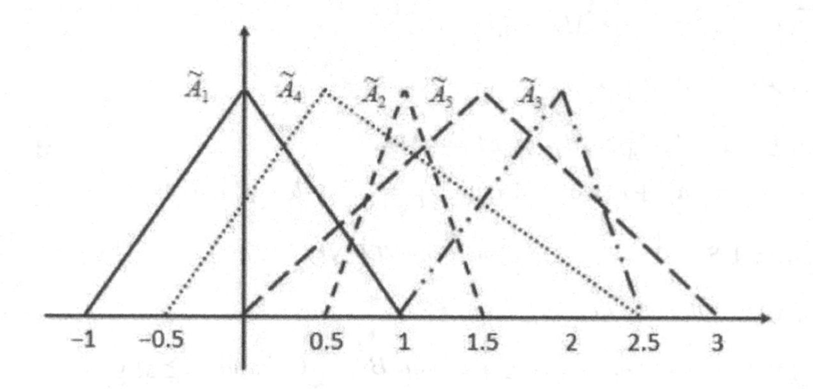

FIGURE 1.3
Membership functions of $\widetilde{A}_i$, $i = 1, 2, ..., 5$ in Example 1.13.

Example 1.13 *Consider five* ***TFNs*** *of* $\widetilde{A}_1 = (-1, 0, 1)_T$, $\widetilde{A}_2 = (0.5, 1, 1.5)_T$, $\widetilde{A}_3 = (1, 2, 2.5)_T$, $\widetilde{A}_4 = (-0.5, 0.5, 2.5)_T$, *and* $\widetilde{A}_5 = (0, 1.5, 3)_T$. *The plots of* $\widetilde{A}_i$, $i = 1, 2, .., 5$ *are shown in Fig. 1.3. The preference degree* $P_d(\widetilde{A}_j \succeq \widetilde{A}_i)$, $i, j = 1, 2, ..., 5$ *are listed in Table 1.1. Accordingly, we can get a sorted result as* $\widetilde{A}_5 \succeq_{P_d} \widetilde{A}_3 \succeq_{P_d} \widetilde{A}_4 \succeq_{P_d} \widetilde{A}_2 \succeq_{P_d} \widetilde{A}_1$.

TABLE 1.1
Preference degrees $P_d(\widetilde{A}_i \succeq \widetilde{A}_j)$ in Example 1.13.

j	1	2	3	4	5
$P_d(\widetilde{A}_1 \succeq \widetilde{A}_j)$	1	0	0	0	0
$P_d(\widetilde{A}_2 \succeq \widetilde{A}_j)$	1	1	0	$\frac{3}{8}$	$\frac{1}{2}$
$P_d(\widetilde{A}_3 \succeq \widetilde{A}_j)$	1	1	1	1	$\frac{3}{8}$
$P_d(\widetilde{A}_4 \succeq \widetilde{A}_j)$	1	$\frac{5}{8}$	0	1	0
$P_d(\widetilde{A}_5 \succeq \widetilde{A}_j)$	1	$\frac{1}{2}$	$\frac{5}{8}$	1	1

Remark 1.3 *Let* $\widetilde{A} = (a^L, a, a^U)_T$ *and* $\widetilde{B} = (b^L, b, b^U)_T$ *be two* ***TFNs****. According to Proposition 1.7, we have* $P_d(\widetilde{A} \succeq \widetilde{B}) = 1$ *if and only if* $a^L \leq b^L$, $a \leq b$ *and* $a^U \leq b^U$.

Further, a different type of ranking for **TFNs** is introduced based on an exact criterion. We employed this criterion to construct a fuzzy hypothesis in Chapter 5.

Definition 1.18 *Let* $\widetilde{A} = (a^L, a, a^U)_T$ *and* $\widetilde{B} = (b^L, b, b^U)_T$ *be two* ***TFNs****. Then, we say that:*

1. $\widetilde{A} \simeq_M \widetilde{B}$*, if* $M_{\widetilde{A}} = M_{\widetilde{B}}$.

2. $\widetilde{A} \succ_M \widetilde{B}$*, if* $M_{\widetilde{A}} > M_{\widetilde{B}}$.

3. $\widetilde{A} \succeq_M \widetilde{B}$*, if* $M_{\widetilde{A}} > M_{\widetilde{B}}$ *or* $M_{\widetilde{A}} = M_{\widetilde{B}}$.

where $M_{\widetilde{A}} = (a^L + a + a^U)/3$ *and* $M_{\widetilde{B}} = (b^L + b + b^U)/3$.

Lemma 1.8 *Let* $\widetilde{A}$*,* $\widetilde{B}$*, and* $\widetilde{C}$ *be three* ***TFN****s. Then:*

1) $\succeq_M$ *is reflexive, i.e.,* $\widetilde{A} \succeq_M \widetilde{A}$.

2) $\succeq_M$ *is transitive, i.e.,* $\widetilde{A} \succeq_M \widetilde{B}$ *and* $\widetilde{B} \succeq_M \widetilde{C}$ *imply* $\widetilde{A} \succeq_M \widetilde{C}$.

3) $\succeq_M$ *is antisymmetric, i.e.,* $\widetilde{A} \succeq_M \widetilde{B}$ *and* $\widetilde{B} \succeq_M \widetilde{A}$ *imply* $\widetilde{A} \simeq_M \widetilde{B}$.

Proof *The proof is left as an exercise.*

1.2.3 Distance measure

In several fields, the necessity to determine the distance that separates two points arises. When there is imprecision on the location of these points, the calculation of the distance has to take this uncertainty into consideration. The methods for measuring the distance between fuzzy sets have become important due to their significant applications in various areas [17, 49, 50, 79, 136, 192, 199]. Particularly, many statistical inferences such as hypothesis test, decision making, regression analysis, and non-parametric inferences rely on distance measures. Therefore, when we speak statistical inferences based on fuzzy data, there is a need to extend the classic distance measures where the observed data are reported by fuzzy quantities.

Definition 1.19 *It is said that d is a distance measure between two fuzzy sets if it meets the following conditions for any fuzzy sets* $\widetilde{A}, \widetilde{B}$*, and* $\widetilde{C}$*:*

1) $d(\widetilde{A}, \widetilde{B}) = 0$ *if and only if* $\widetilde{A} = \widetilde{B}$,

2) $d(\widetilde{A}, \widetilde{B}) = d(\widetilde{B}, \widetilde{A})$,

3) $d(\widetilde{A}, \widetilde{C}) \leq d(\widetilde{A}, \widetilde{B}) + d(\widetilde{B}, \widetilde{C})$.

Here, a non-fuzzy distance measures between two **TFN**s are introduced and discussed.

Lemma 1.9 *Consider two* ***TFN****s* $\widetilde{A} = (a^L, a, a^U)_T$ *and* $\widetilde{B} = (b^L, b, b^U)_T$. *Let:*

1) $d_1(\widetilde{A}, \widetilde{B}) = \frac{|a^L - b^L| + |a - b| + |a^U - b^U|}{3}$,

2) $d_2(\widetilde{A}, \widetilde{B}) = \max\{|a^L - b^L|, |a - b|, |a^U - b^U|\}$,

3) $d_3(\widetilde{A}, \widetilde{B}) = \sqrt{|a^L - b^L|^2 + |a - b|^2 + |a^U - b^U|^2}$.

Then, d_1, d_2, *and* d_3 *are distance measures.*

Proof *The proof is left for the reader.*

Example 1.14 *Consider two **TFNs*** $\widetilde{A} = (-3, -1, 3)_T$ *and* $\widetilde{B} = (-2, 1, 4)_T$. *Then:*

1) $d_1(\widetilde{A}, \widetilde{B}) = \frac{|a^L - b^L| + |a - b| + |a^U - b^U|}{3} = 4/3$,

2) $d_2(\widetilde{A}, \widetilde{B}) = \sqrt{\frac{|a^L - b^L|^2 + |a - b|^2 + |a^U - b^U|^2}{3}} = \sqrt{2}$,

3) $d_3(\widetilde{A}, \widetilde{B}) = \max\{|a^L - b^L|, |a - b|, |a^U - b^U|\} = 2$.

The following definition extends the distance between two **FN**s of $\widetilde{A}$ and $\widetilde{B}$ as a **FN**.

Definition 1.20 *[199] For three fuzzy sets* $\widetilde{A}, \widetilde{B}$, *and* $\widetilde{C}$, *it is said that* $\widetilde{D}$ *is a fuzzy distance measure (**FDM**) if it meets the following conditions:*

1) $\widetilde{D}(\widetilde{A}, \widetilde{B}) = I\{0\}$ *if and only if* $\widetilde{A} = \widetilde{B}$.

2) $\widetilde{D}(\widetilde{A}, \widetilde{B}) = \widetilde{D}(\widetilde{B}, \widetilde{A})$.

3) $\widetilde{D}(\widetilde{A}, \widetilde{C}) \preceq_{P_d} \widetilde{D}(\widetilde{A}, \widetilde{B}) \oplus \widetilde{D}(\widetilde{B}, \widetilde{C})$.

Theorem 1.7 *For two fuzzy sets* $\widetilde{A}$ *and* $\widetilde{B}$ *define:*

$$\widetilde{D}(\widetilde{A}, \widetilde{B})(d) = \sup_{x,y:|x-y|=d} \min\{\widetilde{A}(x), \widetilde{B}(y)\}. \tag{1.25}$$

Then, $\widetilde{D}$ *is a **FDM**.*

Proof *To prove, see [199].*

Example 1.15 *Let* $\mathbb{X} = \{-1, 0, 2, 3, 4\}$, $\widetilde{A} = \{\frac{0.3}{-1}, \frac{0.5}{0}, \frac{0.7}{2}, \frac{1}{3}, \frac{0.8}{4}\}$, *and* $\widetilde{B} = \{\frac{0.8}{-1}, \frac{1}{0}, \frac{0.9}{2}, \frac{0.6}{3}, \frac{0.4}{4}\}$. *Then, the membership functions of* $\widetilde{D}(\widetilde{A}, \widetilde{B})$ *can be evaluated as follows:*

$$\begin{aligned}
\widetilde{D}(\widetilde{A}, \widetilde{B})(0) &= \sup_{x,y:|x-y|=0} \min\{\widetilde{A}(x), \widetilde{B}(y)\} \\
&= \max\Big\{ \min\{\widetilde{A}(-1), \widetilde{B}(-1)\}, \min\{\widetilde{A}(0), \widetilde{B}(0)\} \\
&, \min\{\widetilde{A}(2), \widetilde{B}(2)\}, \min\{\widetilde{A}(3), \widetilde{B}(3)\}, \min\{\widetilde{A}(4), \widetilde{B}(4)\}\Big\} \\
&= \max\Big\{ \min\{0.3, 0.8\}, \min\{0.5, 1\}, \min\{0.7, 0.9\}, \min\{1, 0.6\}, \\
&\min\{0.4, 0.8\}\Big\} = 0.7.
\end{aligned}$$

$$
\begin{aligned}
\widetilde{D}(\widetilde{A},\widetilde{B})(1) &= \sup_{x,y:|x-y|=1} \min\{\widetilde{A}(x),\widetilde{B}(y)\} \\
&= \max\Big\{\min\{\widetilde{A}(-1),\widetilde{B}(0)\},\min\{\widetilde{A}(0),\widetilde{B}(-1)\},\min\{\widetilde{A}(2), \\
&\widetilde{B}(3)\},\min\{\widetilde{A}(3),\widetilde{B}(2)\},\min\{\widetilde{A}(4),\widetilde{B}(3)\},\min\{\widetilde{A}(3),\widetilde{B}(4)\}\Big\} \\
&= \max\Big\{\min\{0.3,1\},\min\{0.8,0.3\},\min\{0.7,0.6\},\min\{0.6,0.7\}, \\
&\min\{1,0.9\},\min\{0.8,0.6\},\min\{0.4,1\}\Big\} = 0.9.
\end{aligned}
$$

Similarly, it can be shown that $\widetilde{D}(\widetilde{A},\widetilde{B})(3)=1$, $\widetilde{D}(\widetilde{A},\widetilde{B})(2)=\widetilde{D}(\widetilde{A},\widetilde{B})(4)=\widetilde{D}(\widetilde{A},\widetilde{B})(5)=0.8$ *and thus:*

$$
\widetilde{D}(\widetilde{A},\widetilde{B})=\{\frac{0.7}{0},\frac{0.9}{1},\frac{0.8}{2},\frac{1}{3},\frac{0.8}{\{4,5\}}\}.
$$

Here, the proposed fuzzy distance measures between two **TFN**s are developed in the fuzzy domain.

Proposition 1.8 *Consider two* ***TFNs*** $\widetilde{A}=(a^L,a,a^U)_T$ *and* $\widetilde{B}=(b^L,b,b^U)_T$. *Let:*

$$
\widetilde{d}_a(\widetilde{A},\widetilde{B})=(d_1(\widetilde{A},\widetilde{B}),d_2(\widetilde{A},\widetilde{B}),d_3(\widetilde{A},\widetilde{B}))_T, \tag{1.26}
$$

where d_1, d_2, *and* d_3 *are as in Lemma 1.9. Then,* $\widetilde{d}_a$ *is* ***FDM***.

Proof *First note that* d_1, d_2, *and* d_3 *are (exact) distance measures and* $d_1(\widetilde{A},\widetilde{B})\le d_2(\widetilde{A},\widetilde{B})\le d_3(\widetilde{A},\widetilde{B})$ *for any* $\widetilde{A},\widetilde{B}\in\mathcal{F}(\mathbb{R})$. *Therefore, it can be checked that*

1) $\widetilde{d}_a(\widetilde{A},\widetilde{B})=I\{0\}$ *if and only if* $\widetilde{A}=\widetilde{B}$,

2) $\widetilde{d}_a(\widetilde{A},\widetilde{B})=\widetilde{d}_a(\widetilde{B},\widetilde{A})$,

and

$$
\begin{aligned}
d_1^L(\widetilde{A},\widetilde{C}) &\le d_1^L(\widetilde{A},\widetilde{B})+d_1^L(\widetilde{B},\widetilde{C}),\\
d_2(\widetilde{A},\widetilde{C}) &\le d_2(\widetilde{A},\widetilde{B})+d_2(\widetilde{B},\widetilde{C}),\\
d_3^U(\widetilde{A},\widetilde{C}) &\le d_3^U(\widetilde{A},\widetilde{B})+d_3^U(\widetilde{B},\widetilde{C}).
\end{aligned}
$$

From Remark 1.3, these conclude that $P_d(\widetilde{d}_a(\widetilde{A},\widetilde{C})\preceq(\widetilde{d}_a(\widetilde{A},\widetilde{B})\oplus\widetilde{d}_a(\widetilde{B},\widetilde{C})))=1$ *or* $\widetilde{d}_a(\widetilde{A},\widetilde{C})\preceq_{P_d}\widetilde{d}_a(\widetilde{A},\widetilde{B})\oplus\widetilde{d}_a(\widetilde{B},\widetilde{C})$. *Therefore,* $\widetilde{d}_a$ *is* ***FDM***.

Example 1.16 *Recall Example 1.14. Therefore, the fuzzy distance between* $\widetilde{A}$ *and* $\widetilde{B}$ *can be evaluated via a* ***TFN*** $\widetilde{d}_a(\widetilde{A},\widetilde{B})=(4/3,\sqrt{2},2)_T$.

1.3 α-values of FNs

Here, the notion of α-values of **FN**s is recalled. We will utilize such a notion to rewrite a common notion of the fuzzy random variable in the next section [78].

Definition 1.21 *The α-values of a **FN** $\widetilde{A}$ is a mapping $\widetilde{A}_\alpha : [0,1] \to \mathbb{R}$ defined by:*

$$\widetilde{A}_\alpha = \begin{cases} \widetilde{A}^L_{2\alpha}, & \alpha \in [0, 0.5], \\ \widetilde{A}^U_{2(1-\alpha)}, & \alpha \in [0.5, 1]. \end{cases} \tag{1.27}$$

Lemma 1.10 *The α-values a **FN** $\widetilde{A}$ is an increasing function of α on $[0,1]$.*

Proof *Since $\widetilde{A}^L_\alpha$ and $\widetilde{A}^U_\alpha$ are increasing and decreasing functions, for any $\alpha_1 < \alpha_2$ in $[0,1]$, we have $\widetilde{A}^L_{2\alpha_1} \leq \widetilde{A}^L_{2\alpha_2}$ for $\alpha \in [0, 0.5]$ and $\widetilde{A}^U_{2(1-\alpha_1)} \leq \widetilde{A}^L_{2(1-\alpha_2)}$ for $\alpha \in [0.5, 1]$. These verify the proof.*

Example 1.17 *Let $\widetilde{A} = (a; l_a, r_a)_{LR}$ be a LR-**FN**. From Definition 1.21, one finds that:*

$$\widetilde{A}_\alpha = \begin{cases} a - l_a L^{-1}(2\alpha), & 0 \leq \alpha \leq 0.5, \\ a + r_a R^{-1}(2(1-\alpha)), & 0.5 \leq \alpha \leq 1. \end{cases}$$

For instance,

1. *If $\widetilde{A} = (a; l_a, r_a)_T$ is a **TFN**, then:*

$$\widetilde{A}_\alpha = \begin{cases} (a - l_a) + 2l_a\alpha, & 0 \leq \alpha \leq 0.5, \\ a + r_a - 2r_a(1-\alpha), & 0.5 \leq \alpha \leq 1. \end{cases}$$

2. *Let $\widetilde{A} = (a; l_a, r_a)_{LR}$ with $L(x) = \sqrt{1 - x^3}$ and $R(x) = 1 - x^5$ then:*

$$\widetilde{A}_\alpha = \begin{cases} a - l_a \sqrt[3]{1 - 4\alpha^2}, & 0 \leq \alpha \leq 0.5, \\ a + r_a \sqrt[5]{2\alpha - 1}, & 0.5 \leq \alpha \leq 1. \end{cases}$$

The relationship between α-values and α-cuts of a **FN** is given below.

Lemma 1.11 *Let $\widetilde{A}[\alpha]$ and $\widetilde{A}_\alpha$ be the α-cuts and α-values of a **FN** $\widetilde{A}$, respectively. Then:*

$$\widetilde{A}[\alpha] = [\widetilde{A}^L_\alpha, \widetilde{A}^U_\alpha] = [\widetilde{A}_{\alpha/2}, \widetilde{A}_{1-\alpha/2}]. \tag{1.28}$$

Proof *The proof is a simple consequence of Eq.* (1.27) *since* $\widetilde{A}_{\alpha/2} = \widetilde{A}^L_\alpha$ *and* $\widetilde{A}_{1-\alpha/2} = \widetilde{A}^U_\alpha$ *for any* $\alpha \in [0,1]$.

Therefore, the membership degree of $\widetilde{A}$ at $x \in \mathbb{R}$ can be written as:

$$\widetilde{A}(x) = \sup\{\alpha \in [0,1] : x \in [\widetilde{A}_{\alpha/2}, \widetilde{A}_{1-\alpha/2}]\}.$$

The following Lemma shows that the interval-valued arithmetic operations of **FN**s can be simplified using their α-values.

Lemma 1.12 *Let* $\lambda \in \mathbb{R}$ *and* $\widetilde{A}$ *and* $\widetilde{B}$ *be two* ***FN****s. Then, for any* $\alpha \in [0,1]$*:*

1. $(\widetilde{A} \oplus \widetilde{B})_\alpha = \widetilde{A}_\alpha + \widetilde{B}_\alpha$.

2. $(\lambda \otimes \widetilde{A})_\alpha = \begin{cases} \lambda \widetilde{A}_\alpha, & \lambda \geq 0, \\ -\lambda \widetilde{A}_{1-\alpha}, & \lambda < 0. \end{cases}$

3. $(\widetilde{A} \otimes \widetilde{B})_\alpha = \begin{cases} \widetilde{A}_\alpha \widetilde{B}_\alpha, & if\ \ \widetilde{A}, \widetilde{B} > 0, \\ \widetilde{A}_{1-\alpha} \widetilde{B}_{1-\alpha}, & if\ \ \widetilde{A}, \widetilde{B} < 0, \\ \widetilde{A}_{1-\alpha} \widetilde{B}_\alpha, & if\ \ \widetilde{A} > 0, \widetilde{B} < 0. \end{cases}$

Proof *From Definition 1.21 and the arithmetic operation of* ***FN****s, it is easy to check that:*

$$\begin{aligned} (\widetilde{A} \oplus \widetilde{B})_\alpha &= \begin{cases} (\widetilde{A} \oplus \widetilde{B})^L_{2\alpha}, & 0 \leq \alpha \leq 0.5, \\ (\widetilde{A} \oplus \widetilde{B})^U_{2(1-\alpha)}, & 0.5 < \alpha \leq 1. \end{cases} \\ &= \begin{cases} \widetilde{A}^L_{2\alpha} + \widetilde{B}^L_{2\alpha}, & 0 \leq \alpha \leq 0.5, \\ \widetilde{A}^U_{2(1-\alpha)} + \widetilde{B}^U_{2(1-\alpha)}, & 0.5 < \alpha \leq 1. \end{cases} \\ &= \widetilde{A}_\alpha + \widetilde{B}_\alpha, \end{aligned} \tag{1.29}$$

and

$$\begin{aligned} (\lambda \otimes \widetilde{A})_\alpha &= \begin{cases} (\lambda \otimes \widetilde{A})^L_{2\alpha}, & 0 \leq \alpha \leq 0.5, \\ (\lambda \otimes \widetilde{A})^U_{2(1-\alpha)}, & 0.5 < \alpha \leq 1. \end{cases} \\ &= \lambda \widetilde{A}_\alpha I(\lambda \geq 0) - \lambda \widetilde{A}_{1-\alpha} I(\lambda < 0). \end{aligned} \tag{1.30}$$

These verify (1) and (2). To prove (3), first assume that $\widetilde{A}$ *and* $\widetilde{B}$ *are two positive* ***FN****s. Since* $(\widetilde{A}^2)_\alpha = (\widetilde{A}_\alpha)^2$ *we get:*

$$\begin{aligned} ((\widetilde{A} \oplus \widetilde{B})^2)_\alpha &= ((\widetilde{A} \oplus \widetilde{B}) \otimes (\widetilde{A} \oplus \widetilde{B}))_\alpha \\ &= (\widetilde{A}_\alpha + \widetilde{B}_\alpha)^2 \\ &= \widetilde{A}^2_\alpha + \widetilde{B}^2_\alpha + 2\widetilde{A}_\alpha \widetilde{B}_\alpha. \end{aligned}$$

By (1) and (2), therefore:

$$\begin{aligned}((\widetilde{A} \oplus \widetilde{B})^2)_\alpha &= (\widetilde{A}^2 \oplus \widetilde{B}^2 \oplus 2\widetilde{A} \otimes \widetilde{B})_\alpha \\ &= \widetilde{A}^2_\alpha + \widetilde{B}^2_\alpha + 2(\widetilde{A} \otimes \widetilde{B})_\alpha.\end{aligned}$$

This shows that $(\widetilde{A} \otimes \widetilde{B})_\alpha = \widetilde{A}_\alpha \widetilde{B}_\alpha$. *Now, if* $\widetilde{A}$ *and* $\widetilde{B}$ *are two negative **FNs**, then we have* $\widetilde{A} = (-1) \otimes \widetilde{A}'$ *and* $\widetilde{B} = (-1) \otimes \widetilde{B}'$ *for two positive **FNs** of* $\widetilde{A}'$ *and* $\widetilde{B}'$. *So, it concludes that:*

$$\begin{aligned}(\widetilde{A} \otimes \widetilde{B})_\alpha &= \Big((-1) \otimes \widetilde{A}' \otimes (-1) \otimes \widetilde{B}'\Big)_\alpha \\ &= (\widetilde{A}' \otimes \widetilde{B}')_\alpha = \widetilde{A}'_\alpha \widetilde{B}'_\alpha = \widetilde{A}_{1-\alpha} \widetilde{B}_{1-\alpha}.\end{aligned}$$

This provides (3). These results can also be verified in cases where $\widetilde{A} > 0$ *and* $\widetilde{B} < 0$.

According to Lemma 1.12, it should be noted that the arithmetic operations of **FN**s can be rewritten according to their α-values.

Lemma 1.13 *Consider a mapping* $\widetilde{A}_\alpha : [0,1] \to \mathbb{R}$ *such that: 1)* $\widetilde{A}_\alpha$ *strictly increasing with respect to* α *for any* $\alpha \in [0,1]$ *and 2)* $\widetilde{A}_{0.5}$ *is a constant number. Then,* $\{\widetilde{A}_\alpha\}_{\alpha \in [0,1]}$ *provides a **FN*** $\widetilde{A}$ *with the following* α*-cut:*

$$\widetilde{A}[\alpha] = [\widetilde{A}_{\alpha/2}, \widetilde{A}_{1-\alpha/2}]. \tag{1.31}$$

Proof *Let* $\widetilde{A}_\alpha : [0,1] \to \mathbb{R}$ *be a mapping such that: 1)* $\widetilde{A}_\alpha$ *strictly increasing with respect to* α *for any* $\alpha \in [0,1]$ *and 2)* $\widetilde{A}_{0.5}$ *is a constant number. It is easy to verify that* $[\widetilde{A}_{1-\alpha_2/2}, \widetilde{A}_{\alpha_2/2}] \subseteq [\widetilde{A}_{1-\alpha_1/2}, \widetilde{A}_{\alpha_1/2}]$ *for any* $\alpha_1 < \alpha_2$. *Therefore,* $\widetilde{A}[\alpha] = [\widetilde{A}_{1-\alpha/2}, \widetilde{A}_{\alpha/2}]$ *construct a sequence of* α*-cuts of a **FN**.*

This is the reason why we focused on α-values since a **FN** of $\widetilde{A}$ can be traceable via its α-values ($\{\widetilde{A}_\alpha\}$) instead of its α-cuts ($\{\widetilde{A}[\alpha]\}$).

Lemma 1.14 *Let* $\widetilde{A} \in \mathcal{F}(\mathbb{R})$ *and* g *be a continuous and strictly monotone real-valued function. Then, for every* $\alpha \in [0,1]$

$$(g(\widetilde{A}))_\alpha = \begin{cases} g(\widetilde{A}_\alpha), & g \text{ is increasing}, \\ g(\widetilde{A}_{1-\alpha}), & g \text{ is decreasing}. \end{cases} \tag{1.32}$$

Proof *Without loss of generality, assume that* g *is a strictly increasing function. Regarding the extension principle and Definition 1.21, first note that* $g(\widetilde{A})(y) = \widetilde{A}(g^{-1}(y))$ *and therefore:*

$$\begin{aligned}g(\widetilde{A})[\alpha] &= g(\widetilde{A}[\alpha]) \\ &= [g(\widetilde{A}^L_\alpha), g(\widetilde{A}^U_\alpha)] \\ &= [g(\widetilde{A}_{\alpha/2}), g(\widetilde{A}_{1-\alpha/2})].\end{aligned}$$

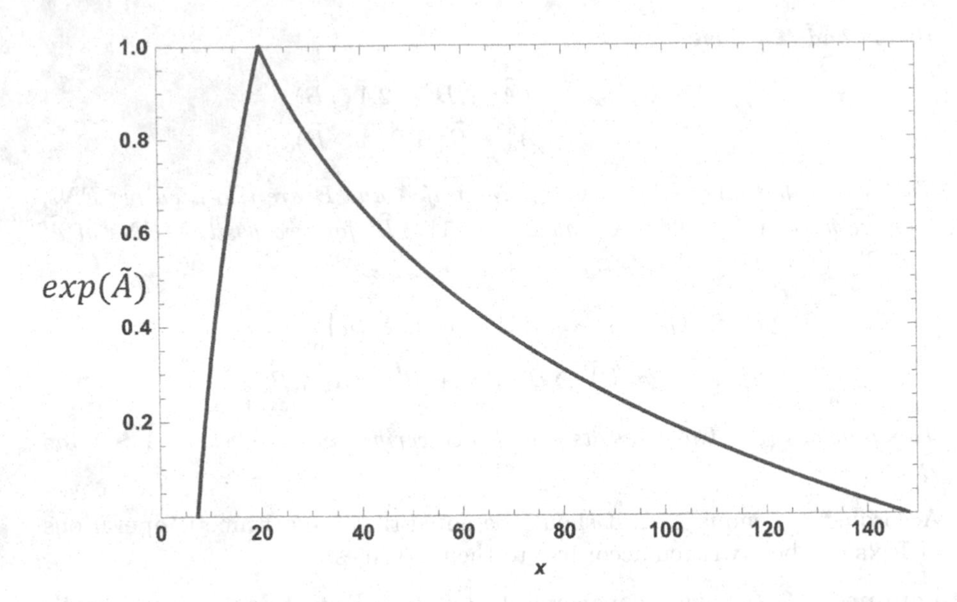

FIGURE 1.4
Membership function of $\exp(\widetilde{A})$ in Example 1.18.

Now, by the relationship between α-values and α-cuts given in Definition 1.21, it can be seen that $(g(\widetilde{A}))_\alpha = g(\widetilde{A}_\alpha)$ for every $\alpha \in [0,1]$.

Remark 1.4 *By the relationship between α-values and α-cuts of **FNs**, note that the preference degree that '$\widetilde{A}$ is greater than $\widetilde{B}$' can be rewritten as follows:*

$$P_d(\widetilde{A} \succ \widetilde{B}) = \begin{cases} 0.5, & \widetilde{A} = \widetilde{B}, \\ \frac{\Delta_{\widetilde{A}\widetilde{B}}}{\Delta_{\widetilde{A}\widetilde{B}} + \Delta_{\widetilde{B}\widetilde{A}}}, & \widetilde{A} \neq \widetilde{B}. \end{cases}$$

where

$$\Delta_{\widetilde{A}\widetilde{B}} = \int_{\{\alpha:\widetilde{A}_\alpha \geq \widetilde{B}_\alpha\}} (\widetilde{A}_\alpha - \widetilde{B}_\alpha) d\alpha. \tag{1.33}$$

Therefore $P_d(\widetilde{A} \succeq \widetilde{B}) = 1$ if and only if $\widetilde{A}_\alpha \geq \widetilde{B}_\alpha$ for any $\alpha \in [0,1]$.

From Lemma 1.14, we can observe that the conventional function of a **FN** (based on the extension principle) can also be rewritten based on the α-values of the **FN**s.

Example 1.18 *Let $g(x) = \exp(x)$ and $\widetilde{A} = (a; l_a, r_a)_T$ be a **TFN**. Therefore, the α-cuts of $g(\widetilde{A})$ can be evaluated as:*

$$(g(\widetilde{A}))_\alpha = \begin{cases} \exp((a - l_a) + 2l_a\alpha), & 0 \leq \alpha \leq 0.5, \\ \exp(a + r_a - 2r_a(1-\alpha)), & 0.5 < \alpha \leq 1. \end{cases}$$

Hence,

$$g(\widetilde{A})(x) = \sup\{\alpha \in [0,1] : x \in [\exp((a - l_a) + l_a\alpha), \exp(a + r_a - \alpha r_a)]\}.$$

For instance, assume that $\widetilde{A} = (3; 1, 2)_T$. *The membership function of* $g(\widetilde{A})$ *is drawn in Fig. 1.4.*

Theorem 1.8 *For two* ***FNs*** $\widetilde{A}$ *and* $\widetilde{B}$ *define:*

$$d_p(\widetilde{A}, \widetilde{B}) = \begin{cases} (\int_0^1 g_p(\alpha)|\widetilde{A}_\alpha - \widetilde{B}_\alpha|^p d\alpha)^{1/p}, & p \geq 1, \\ \sup_{\alpha \in [0,1]} |\widetilde{A}_\alpha - \widetilde{B}_\alpha|, & p = \infty, \end{cases} \tag{1.34}$$

where

$$g_p(\alpha) = \begin{cases} (p+1)2^p\alpha^2, & 0 \leq \alpha \leq 0.5, \\ (p+1)2^p(1-\alpha)^2, & 0.5 \leq \alpha \leq 1. \end{cases} \tag{1.35}$$

Then d_p, $p \geq 1$ *is a distance measure.*

Proof *Having a* ***FN*** *of* $\widetilde{A}$, *it should be noted that* $\widetilde{A}_\alpha : [0,1] \to \mathbb{R}$ *can be regarded as a sequence of the continuous and bounded support function. Since* g *is a probability density function on* $[0,1]$, *then* d_p, $1 \leq p < \infty$ *can be regarded as the expectation of* $|\widetilde{A}_\alpha - \widetilde{B}_\alpha|^p$ *with respect to* g_p *as* $d_p(\widetilde{A}, \widetilde{B}) = (E_g(|\widetilde{A}_\alpha - \widetilde{B}_\alpha|^p))^{1/p}$. *To prove that* d_p, $p \geq 1$ *is a distance measure, it is enough to show that* d_p *satisfies the following conditions:*

1) $d_p(\widetilde{A}, \widetilde{B}) = 0$ *if and only if* $\widetilde{A} = \widetilde{B}$,

2) $d_p(\widetilde{A}, \widetilde{B}) = d_p(\widetilde{B}, \widetilde{A})$,

3) $d_p(\widetilde{A}, \widetilde{C}) \leq d_p(\widetilde{A}, \widetilde{B}) + d_p(\widetilde{B}, \widetilde{C})$,

for any ***FNs*** $\widetilde{A}$, $\widetilde{B}$, *and* $\widetilde{C}$. *The parts (1) and (2) can be easily verified. Next, consider part (3). To check this, note that:*

$$\begin{aligned} d_p(\widetilde{A}, \widetilde{C}) &= (E_g(|(\widetilde{A}_\alpha - \widetilde{B}_\alpha) + (\widetilde{B}_\alpha - \widetilde{C}_\alpha)|^p)^{1/p} \\ &\leq d_p(\widetilde{A}, \widetilde{B}) + d_p(\widetilde{B}, \widetilde{C}). \end{aligned}$$

The last inequality follows from Minkowski inequality [31]. Thus, d_p *is a distance measure for* $1 \leq p < \infty$. *Similarly, it can be shown that* d_∞ *is also a distance measure.*

Our focus on the density function of g since it may serve as a weighted function emphasizing on the values of α near to $|\widetilde{A}[1] - \widetilde{B}[1]|^p$ more than other values $|a - b|^p$ where $a \in \widetilde{A}[0]$ and $b \in \widetilde{B}[0]$.

Example 1.19 *Let $\widetilde{A} = (a; l_a, r_a)_T$ and $\widetilde{B} = (b; l_b, r_b)_T$ be two* ***TFNs****. Then, for $1 \leq p < \infty$:*

$$d_2^2(\widetilde{A}, \widetilde{B}) = \int_0^1 g_2(\alpha)(\widetilde{A}_\alpha - \widetilde{B}_\alpha)^2 d\alpha$$
$$= \int_0^{0.5} 12\alpha^2(a - (1-2\alpha)l_a - b + (1-2\alpha)l_b)^2 d\alpha$$
$$+ \int_{0.5}^1 12(1-\alpha)^2(a + (1-2\alpha)r_a - b - (1-2\alpha)r_b)^2 d\alpha$$
$$= 0.05\Big(20(a-b)^2 + (l_a - l_b)^2 + (r_a - r_b)^2$$
$$+ 5(a-b)(r_a - l_a + l_b - r_b)\Big).$$

In special cases that $\widetilde{A}$ and $\widetilde{B}$ are two ***STFNs****, then:*

$$d_2^2(\widetilde{A}, \widetilde{B}) = (a-b)^2 + 0.1(l_a - l_b)^2.$$

Example 1.20 *Let $\widetilde{A} = (a; l_a)_T$ and $\widetilde{B} = (b; l_b)_T$ be two* ***STFNs*** *with $a > b$ and $l_a > l_b$. Then:*

$$d_\infty(\widetilde{A}, \widetilde{B}) = \sup_{\alpha \in [0,1]} |\widetilde{A}_\alpha - \widetilde{B}_\alpha|$$
$$= \max\{ \sup_{\{\alpha : \widetilde{A}_\alpha \geq \widetilde{B}_\alpha\}} (\widetilde{A}_\alpha - \widetilde{B}_\alpha), \sup_{\{\alpha : \widetilde{B}_\alpha \geq \widetilde{A}_\alpha\}} (\widetilde{B}_\alpha - \widetilde{A}_\alpha)\}.$$

But

$$\sup_{\{\alpha : \widetilde{A}_\alpha \geq \widetilde{B}_\alpha\}} (\widetilde{A}_\alpha - \widetilde{B}_\alpha) = \sup_{\{\alpha : \max\{0, 0.5(1 - \frac{a-b}{l_a - l_b})\} \leq \alpha \leq 1\}} (a - b + (1-2\alpha)(l_b - l_a))$$
$$= a - b - (l_b - l_a).$$

Similarly, we get:

$$\sup_{\{\alpha : \widetilde{B}_\alpha \geq \widetilde{A}_\alpha\}} (\widetilde{B}_\alpha - \widetilde{A}_\alpha) = \sup_{\{\alpha : 0 \leq \alpha \leq \max\{0, 0.5(1 - \frac{a-b}{l_a - l_b})\}\}} (b - a + (1-2\alpha)(l_a - l_b))$$
$$= b - a + (l_a - l_b).$$

That concludes that:

$$d_\infty(\widetilde{A}, \widetilde{B}) = \max\{a - b - (l_b - l_a), b - a + (l_a - l_b)\} = a - b - (l_b - l_a).$$

Lemma 1.15 *If $\{\widetilde{A}_n\}_{n \in \mathbb{N}}$ is a countable collection of* ***FNs****, then $\widetilde{A}_n \to \widetilde{A}$ if and only if $d_\infty(\widetilde{A}_n, \widetilde{A}) = 0$.*

Proof *The proof is left as an exercise.*

Lemma 1.16 *Let $\widetilde{A}$ be a **FN** and f_1 and f_2 be two continuous functions. If for all $x \in \mathbb{R}$, $f_1(x) < f_2(x)$ then $P_d(f_1(\widetilde{A}) \preceq f_2(\widetilde{A})) = 1$.*

Proof *Since f_1 is a continuous function, according to the extension principal, we have $f_1(\widetilde{A})[\alpha] = f_1(\widetilde{A}[\alpha]) = f_1(x : x \in \widetilde{A}[\alpha])$ for any $\alpha \in [0,1]$. It simply concludes that $(f_1(\widetilde{A}))_\alpha \leq (f_2(\widetilde{A}))_\alpha$ for any $\alpha \in [0,1]$ which completes the proof by the last part of Remark 1.4.*

1.3.1 A generalized difference for FNs

It should be noted that the conventional difference ($\ominus$) between two **FNs** $\widetilde{A}$ and $\widetilde{B}$ is based on the difference between their α-cuts. It takes into account all the possible combinations of difference between two elements a and b, one from the α-level of $\widetilde{A}$ and the other from the α-level of $\widetilde{B}$. Consequently, the result is always greater (in diameter) than any of the sets involved in the operation. For a given **FN** of $\widetilde{A}$, it is observed that

$$(\widetilde{A} \ominus \widetilde{A})[\alpha] = [\widetilde{A}^L - \widetilde{A}^U, \widetilde{A}^U - \widetilde{A}^L].$$

Thus, the subtracting a **FN** from itself is never the exact number zero. In this regard, some studies were conducted to overcome such shortcoming [40, 161, 170, 186]. Here, generalized difference between $\widetilde{A}$ and $\widetilde{B}$ is given and discussed.

Definition 1.22 *[78] Let $\widetilde{A}, \widetilde{B} \in \mathcal{F}(\mathbb{R})$. The generalized difference between $\widetilde{A}$ and $\widetilde{B}$ is defined as a **FN** $\widetilde{A} \ominus_G \widetilde{B}$ with the following α-values:*

$$(\widetilde{A} \ominus_G \widetilde{B})_\alpha = \begin{cases} \inf_{\beta \in [\alpha, 1-\alpha]} (\widetilde{A}_\beta - \widetilde{B}_\beta), & 0.0 \leq \alpha \leq 0.50, \\ \sup_{\beta \in [1-\alpha, \alpha]} (\widetilde{A}_\beta - \widetilde{B}_\beta), & 0.50 \leq \alpha \leq 1.0. \end{cases} \tag{1.36}$$

Theorem 1.9 *For two **FNs** $\widetilde{A}$ and $\widetilde{B}$, $\ominus_G$ meets the following properties:*

1) $\widetilde{A} \ominus_G \widetilde{A} = I\{0\}$.

2) $\widetilde{A} \ominus_G \widetilde{B} = (-1) \otimes (\widetilde{B} \ominus_G \widetilde{A}) = ((-1) \otimes \widetilde{B}) \ominus_G ((-1) \otimes \widetilde{A})$.

3) $(\widetilde{A} \oplus \widetilde{B}) \ominus_G \widetilde{B} = \widetilde{A}$.

4) $\widetilde{A} \ominus_G \widetilde{B} = \widetilde{B} \ominus_G \widetilde{A}$ iff $\widetilde{A} = \widetilde{B}$.

5) $(\widetilde{A} \oplus \widetilde{C}) \ominus_G (\widetilde{B} \oplus \widetilde{C}) = \widetilde{A} \ominus_G \widetilde{B}$.

6) $I\{0\} \ominus_G (\widetilde{A} \ominus \widetilde{B}) = \widetilde{B} \ominus_G \widetilde{A}$.

Proof *Note that:*

$$(\widetilde{A} \ominus_G \widetilde{A})_\alpha = \begin{cases} \inf_{\beta \in [\alpha, 1-\alpha]} (\widetilde{A}_\beta - \widetilde{A}_\beta) = 0, & 0 \leq \alpha \leq 0.5, \\ \sup_{\beta \in [1-\alpha, \alpha]} (\widetilde{A}_\beta - \widetilde{A}_\beta) = 0, & 0.5 \leq \alpha \leq 1, \end{cases}$$

which means $\widetilde{A} \ominus_G \widetilde{A} = I\{0\}$ *and thus part (1) is verified. To establish (2), for two* ***FNs*** $\widetilde{A}$ *and* $\widetilde{B}$, *note that:*

$$((-1) \otimes (\widetilde{B} \ominus_G \widetilde{A}))_\alpha = -(\widetilde{B} \ominus_G \widetilde{A}))_{1-\alpha}$$

$$= \begin{cases} -\inf_{\beta \in [1-\alpha, \alpha]}(\widetilde{B}_\beta - \widetilde{A}_\beta), & 0 \leq 1-\alpha \leq 0.5, \\ -\sup_{\beta \in [\alpha, 1-\alpha]}(\widetilde{B}_\beta - \widetilde{A}_\beta), & 0.5 \leq 1-\alpha \leq 1, \end{cases}$$

$$= \begin{cases} \sup_{\beta \in [1-\alpha, \alpha]}(\widetilde{B}_\beta - \widetilde{A}_\beta), & 0.5 \leq \alpha \leq 1, \\ \inf_{\beta \in [\alpha, 1-\alpha]}(\widetilde{B}_\beta - \widetilde{A}_\beta), & 0 \leq \alpha \leq 0.5, \end{cases}$$

which is equal to $(\widetilde{A} \ominus_G \widetilde{B})_\alpha$ *and therefore* $(-1) \otimes (\widetilde{B} \ominus_G \widetilde{A}) = \widetilde{A} \ominus_G \widetilde{B}$. *In addition:*

$$((-1) \otimes \widetilde{B}) \ominus_G ((-1) \otimes \widetilde{A})_\alpha = \begin{cases} \inf_{\beta \in [\alpha, 1-\alpha]}(-\widetilde{B}_{1-\beta} + \widetilde{A}_{1-\beta}), & 0 \leq \alpha \leq 0.5, \\ \sup_{\beta \in [1-\alpha, \alpha]}(-\widetilde{B}_{1-\beta} + \widetilde{A}_{1-\beta}), & 0.5 \leq \alpha \leq 1, \end{cases}$$

$$= \begin{cases} \inf_{\beta \in [\alpha, 1-\alpha]}(\widetilde{A}_\beta - \widetilde{B}_\beta), & 0 \leq \alpha \leq 0.5, \\ \sup_{\beta \in [1-\alpha, \alpha]}(\widetilde{A}_\beta - \widetilde{B}_\beta), & 0.5 \leq \alpha \leq 1, \end{cases}$$

which is equal to $(\widetilde{A} \ominus_G \widetilde{B})_\alpha$ *and therefore* $((-1) \otimes \widetilde{B}) \ominus_G ((-1) \otimes \widetilde{A}) = \widetilde{A} \ominus_G \widetilde{B}$. *It is readily seen that to to prove assertion (3):*

$$((\widetilde{A} \oplus \widetilde{B}) \ominus_G \widetilde{B})_\alpha$$

$$= \begin{cases} \inf_{\beta \in [\alpha, 1-\alpha]}((\widetilde{A} \oplus \widetilde{B})_\beta - \widetilde{B}_\beta), & 0 \leq \alpha \leq 0.5, \\ \sup_{\beta \in [1-\alpha, \alpha]}((\widetilde{A} \oplus \widetilde{B})_\beta - \widetilde{B}_\beta), & 0.5 \leq \alpha \leq 1, \end{cases}$$

$$= \begin{cases} \inf_{\beta \in [\alpha, 1-\alpha]}((\widetilde{A}_\beta + \widetilde{B}_\beta) - \widetilde{B}_\beta), & 0 \leq \alpha \leq 0.5, \\ \sup_{\beta \in [1-\alpha, \alpha]}((\widetilde{A}_\beta + \widetilde{B}_\beta) - \widetilde{B}_\beta), & 0.5 \leq \alpha \leq 1, \end{cases}$$

$$= \widetilde{A}_\alpha.$$

Thus, $(\widetilde{A} \oplus \widetilde{B}) \ominus_G \widetilde{B} = \widetilde{A}$. *For providing (4), first assume that* $\widetilde{A} \ominus_G \widetilde{B} = \widetilde{B} \ominus_G \widetilde{A}$. *That is* $(\widetilde{A} \ominus_G \widetilde{B})_\alpha = (\widetilde{B} \ominus_G \widetilde{A})_\alpha$ *for any* $\alpha \in [0, 1]$. *It means that*

$$\inf_{\beta \in [\alpha, 1-\alpha]}(\widetilde{A}_\beta - \widetilde{B}_\beta) = \inf_{\beta \in [\alpha, 1-\alpha]} -(\widetilde{A}_\beta - \widetilde{B}_\beta),$$

for $\alpha \in [0, 0.5]$ *and*

$$\sup_{\beta \in [1-\alpha, \alpha]}(\widetilde{A}_\beta - \widetilde{B}_\beta) = \sup_{\beta \in [1-\alpha, \alpha]} -(\widetilde{A}_\beta - \widetilde{B}_\beta),$$

for $\alpha \in [0.5, 1]$ *which conclude that* $\widetilde{A}_\alpha = \widetilde{B}_\alpha$ *for any* $\alpha \in [0, 1]$ *or* $\widetilde{A} = \widetilde{B}$. *Conversely, assume that* $\widetilde{A} = \widetilde{B}$. *It is easily seen that* $\widetilde{A} \ominus_G \widetilde{B} = \widetilde{B} \ominus_G \widetilde{A}$. *To prove (5), note that:*

$$((\widetilde{A} \oplus \widetilde{C}) \ominus_G (\widetilde{B} \oplus \widetilde{C}))_\alpha =$$

$$\begin{cases} \inf_{\beta\in[\alpha,1-\alpha]}((\widetilde{A}\oplus\widetilde{C})_\beta - (\widetilde{B}\oplus\widetilde{C})_\beta), & 0\le\alpha\le 0.5, \\ \sup_{\beta\in[1-\alpha,\alpha]}((\widetilde{A}\oplus\widetilde{C})_\beta - (\widetilde{B}\oplus\widetilde{C})_\beta), & 0.5\le\alpha\le 1, \end{cases}$$

$$= \begin{cases} \inf_{\beta\in[\alpha,1-\alpha]}((\widetilde{A}_\beta + \widetilde{C}_\beta) - (\widetilde{B}_\beta + \widetilde{C}_\beta)), & 0\le\alpha\le 0.5, \\ \sup_{\beta\in[1-\alpha,\alpha]}((\widetilde{A}_\beta + \widetilde{C}_\beta) - (\widetilde{B}_\beta + \widetilde{C}_\beta)), & 0.5\le\alpha\le 1, \end{cases}$$

which is equal to $(\widetilde{A}\ominus_G\widetilde{B})_\alpha$. Finally, to prove 6), we get:

$$(I\{0\}\ominus_G(\widetilde{A}\ominus\widetilde{B}))_\alpha =$$

$$\begin{cases} \inf_{\beta\in[\alpha,1-\alpha]}(0-(\widetilde{A}_\beta - \widetilde{B}_{1-\beta})), & 0\le\alpha\le 0.5, \\ \sup_{\beta\in[1-\alpha,\alpha]}(0-(\widetilde{A}_\beta - \widetilde{B}_{1-\beta})), & 0.5\le\alpha\le 1, \end{cases}$$

$$= \begin{cases} \inf_{\beta\in[\alpha,1-\alpha]}(\widetilde{B}_\beta - \widetilde{A}_{1-\beta}), & 0\le\alpha\le 0.5, \\ \sup_{\beta\in[1-\alpha,\alpha]}(\widetilde{B}_\beta - \widetilde{A}_{1-\beta}), & 0.5\le\alpha\le 1, \end{cases}$$

which is equal to $\widetilde{B}\ominus_G\widetilde{A}$. This complete the proof.

Remark 1.5 *Let $\widetilde{A}$ be a **FN** and k be a constant number. Then, it is easy to show that $\widetilde{A}\ominus_G k = \widetilde{A}\ominus k$.*

Proposition 1.9 *For two **TFNs** $\widetilde{A}=(a;l_a,r_a)_T$ and $\widetilde{B}=(b;l_b,r_b)_T$:*

$$\widetilde{A}\ominus_G\widetilde{B} = (a-b;|l_a-l_b|,|r_a-r_b|)_T. \tag{1.37}$$

Proof *Note that:*

$$(\widetilde{A}\ominus_G\widetilde{B})_\alpha =$$

$$\begin{cases} \inf_{\beta\in[\alpha,1-\alpha]}(a-(1-2\beta)l_a - b + (1-2\beta)l_b), & 0\le\alpha\le 0.5, \\ \sup_{\beta\in[1-\alpha,\alpha]}(a-(1-2\beta)r_a - b + (1-2\beta)r_b), & 0.5\le\alpha\le 1, \end{cases}$$

$$= \begin{cases} a-b+\inf_{\beta\in[\alpha,1-\alpha]}((l_b-l_a)(1-2\beta)), & 0\le\alpha\le 0.5, \\ a-b+\sup_{\beta\in[1-\alpha,\alpha]}((r_b-r_a)(1-2\beta)), & 0.5\le\alpha\le 1, \end{cases}$$

$$= \begin{cases} a-b-(l_b-l_a)I(l_b\ge l_a)(1-2\alpha)+ & \\ (l_b-l_a)I(l_b<l_a)(1-2\alpha), & 0\le\alpha\le 0.5, \\ a-b-(r_b-r_a)I(r_b\ge r_a)(1-2\alpha)+ & \\ (r_b-r_a)I(r_b<r_a)(1-2\alpha), & 0.5\le\alpha\le 1, \end{cases}$$

$$= \begin{cases} a-b-(1-2\alpha)|l_a-l_b|, & 0\le\alpha\le 0.5, \\ a-b-(1-2\alpha)|r_a-r_b|, & 0.5\le\alpha\le 1. \end{cases}$$

*According to α-values of a **TFN**, this simply concludes that $\widetilde{A}\ominus_G\widetilde{B} = (a-b;|l_a-l_b|,|r_a-r_b|)_T$.*

1.3.2 Maximum and minimum of the two FNs

Max and Min are the lattice operations to be used in the ordering of **FN**s. Here a notion of maximum and minimum of two **FN**s are extended based on α-values of **FN**s [71].

Definition 1.23 *Let $\widetilde{A}$ and $\widetilde{B}$ be two **FN**s. The α-values of maximum and minimum of the two **FN**s $\widetilde{A}$ and $\widetilde{B}$ are defined as follows:*

$$(\widetilde{\max}(\widetilde{A}, \widetilde{B}))_\alpha = \max(\widetilde{A}_\alpha, \widetilde{B}_\alpha), \tag{1.38}$$

$$(\widetilde{\min}(\widetilde{A}, \widetilde{B}))_\alpha = \min(\widetilde{A}_\alpha, \widetilde{B}_\alpha). \tag{1.39}$$

Proposition 1.10 *The maximum and minimum of two **FN**s introduced in Definition 1.23 construct two **FN**s.*

Proof *Let $\widetilde{A}$ and $\widetilde{B}$ be two **FN**s. According to the relationship between α-values and α-cuts of **FN**s, we have*

$$\widetilde{\max}(\widetilde{A}, \widetilde{B})[\alpha] = \left[\max(\widetilde{A}_{\alpha/2}, \widetilde{B}_{\alpha/2}), \max(\widetilde{A}_{1-\alpha/2}, \widetilde{B}_{1-\alpha/2})\right],$$

$$\widetilde{\min}(\widetilde{A}, \widetilde{B})[\alpha] = \left[\min(\widetilde{A}_{\alpha/2}, \widetilde{B}_{\alpha/2}), \min(\widetilde{A}_{1-\alpha/2}, \widetilde{B}_{1-\alpha/2})\right].$$

*Since $\widetilde{A}$ and $\widetilde{B}$ are two **FN**s, it can be seen that 1) $\max(\widetilde{A}_{\alpha/2}, \widetilde{B}_{\alpha/2})$ and $\min(\widetilde{A}_{\alpha/2}, \widetilde{B}_{\alpha/2})$ are increasing function of α, and 2) $\max(\widetilde{A}_{1-\alpha/2}, \widetilde{B}_{1-\alpha/2})$ and $\min(\widetilde{A}_{1-\alpha/2}, \widetilde{B}_{1-\alpha/2})$ are decreasing function of α. Also, $\widetilde{\max}(\widetilde{A}, \widetilde{B})[1] = \max\{\widetilde{A}[1], \widetilde{B}[1]\}$ and $\widetilde{\min}(\widetilde{A}, \widetilde{B})[1] = \min\{\widetilde{A}[1], \widetilde{B}[1]\}$. Therefore, 1) $\widetilde{\max}(\widetilde{A}, \widetilde{B})[\alpha]$ and $\widetilde{\min}(\widetilde{A}, \widetilde{B})[\alpha]$ are two non-empty, compact, and decreasing intervals, and 2) there are two unique constants of x_1 and x_2 such that $\widetilde{\min}(\widetilde{A}, \widetilde{B})[1] = x_1$ and $\max\{\widetilde{A}[1], \widetilde{B}[1]\} = x_2$. According to the definition of a **FN**, $\widetilde{\max}(\widetilde{A}, \widetilde{B})$ and $\widetilde{\min}(\widetilde{A}, \widetilde{B})$ are then two **FN**s.*

Example 1.21 *Fig. 1.5 shows the membership functions of two **TFN**s of $\widetilde{A}$ and $\widetilde{B}$. Then, it can be checked that:*

$$\widetilde{\min}(\widetilde{A}, \widetilde{B})[\alpha] = [\min(\widetilde{A}^L_\alpha, \widetilde{B}^L_\alpha), \min(\widetilde{A}^U_\alpha, \widetilde{B}^U_\alpha)]$$

$$= \begin{cases} [a-(1-\alpha)l_a, a+(1-\alpha)r_a], & 0 \le \alpha \le \frac{(b-l_b+a-l_a)}{(l_b-l_a)}, \\ [b-(1-\alpha)l_b, a+(1-\alpha)r_a], & \frac{(b-l_b+a-l_a)}{(l_b-l_a)} \le \alpha \le \frac{a+r_a-(b+r_b)}{r_a-r_b}, \\ [b-(1-\alpha)l_b, b+(1-\alpha)r_b], & \frac{a+r_a-(b+r_b}{r_a-r_b)} \le \alpha \le 1, \end{cases}$$

and

$$\widetilde{\max}(\widetilde{A}, \widetilde{B})[\alpha] = [\max(\widetilde{A}^L_\alpha, \widetilde{B}^L_\alpha), \max(\widetilde{A}^U_\alpha, \widetilde{B}^U_\alpha)]$$

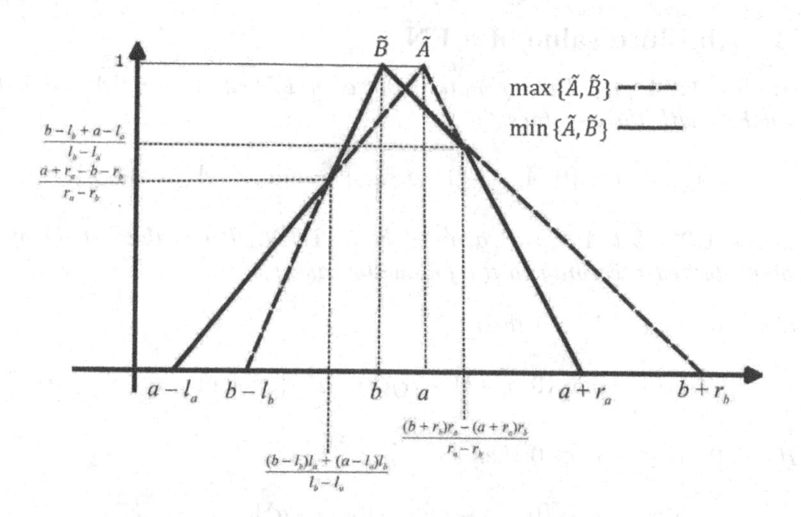

FIGURE 1.5
Membership functions of $\widetilde{\min}(\widetilde{A},\widetilde{B})$ and $\widetilde{\max}(\widetilde{A},\widetilde{B})$ in Example 1.21.

$$
= \begin{cases} [b-(1-\alpha)l_b, b+(1-\alpha)r_b], & 0 \le \alpha \le \frac{(b-l_b+a-l_a)}{(l_b-l_a)}, \\ [a-(1-\alpha)l_a, b+(1-\alpha)r_b], & \frac{(b-l_b+a-l_a)}{(l_b-l_a)} \le \alpha \le \frac{a+r_a-(b+r_b}{r_a-r_b}, \\ [a-(1-\alpha)l_a, a+(1-\alpha)r_a], & \frac{a+r_a-(b+r_b}{r_a-r_b} \le \alpha \le 1. \end{cases}
$$

Therefore, the membership functions of $\widetilde{\min}(\widetilde{A},\widetilde{B})$ and $\widetilde{\max}(\widetilde{A},\widetilde{B})$ are:

$$
\widetilde{\min}(\widetilde{A},\widetilde{B})(x) = \begin{cases} \frac{x-(a-l_a)}{l_a}, & a-l_a x \le \frac{(b-l_b)l_a+(a-l_a)l_b}{l_b-l_a}, \\ \frac{x-(b-l_b)}{l_b}, & \frac{(b-l_b)l_a+(a-l_a)l_b}{l_b-l_a} \le x \le b, \\ \frac{b+r_b-x}{r_b}, & b \le x \le \frac{(b+r_b)r_a-(a+r_a)r_b}{r_b-r_a}, \\ \frac{a+r_a-x}{r_a}, & \frac{(b+r_b)r_a-(a+r_a)r_b}{r_b-r_a} \le x \le b+r_b, \end{cases}
$$

and

$$
\widetilde{\max}(\widetilde{A},\widetilde{B})(x) = \begin{cases} \frac{x-(b-l_b)}{l_b}, & b-l_b \le x \le \frac{(b-l_b)l_a+(a-l_a)l_b}{l_b-l_a}, \\ \frac{x-(a-l_a)}{l_a}, & \frac{(b-l_b)l_a+(a-l_a)l_b}{l_b-l_a} \le x \le a, \\ \frac{a+r_a-x}{r_a}, & a \le x \le \frac{(b+r_b)r_a-(a+r_a)r_b}{r_b-r_a}, \\ \frac{b+r_b-x}{r_b}, & \frac{(b+r_b)r_a-(a+r_a)r_b}{r_b-r_a} \le x \le a+r_a. \end{cases}
$$

The membership functions of $\widetilde{\min}(\widetilde{A},\widetilde{B})$ and $\widetilde{\max}(\widetilde{A},\widetilde{B})$ are also shown in Fig. 1.5.

1.3.3 Absolute value of a FN

Definition 1.24 *[78] The absolute value of a* ***FN*** *of $\widetilde{A}$, say $||\widetilde{A}||$, is defined to be a* ***FN*** *with the α-values:*

$$||\widetilde{A}||_\alpha = \max\{0, \widetilde{A}_\alpha, ((-1) \otimes \widetilde{A})_\alpha\} = \max\{0, \widetilde{A}_\alpha, -\widetilde{A}_{1-\alpha}\}.$$

Example 1.22 *Let $\widetilde{A} = (a^L, a, a^U)_T$ be a* ***TFN****. Then, the α-cuts of $||\widetilde{A}||$ can be evaluated according to the following cases:*

1. *If $a > 0$ and $a - l_a < 0$ then*

$$||\widetilde{A}||[\alpha] = [\max\{0, a - (1-\alpha)(a - a^L)\}, a + (1-\alpha)(a^U - a)].$$

2. *If $a < 0$ and $a + r_a > 0$ then*

$$||\widetilde{A}||[\alpha] = [\max\{0, -a - (1-\alpha)(a^U - a)\}, \max\{-a + (1-\alpha)(a - a^L), a + (1-\alpha)(a^U - a)\}].$$

3. *If $a = 0$ then*

$$||\widetilde{A}||[\alpha] = [0, \max\{(1-\alpha)(a - a^L), (1-\alpha)(a^U - a)\}].$$

4. *If $a + r_a < 0$ then*

$$||\widetilde{A}||[\alpha] = [\max\{0, -a - (1-\alpha)(a^U - a)\}, \max\{-a + (1-\alpha)(a - a^L), a + (1-\alpha)(a^U - a)\}].$$

Therefore, the membership functions of $||\widetilde{A}||$ in each case can be evaluated as follows:

1. *If $a > 0$ and $a - l_a < 0$ then:*

$$||\widetilde{A}||(x) = \begin{cases} \frac{x - a^L}{a - a^L}, & 0 \le x \le a, \\ \frac{a^U - x}{a^U - a}, & a \le x \le a^U. \\ 0, & x \in \mathbb{R} - [0, a^U]. \end{cases}$$

2. *If $a < 0$ and $a + r_a > 0$ then:*

$$||\widetilde{A}||(x) = \begin{cases} 0, & a^L \le x \le 0, \\ \frac{x - a^U}{a^U - a}, & 0 \le x \le a, \\ \frac{a^L - x}{a - a^L}, & a \le x \le a^L. \\ 0, & x > a^L. \end{cases}$$

3. If $a = 0$ then:

$$||\widetilde{A}||(x) = \begin{cases} 0, & x \leq 0, \\ \frac{a^U - x}{a^U - a}, & 0 < x \leq a^U. \\ 0, & x > a^U. \end{cases}$$

4. If $a + r_a < 0$ then $||\widetilde{A}|| = (-a^U, -a, -a^L)_T$.

Lemma 1.17 *If $\widetilde{A}$ be a **FN**, then $P_d(\widetilde{A} \preceq ||\widetilde{A}||) = 1$.*

Proof *For proving, it is enough to put $f_1(x) = x$ and $f_2(x) = |x|$ in Lemma 1.16.*

1.4 Exercise

Exercise 1.1 *For two fuzzy sets of $\widetilde{A}$ and $\widetilde{B}$, prove that $\widetilde{A} \triangle \widetilde{B} = \widetilde{A}^c \triangle \widetilde{B}^c$.*

Exercise 1.2 *For two fuzzy sets of $\widetilde{A}$ and $\widetilde{B}$, prove that $\widetilde{A} - \widetilde{B} = \widetilde{B}^c - \widetilde{A}^c$.*

Exercise 1.3 *For a fuzzy set of $\widetilde{A}$, prove that $(\widetilde{A} \cup \widetilde{A}^c)(x) \geq 0.5$ and $(\widetilde{A} \cap \widetilde{A}^c)(x) \leq 0.5$ for any $x \in \mathbb{R}$.*

Exercise 1.4 *For two fuzzy sets of $\widetilde{A}$ and $\widetilde{B}$, prove that $\widetilde{A} \cap (\widetilde{B} \cup \widetilde{A}) = \widetilde{A}$ and $\widetilde{A} \cup (\widetilde{B} \cap \widetilde{A}) = \widetilde{A}$.*

Exercise 1.5 *Prove that $(\widetilde{A}^c)^c = \widetilde{A}$.*

Exercise 1.6 *Let $\{\widetilde{A}_i\}_{i=1}^n$ be a countable fuzzy sets. For a given fuzzy set of $\widetilde{A}$, show that*

1. $\widetilde{A} - \cup_{i=1}^n \widetilde{A}_i = \cap_{i=1}^n (\widetilde{A} - \widetilde{A}_i)$.
2. $\widetilde{A} - \cap_{i=1}^n \widetilde{A}_i = \cup_{i=1}^n (\widetilde{A} - \widetilde{A}_i)$.

Exercise 1.7 *Consider two fuzzy sets of $\widetilde{A}$ and $\widetilde{B}$ on $\mathbb{X} = \{-2, -1, 0, 1, 2, 3\}$ with the membership functions $\widetilde{A} = \{\frac{0.1}{-2}, \frac{0.4}{-1}, \frac{0.6}{0}, \frac{1}{1}, \frac{0.9}{2}, \frac{0.6}{3}\}$ and $\widetilde{B} = \{\frac{0.5}{-2}, \frac{0.7}{-1}, \frac{1}{0}, \frac{0.8}{1}, \frac{0.6}{2}, \frac{0.3}{3}\}$. Compute $\widetilde{A} \cap \widetilde{B}$, $\widetilde{A} \cup \widetilde{B}$, $\widetilde{B} \triangle \widetilde{A}$, and $\widetilde{A}^c - \widetilde{B}$.*

Exercise 1.8 *For two fuzzy sets of $\widetilde{A}$ and $\widetilde{B}$, prove that $|\widetilde{A} \triangle \widetilde{B}| = |\widetilde{A}| + |\widetilde{B}| - 2|\widetilde{A} \cap \widetilde{B}|$.*

Exercise 1.9 *For two fuzzy sets of $\widetilde{A}$ and $\widetilde{B}$, prove that $\widetilde{A} \times \widetilde{B} = \emptyset$ if and only if $\widetilde{A} = \emptyset$ or $\widetilde{B} = \emptyset$.*

Exercise 1.10 *Let $g : \mathbb{X} \longrightarrow \mathbb{Y}$ be an arbitrary real-valued function. For any $\widetilde{A}_i \in \mathbb{F}(\mathbb{X})$ prove that $g(\cup_{i\in I}\widetilde{A}_i) = \cup_{i\in I} g(\widetilde{A}_i)$ and $g(\cap_{i\in I}\widetilde{A}_i) \subseteq \cap_{i\in I} g(\widetilde{A}_i)$.*

Exercise 1.11 *Let $\mathbb{X} = \{1, 2, ..., 10\}$ and $\widetilde{A}_n(x) = (1 + x/n)^{-n}$ for $x = 1, 2, ..., 10$ and $n \geq 1$. Find $\liminf \widetilde{A}_n$ and $\limsup \widetilde{A}_n$.*

Exercise 1.12 *For two **TFNs** of $\widetilde{A} = (3; 2, 3)_T$ and $\widetilde{B} = (4; 3, 2)_T$, find $\widetilde{\max}\{\widetilde{A}, \widetilde{B}\}$ and $\widetilde{\min}\{\widetilde{A}, \widetilde{B}\}$.*

Exercise 1.13 *For three fuzzy sets of $\widetilde{A}$, $\widetilde{B}$, and $\widetilde{C}$ prove or disprove that $(\widetilde{A} \cap \widetilde{B}) - \widetilde{C} = (\widetilde{A} - \widetilde{C}) \cap (\widetilde{B} - \widetilde{C})$.*

Exercise 1.14 *For two fuzzy sets of $\widetilde{A}$ and $\widetilde{B}$, prove that $\widetilde{A} - \widetilde{B} = \widetilde{A} - (\widetilde{A} \cap \widetilde{B})$.*

Exercise 1.15 *For two fuzzy sets of $\widetilde{A}$ and $\widetilde{B}$, prove that $\widetilde{A} - (\widetilde{A} - \widetilde{B}) = \widetilde{B}$ if and only if $\widetilde{B} \subseteq \widetilde{A}$.*

Exercise 1.16 *For two fuzzy sets of $\widetilde{A}$ and $\widetilde{B}$, prove that $\widetilde{A} - \widetilde{B} = \widetilde{B} - \widetilde{A}$ if and only if $\widetilde{A} = \widetilde{B}$.*

Exercise 1.17 *For two fuzzy sets of $\widetilde{A}$ and $\widetilde{B}$, prove that $\widetilde{A} - \widetilde{B} = \emptyset$ if and only if $\widetilde{A} \subseteq \widetilde{B}$.*

Exercise 1.18 *For two fuzzy sets of $\widetilde{A}$ and $\widetilde{B}$, prove that $\widetilde{A} \cap \widetilde{B} = \emptyset$ if and only if $\widetilde{A} - \widetilde{B} = \widetilde{A}$ if and only if $\widetilde{B} - \widetilde{A} = \widetilde{B}$.*

Exercise 1.19 *For two fuzzy sets of $\widetilde{A}$ and $\widetilde{B}$, prove that $\widetilde{A} \cup \widetilde{B} = (\widetilde{A} \triangle \widetilde{B}) \triangle (\widetilde{A} \cap \widetilde{B})$.*

Exercise 1.20 *For two fuzzy sets of $\widetilde{A}$ and $\widetilde{B}$, prove that $\widetilde{A} \subseteq \widetilde{B}$ if and only if $\widetilde{A} \triangle \widetilde{B} = \widetilde{B} - \widetilde{A}$.*

Exercise 1.21 *For two fuzzy sets of $\widetilde{A}$ and $\widetilde{B}$, prove that $\widetilde{A} \subseteq \widetilde{B}$ if and only if $\widetilde{A} \triangle (\widetilde{B} - \widetilde{A}) = \widetilde{B}$.*

Exercise 1.22 *For two fuzzy sets of $\widetilde{A}$ and $\widetilde{B}$, prove that $\widetilde{A} \triangle \widetilde{B} = \emptyset$ if and only if $\widetilde{A} = \widetilde{B}$.*

Exercise 1.23 *Let $\mathbb{X}_1 = \mathbb{X}_2 = \{1, 2, 3, 4\}$, $\widetilde{A}_1 = \{\frac{0.5}{1}, \frac{1}{2}, \frac{0.9}{3}, \frac{0.7}{4}\}$, and $\widetilde{A}_2 = \{\frac{0.4}{2}, \frac{1}{3}, \frac{0.8}{4}\}$. Compute $\widetilde{A}_1 \times \widetilde{A}_2$, $\widetilde{A}_2^c \times \widetilde{A}_1$, $(\widetilde{A}_1 \cap \widetilde{A}_2) \times (\widetilde{A}_1 - \widetilde{A}_2)$.*

Exercise 1.24 *Let $\{\widetilde{A}_n\}_{n\geq 1}$ be a countable fuzzy sets. For a given fuzzy set $\widetilde{A}$, prove that:*

a) $\widetilde{A} - \liminf \widetilde{A}_n = \limsup(\widetilde{A} - \widetilde{A}_n)$.

b) $\widetilde{A} - \limsup \widetilde{A}_n = \liminf(\widetilde{A} - \widetilde{A}_n)$.

Exercise 1.25 *For three fuzzy sets of $\widetilde{A}$, $\widetilde{B}$, and $\widetilde{C}$ prove that $(\widetilde{A}-\widetilde{B})\times\widetilde{C} = (\widetilde{A}\times\widetilde{C})-(\widetilde{B}\times\widetilde{C})$ is not true in general.*

Exercise 1.26 *For two countable fuzzy sets $\{\widetilde{A}_n\}_{n\geq 1}$ and $\{\widetilde{B}_n\}_{n\geq 1}$, prove that*

1. $\limsup(\widetilde{A}_n \cup \widetilde{B}_n) = \limsup \widetilde{A}_n \cup \limsup \widetilde{B}_n$.
2. $\liminf(\widetilde{A}_n \cap \widetilde{B}_n) = \liminf \widetilde{A}_n \cap \liminf \widetilde{B}_n$.

Exercise 1.27 *Consider a LR-**FN** $\widetilde{A} = (1; 4, 2)_{LR}$ with $L(x) = (1-x)^2$ and $R(x) = 1 - x^3$. Plot the membership function of $||\widetilde{A}||$.*

Exercise 1.28 *Prove Proposition 1.1.*

Exercise 1.29 *Prove Proposition 1.6.*

Exercise 1.30 *Prove Lemma 1.8.*

Exercise 1.31 *Prove Lemma 1.9.*

Exercise 1.32 *Prove Lemma 1.15.*

1.5 Glossary

Fuzzy similarity measure: A fuzzy-valued function to quantify the similarity between two fuzzy number.

Fuzzy distance: A generalization of distance measure between two fuzzy sets.

Fuzzy function: Generalized real function whose values are fuzzy numbers.

Membership function: Mathematical description of a fuzzy set.

Preference degree: A criterion to measure the degree to which a fuzzy number is greater than another one.

2

Probability of a Fuzzy Event

In the models of decision-making under risk, the probability space are often considered. In such ways, we are able to assign probabilities to some precisely defined random events. However, in many real applications, imprecision regarding probability of an event may essentially arise from fuzziness due to partial information such as expert opinions or sparse data sets. In such cases, it is better to describe such events as fuzzy quantities. For instance, consider an urn contains 20 balls $b_1, b_2, ..., b_{20}$ of various sizes. We wish to evaluate the probability that a ball drawn at random is large. However, the concept 'large' is imprecise and it can be demonstrated via fuzzy set on $\mathbb{X} = \{b_1, b_2, ..., b_{20}\}$. Therefore, to evaluate the probability of the event 'large balls', there is a need to extend the conventional probability of an event.

Theoretical probabilistic methods are successfully used in scientific research for modeling many aspects of imprecision using the terms of randomness. There are many situations that events of a random experiment are fuzzy quantities instead of exact ones. For instance 'to get a big yellow apple' in randomly choosing an apple from a basket. For such cases, when we speak about the connections between probabilities and fuzzy events we have two mathematical tools to evaluate the probability of 'to get a big yellow apple'; 1) classical probabilities of 'to get a big yellow apple', and 2) fuzzy probabilities of 'to get a big yellow apple'. Therefore, combining fuzzy set and probability theory can describe human's perception of random experiment to model 'chance' in various applied fields. The probability of a fuzzy event was first introduced by Zadeh [218] to describe the situations with non-sharp boundaries and partial availability or imprecise information (e.g., natural language) which might occur in random experiments. This probability shows the expected value of the membership function of a fuzzy event by providing a flexible criterion to combine fuzziness with randomness. Since then, many studies have reported probability measures based on fuzzy information. Such probability measures can be classified as (1) exact [10, 47, 133, 162, 163, 184, 189, 191, 202, 201] and (2) fuzzy quantities [67, 88, 112, 222, 212, 211] under imprecise information including exact/fuzzy events.

In this chapter, a modified version of exact and fuzzy-valued probability measures of a fuzzy event were proposed as some extensions of Zadeh's probability. For this purpose, two scenario can be addressed. The conventional population is adopted with exact or fuzzy parameters. In practical studies, however, fuzzy parameters may involve in a random experiment. Sometimes,

DOI: 10.1201/9781003248644-2

the distributional parameters involve more information by estimating from expert opinion(s) such as linguistic variables or 'about a number'. For instance, the equipment failure rate is generally used to quantify the probability of failure of a specific piece of equipment. They are statistical values with a lot of subjectivity and are also used to evaluate equipment of similar types. Although equipment status in the future is not predictable, the failure rate of this equipment can be reported (from experience) as some fuzzy quantities.

2.1 Non-fuzzy Probability of a Fuzzy Event

In statistical inference, the data set is viewed as a realization or observation of a random variable of X defined on a probability space $(\Omega, \mathcal{A}, P)$. Now let $X : \Omega \to \mathbb{R}$ be a random variable, where $\mathbb{R}$ is equipped with the σ-algebra $\mathcal{B}(\mathbb{R})$, the set of all Borel subsets of $\mathbb{R}$. The cumulative distribution function (**CDF**) of X is denoted by F_X.

Definition 2.1 *[100] Assuming $X : \Omega \to \mathbb{R}$ is a random variable, the mapping of $\widetilde{A} : \mathcal{B}(\mathbb{R}) \to \mathcal{F}(\mathbb{R})$ is called a fuzzy event (**FE**) if $\widetilde{A}$ is a measurable function. The set of such **FE**s will be denoted by $\mathbb{F}_m(\mathbb{R})$. Moreover, the triple $(\mathcal{B}(\mathbb{R}), \mathbb{F}_m(\mathbb{R}))$ is called a fuzzy measurable space.*

Definition 2.2 *[100] Let $X : \Omega \to \mathbb{R}$ be a random variable. The mapping of $P_X : \mathbb{F}_m(\mathbb{R}) \to [0, 1]$ is called a probability measure if it satisfies the following conditions:*

(i) $P_X(I(\mathbb{R})) = 1$,

(ii) *If $\widetilde{A}, \widetilde{B} \in \mathbb{F}_m(\mathbb{R})$ and $\widetilde{A} \cap \widetilde{B} = \emptyset$ then $P_X(\widetilde{A} \cup \widetilde{B}) = P_X(\widetilde{A}) + P_X(\widetilde{B})$.*

Theorem 2.1 *Let $X : \Omega \to \mathbb{R}$ be a random variable and $\widetilde{A} \in \mathbb{F}_m(\mathbb{R})$. Then, $P_X(\widetilde{A}) = E_X(\widetilde{A}) = \int \widetilde{A} dF_X$ is a probability measure.*

Proof *It is shown that P_X satisfies the conditions of Definition 2.2. To establish (i) note that $P_X(I(\mathbb{R})) = \int_{\mathbb{R}} dF_X = 1$. To provide (ii), for $\widetilde{A}, \widetilde{B} \in \mathbb{F}_m(\mathbb{R})$ with $\widetilde{A} \cap \widetilde{B} = \emptyset$, one has:*

$$\begin{aligned}
P_X(\widetilde{A} \cup \widetilde{B}) &= \int_{\mathbb{R}} \max\{\widetilde{A}(x), \widetilde{B}(x)\} dF_X(x) \\
&= \int_{\mathbb{R}} (\widetilde{A}(x) + \widetilde{B}(x) - \min\{\widetilde{A}(x), \widetilde{B}(x)\}) dF_X(x) \\
&= \int_{\mathbb{R}} (\widetilde{A}(x) + \widetilde{B}(x)) dF_X(x) \\
&= P_X(\widetilde{A}) + P_X(\widetilde{B}).
\end{aligned}$$

Hence, the proof is completed.

Example 2.1 *When one drives a car on a highway, it is hard to say that the speed is exactly 70 km/h, for instance. For such cases, it is better to say that 'the speed is around 70 km/h'. Assume that the membership function of 'the speed is around 70 km/h' can be modeled by a* ***STFN*** $\widetilde{A} = (70; 4)_T$. *If the speed of a car is a random variable with uniform distribution on* $(25, 105)$ *then:*

$$\begin{aligned} P_X(\widetilde{A}) &= \int_{66}^{70} \frac{1}{80}(\frac{x-66}{4})dx + \int_{70}^{74} \frac{1}{80}(\frac{74-x}{4})dx \\ &= 0.025 + 0.025 = 0.05. \end{aligned}$$

Example 2.2 *Consider a random experiment that a person may be in target shooting. Define:*

$$X = \begin{cases} 0, & \textit{missing the target}, \\ 1, & \textit{hitting the } i^{th} \textit{ ring},\ i = 1, 2, \ldots, 10, \end{cases}$$

with probability density of $P_X(X = 0) = 1/6$ *and* $P_X(X = i) = 1/12$, $i = 1, 2, \ldots, 10$. *Assume that* $\widetilde{A} = \{\frac{0.5}{6}, \frac{0.9}{7}, \frac{1}{8}, \frac{0.6}{9}, \frac{0.4}{10}\}$ *represents a* ***FE*** *to describe the notion of 'more or less hitting result'. Hence, we have:*

$$\begin{aligned} P_X(\widetilde{A}) &= \sum_{x=0}^{10} \widetilde{A}(x) P_X(X = x) \\ &= \widetilde{A}(6)P_X(X = 6) + \widetilde{A}(7)P_X(X = 7) + \widetilde{A}(8)P_X(X = 8) \\ &+ \widetilde{A}(9)P_X(X = 9) + \widetilde{A}(10)P_X(X = 10) \\ &= 0.0417 + 0.075 + 0.083 + 0.05 + 0.033 = 0.2827. \end{aligned}$$

Next, some basic results of P_X are explored in a fuzzy domain.

Proposition 2.1 *For two* ***FEs*** *of* $\widetilde{A}$ *and* $\widetilde{B}$, *the probability measure of* P_X *meets the following main properties:*

(1) *If* $\widetilde{A} \subseteq \widetilde{B}$ *then* $P_X(\widetilde{A}) \leq P_X(\widetilde{B})$.

(2) $P_X(\widetilde{A}^c) = 1 - P_X(\widetilde{A})$.

(3) $P_X(\widetilde{A} \cup \widetilde{B}) = P_X(\widetilde{A}) + P_X(\widetilde{B}) - P_X(\widetilde{A} \cap \widetilde{B})$.

(4) $P_X(\widetilde{A} - \widetilde{B}) = P_X(\widetilde{A}) - P_X(\widetilde{A} \cap \widetilde{B})$.

(5) $P_X(\widetilde{A} \triangle \widetilde{B}) = P_X(\widetilde{A}) + P_X(\widetilde{B}) - 2P_X(\widetilde{A} \cap \widetilde{B})$.

Proof *To prove (1), take* $\widetilde{A} \subseteq \widetilde{B}$, $P_X(\widetilde{A}) = E_X(\widetilde{A}) \leq E_X(\widetilde{B}) = P_X(\widetilde{B})$. *For (2), note that* $P_X(\widetilde{A}^c) = E_X(\widetilde{A}^c) = E_X(1 - \widetilde{A}) = 1 - P_X(\widetilde{A})$. *If* $\widetilde{A}$ *and*

$\widetilde{B}$ *are two* **FEs** *then* $P_X(\widetilde{A} \cup \widetilde{B}) = E_X(\widetilde{A} \cup \widetilde{B}) = E_X(\widetilde{A} + \widetilde{B} - \widetilde{A} \cap \widetilde{B}) = P_X(\widetilde{A}) + P_X(\widetilde{B}) - P_X(\widetilde{A} \cap \widetilde{B})$ *which proves (3). Next:*

$$\begin{aligned} P_X(\widetilde{A} - \widetilde{B}) &= \int_{\{\widetilde{A} \geq \widetilde{B}\}} \max\{0, \widetilde{A} - \widetilde{B}\} dF_X + \int_{\{\widetilde{A} < \widetilde{B}\}} \max\{0, \widetilde{A} - \widetilde{B}\} dF_X \\ &= \int (\widetilde{A} - \min\{\widetilde{A}, \widetilde{B}\}) dF_X = P_X(\widetilde{A}) - P_X(\widetilde{A} \cap \widetilde{B}), \end{aligned}$$

which proves (4). Further, it is easy to check that $P_X(\widetilde{A} \triangle \widetilde{B}) = P_X(\widetilde{A} \cup \widetilde{B}) - P_X(\widetilde{B} \cap \widetilde{A}) = P_X(\widetilde{A}) + P_X(\widetilde{B}) - 2P_X(\widetilde{A} \cap \widetilde{B})$*, and hence (5) is verified.*

Example 2.3 *Let* X *be a uniform discrete random variable on* $\{1, 2, \ldots, 10\}$*. Suppose that the fuzzy sets* $\widetilde{A}$ *and* $\widetilde{B}$ *describe the notions of 'small' and '5' with the following membership functions:*

$$\widetilde{A} = \{\frac{1}{1}, \frac{0.8}{2}, \frac{0.6}{3}, \frac{0.3}{4}, \frac{0.1}{5}\}, \quad \widetilde{B} = \{\frac{0.3}{3}, \frac{0.8}{4}, \frac{1}{5}, \frac{0.8}{6}, \frac{0.3}{7}\}.$$

According to Proposition 2.1, it can be easily verified that:

(1) $P_X(\widetilde{A}) = E_X(\widetilde{A}) = 0.28$ *and* $P_X(\widetilde{B}) = E_X(\widetilde{B}) = 0.32$.

(2) $\widetilde{A} \cap \widetilde{B} = \{\frac{0.4}{3}, \frac{0.3}{4}, \frac{0.1}{5}\}$ *and therefore* $P_X(\widetilde{A} \cap \widetilde{B}) = 0.08$.

(3) $P_X(\widetilde{A} \cup \widetilde{B}) = P_X(\widetilde{A}) + P_X(\widetilde{B}) - P_X(\widetilde{A} \cap \widetilde{B}) = 0.52$.

(4) $P_X(\widetilde{A} - \widetilde{B}) = P_X(\widetilde{A}) - P_X(\widetilde{A} \cap \widetilde{B}) = 0.20$.

(5) $P_X(\widetilde{A} \triangle \widetilde{B}) = P_X(\widetilde{A}) + P_X(\widetilde{B}) - 2P_X(\widetilde{A} \cap \widetilde{B}) = 0.44$.

Proposition 2.2 *For a finite mutually disjoint collection of* **FEs** $\{\widetilde{A}_k\}_{k=1}^n$, $P_X(\bigcup_{k=1}^n \widetilde{A}_k) = \sum_{k=1}^n P_X(\widetilde{A}_k)$.

Proof *The proof can be shown by mathematical induction. For* $n = 2$*, it holds applying (3) in Proposition 2.1. Now, for* $n = m$ *assume* $P_X(\bigcup_{k=1}^m \widetilde{A}_k) = \sum_{k=1}^n P_X(\widetilde{A}_k)$*. Since* $\widetilde{A}_i \cap \widetilde{A}_j = \emptyset$ *for* $i \neq j$*:*

$$P_X(\bigcup_{k=1}^{n+1} \widetilde{A}_k) = P_X(\bigcup_{k=1}^{n} \widetilde{A}_k) + P_X(\widetilde{A}_{n+1}) = \sum_{k=1}^{n} P_X(\widetilde{A}_k) + P_X(\widetilde{A}_{n+1}).$$

Therefore the result is true for every integer $n \geq 1$ *which completes the proof.*

Example 2.4 *Let* X *be a uniform random variable distributed on* $[0, 5/8]$ *and* $\widetilde{A}_n = 0.1 \oplus (0.9 \otimes (2^{-n}; 2^{n+2})_T)$*. It is not difficult to show that* $P_X(\widetilde{A}_k) = \frac{11}{25} 2^{-k}$*. Since* $\widetilde{A}_n \cap \widetilde{A}_{n'} = \emptyset$ *for any* $n \neq n'$*, from Proposition 2.2, it follows that* $P_X(\bigcup_{k=1}^n \widetilde{A}_k) = \sum_{k=1}^n P_X(\widetilde{A}_k) = \sum_{k=1}^n \frac{11}{25} 2^{-k} = \frac{22}{25}(1 - 2^{-n})$.

An important continuity property of the probability measure of P_X is explored as follows.

Proposition 2.3 *For a countable sequence of* ***FEs*** $\{\widetilde{A}_n\}_{n=1}^{\infty}$:

(1) *if* $\widetilde{A}_n \uparrow \widetilde{A}$ *then* $P_X(\widetilde{A}_n) \uparrow P_X(\widetilde{A})$.

(2) *if* $\widetilde{A}_n \downarrow \widetilde{A}$ *then* $P_X(\widetilde{A}_n) \downarrow P_X(\widetilde{A})$.

Proof *First note that* $\widetilde{A}$ *is a* ***FE***. *Considering the monotone convergence theorem (M.C.T) [13] and the fact that* $\widetilde{A}_n \downarrow \widetilde{A}$, $P_X(\widetilde{A}_n) = E_X(\widetilde{A}_n) \downarrow E_X(\widetilde{A}) = P_X(\widetilde{A})$ *as* $n \to \infty$, *the first part (1) will be proven. The second part can be verified similarly.*

Example 2.5 *Suppose that X is a random variable distributed uniformly over* $(0, 1)$. *Let* $\widetilde{A}_n = (0.5; 0.2 + 2^{-n}, 0.3 + 3^{-n})_T$ *and* $\widetilde{A} = (0.5; 0.2, 0.3)_T$. *Since* $\widetilde{A}_n \downarrow \widetilde{A}$, *by Proposition 2.3,* $\lim_{n\to\infty} P_X(\widetilde{A}_n) = P_X(\widetilde{A}) = E_X(\widetilde{A}) = 0.25$. *To check this via a direct computation:*

$$\begin{aligned} P_X(\widetilde{A}_n) &= E_X(\widetilde{A}_n) \\ &= \frac{0.5(0.025 - (0.3 + 2^{-n})^2) + (-0.3 - 2^{-n})(0.2 - 2^{-n})}{0.2 + 2^{-n}} \\ &\quad - \frac{0.5(-0.25 + (0.8 + 3^{-n})^2)}{0.3 + 3^{-n}} + 0.8 + 3^{-n}. \end{aligned}$$

This yields $\lim_{n\to\infty} P_X(\widetilde{A}_n) = 0.25 = P_X(\widetilde{A})$.

Proposition 2.4 *Let* $\{\widetilde{A}_n\}_{n=1}^{\infty}$ *be a countable collection of* ***FEs***. *Then:*

(1) *Suppose h is a strictly decreasing function on* $\mathbb{R}$. *If* $\widetilde{A}_n \downarrow \widetilde{A}$ *then* $P_X(h(\widetilde{A}_n)) \downarrow P_X(h(\widetilde{A}))$.

(2) *Assume that h is a strictly increasing function on* $\mathbb{R}$. *If* $\widetilde{A}_n \uparrow \widetilde{A}$ *then* $P_X(h(\widetilde{A}_n)) \uparrow P_X(h(\widetilde{A}))$.

Proof *Since* $h(\widetilde{A}_n) \downarrow h(\widetilde{A})$, $P_X(h(\widetilde{A}_n)) \downarrow P_X(h(\widetilde{A}))$ *by Proposition 2.3. This verifies (1). The item (2) can be also verified similar to that of (1).*

Proposition 2.5 *Let* $\widetilde{A}$ *be a* ***FE***.

(1) *Consider a sequence of strictly decreasing functions* $\{h_n\}_{n=1}^{\infty}$ *with* $\lim_{n\to\infty} h_n(x) = h(x)$ *for all x. Then* $P_X(h_n(\widetilde{A})) \downarrow P_X(h(\widetilde{A}))$.

(2) *Consider a sequence of strictly increasing functions* $\{h_n\}_{n=1}^{\infty}$ *with* $\lim_{n\to\infty} h_n(x) = h(x)$ *for all x. Then* $P_X(h_n(\widetilde{A})) \uparrow P_X(h(\widetilde{A}))$.

Proof *We only prove (1). For this purpose, by hypothesis, we simply have* $h_n(\widetilde{A})) \downarrow h(\widetilde{A})$ *and hence* $P_X(h_n(\widetilde{A})) \downarrow P_X(h(\widetilde{A}))$ *by Proposition 2.3.*

Example 2.6 *Let $X \sim U(0,5)$, $h_n(x) = (1+1/n)x$ for $x \geq 0$ and $\widetilde{A} = (2;1,2)_T$. Since $h_n(x) \downarrow h(x) = x$ and $\widetilde{A}$ has a continuous membership function, it follows that $P_X(h_n(\widetilde{A})) \downarrow P_X(h(\widetilde{A}))$ according to Proposition 2.5. To check this using numerical evaluations, one can verify that:*

$$h_n(\widetilde{A})(y) = \widetilde{A}(\frac{ny}{n+1}) = \begin{cases} \frac{ny}{n+1} - 1, & 1+\frac{1}{n} \leq y \leq 2+\frac{1}{n}, \\ \frac{4n+1-ny}{2(n+1)}, & 2+\frac{1}{n} \leq x \leq 4+\frac{1}{n}, \\ 0, & x \in \mathbb{R} - [1+\frac{1}{n}, 4+\frac{1}{n}], \end{cases}$$

and $h(\widetilde{A}) = \widetilde{A}$. Further, it can be shown that $P_X(h(\widetilde{A})) = \int_0^5 \frac{1}{5}\widetilde{A}(x)dx = 0.3$ and

$$\begin{aligned} P_X(h_n(\widetilde{A})) &= \int_0^5 \frac{1}{5} h_n(\widetilde{A})(y)dy \\ &= \frac{1}{5}(\int_{1+1/n}^{2+1/n} (\frac{ny}{n+1} - 1)dy + \int_{2+1/n}^{4+1/n} (\frac{4n+1-ny}{2(n+1)})dy) \\ &= \frac{1}{5}\Big[\frac{n}{2(n+1)}((2+1/n)^2 - (1+1/n)^2) - 1 \\ &+ \frac{8n+2}{2n+1} - \frac{n}{4(n+1)}((4+1/n)^2 - (2+1/n)^2)\Big]. \end{aligned}$$

Taking limit on both of above equations results in $\lim_{n\to\infty} P_X(h_n(\widetilde{A})) = \frac{1}{5}(\frac{4-1}{2} - 1 + 4 - \frac{16-4}{4}) = 0.3$ and hence $\lim_{n\to\infty} P_X(h_n(\widetilde{A})) = P_X(h(\widetilde{A}))$.

Proposition 2.6 *Let $\{\widetilde{A}_k\}_{k=1}^n$ be a finite disjoint collection of* **FE***s. If $\bigcup_{k=1}^n \widetilde{A}_k \subseteq \widetilde{A}$ and $\bigcup_{k=1}^n \widetilde{A}_k \to \widetilde{A}$ for all $n = 1,2,....$ Then $P_X(\bigcup_{k=1}^\infty \widetilde{A}_k) = \sum_{k=1}^\infty P_X(\widetilde{A}_k)$.*

Proof *Let $\widetilde{S}_m(n) = \bigcup_{k=n+1}^m \widetilde{A}_k$ where n is a fixed integer, $m = n+1, n+2, \ldots$, and $\{\widetilde{S}_{n+k}(n)\}_{k=1}^\infty$ is an increasing collection of* **FE***s. Define $\widetilde{Q}_n = \lim_{m\to\infty} \widetilde{S}_m(n)$ and fix $m \geq n+1$. Since $\widetilde{S}_m(n) \supseteq \widetilde{S}_m(n+1)$ $\forall n \geq 1$, it follows that $\widetilde{S}_m(n) \downarrow \emptyset$. Now, let $\bigcup_{k=1}^m \widetilde{A}_k = (\bigcup_{k=1}^n \widetilde{A}_k) \cup \widetilde{S}_m(n)$ for $m = n+1, n+2, ...$ It is easy to show that $(\bigcup_{k=1}^n \widetilde{A}_k) \cup \widetilde{Q}_n \downarrow \emptyset$. Since $\widetilde{A}_i \cap \widetilde{A}_j = \emptyset$ $\forall i \neq j$, from proposition 2.2, we will have $P_X(\bigcup_{k=1}^\infty \widetilde{A}_k) = \sum_{k=1}^n P_X(\widetilde{A}_k) + P_X(\widetilde{Q}_n)$ for any n. This simply concludes that $P_X(\bigcup_{k=1}^\infty \widetilde{A}_k) = \sum_{k=1}^\infty P_X(\widetilde{A}_k)$.*

Next, an extended Fatou's lemma is explored.

Proposition 2.7 *Let $\{\widetilde{A}_k\}_{k=1}^\infty$ be a collection of* **FE***s which converges to $\widetilde{A} \in \mathbb{F}_m(\mathbb{R})$. For any x, if $\widetilde{A}(x) = \liminf_{n\to\infty} \widetilde{A}_n(x)$, then $P_X(\widetilde{A}) \leq \liminf_{n\to\infty} P_X(\widetilde{A}_n)$.*

Proof *Writing* $\widetilde{h}_n = \bigcap_{k\geq n} \widetilde{A}_k$, *simply concludes that* $\lim_{n\to\infty} E_X(\widetilde{h}_n) \leq \liminf_{n\to\infty} E_X(\widetilde{A}_n)$. *By proposition 2.3,* $\widetilde{h}_n \uparrow \widetilde{A}$ *implies* $P_X(\widetilde{h}_n) \uparrow P_X(\widetilde{A})$*; hence:*

$$P_X(\widetilde{A}) = E_X(\widetilde{A}) = lim_{n\to\infty} E_X(\widetilde{h}_n) \leq \liminf_{n\to\infty} E_X(\widetilde{A}_n) = \liminf_{n\to\infty} P_X(\widetilde{A}_n).$$

The following is an extension of Borel-Cantelli Lemma in the fuzzy domain.

Lemma 2.1 *Let* $\{\widetilde{A}_k\}_{k=1}^{\infty}$ *be a countable collection of* ***FEs****. If* $\sum_{n=1}^{\infty} P_X(\widetilde{A}_n) < \infty$ *then* $P_X(\limsup_{n\to\infty} \widetilde{A}_n) = 0$.

Proof *By hypothesis, since* $\sum_{n=1}^{\infty} P_X(\widetilde{A}_n) < \infty$*:*

$$\begin{aligned} P_X(\limsup_{n\to\infty} \widetilde{A}_n) &= \int \lim_{n\to\infty} (\cup_{k\geq n} \widetilde{A}_k)(x) dF_X(x) \\ &= \lim_{n\to\infty} P_X(\cup_{k=n}^{\infty} \widetilde{A}_k) \\ &\leq \lim_{n\to\infty} \sum_{k=n}^{\infty} P_X(\widetilde{A}_k) = 0. \end{aligned}$$

Thus, the proof is complete.

Remark 2.1 *Assume that* X *is a random variable with a parametric density function of* $f_X^{\boldsymbol{\theta}}$, $\boldsymbol{\theta} = (\theta_1, \theta_2, \ldots, \theta_p)^\top \in \mathbb{R}^p$. *As remarked in chapter one, some of* θ_j *values may not be exact values. For instance, assume that* X *denotes the expected time (in hours) to do a certain task, with uniform distribution on* $(0, \theta)$. *However, due to uncertainty related to the exact estimation, it is hard to imagine that* θ *has an exact mean of 10 hours. For such a case, it is better to say that* θ *has a mean of about 10 hours, making it well modelable which can be modeled by a* ***TFN*** *as 'about 10 hours'. Now, consider the cases that* X *is a random variable with the density function of* $f_X^{\boldsymbol{\theta}}$, $\boldsymbol{\theta} = (\theta_1, \theta_2, \ldots, \theta_p)^\top \in \mathbb{R}^p$ *in which some of the parameters are reported as '*θ_i *is about* $\widetilde{\theta}_i$*'. Without loose of generality, assume that* $\widetilde{\boldsymbol{\theta}} = (\widetilde{\boldsymbol{\theta}}_0, \boldsymbol{\theta}_1)^\top$ *where* $\widetilde{\boldsymbol{\theta}} = (\widetilde{\theta}_1^0, \widetilde{\theta}_2^0, \ldots, \widetilde{\theta}_k^0)^\top$ *and* $\boldsymbol{\theta}_1 = (\theta_{k+1}, \ldots, \theta_p)^\top$ *with* $k \leq p$. *For such cases, Zadeh's probability can be extended as follows.*

Definition 2.3 *Let* X *be a random variable with a density function of* $f_X^{\boldsymbol{\theta}}$, $\boldsymbol{\theta} = (\theta_1, \theta_2, \ldots, \theta_p)^\top \in \mathbb{R}^p$. *The Zadeh's fuzzy probability with fuzzy parameters of* $\widetilde{\boldsymbol{\theta}}$ *can be defined as follows:*

$$\widetilde{P}_X^{\widetilde{\boldsymbol{\theta}}}(\widetilde{A}) = E_W\left[E_{\pi_{\widetilde{\boldsymbol{\theta}}_0}}\left[P_X^{\widetilde{\boldsymbol{\theta}}_0}(\widetilde{A}[W])\right]\right], \tag{2.1}$$

where W *is a random variable uniformly distributed on* $(0, 1)$ $(W \sim U(0, 1))$ *and*

$$\pi_{\widetilde{\boldsymbol{\theta}}_0}(\boldsymbol{z}) = \frac{\prod_{i=1}^{k} \widetilde{\theta}_i^0(z_i)}{\int \prod_{i=1}^{k} \widetilde{\theta}_i^0(z_i) d\boldsymbol{z}},$$

can be considered as the joint density function of $(\widetilde{\theta}_1^0, \widetilde{\theta}_2^0, \ldots, \widetilde{\theta}_k^0)^\top$ *at* $\boldsymbol{z} = (z_1, z_2, ..., z_k)^\top$.

Based on the aforementioned notations, $\widetilde{P}_X^{\widetilde{\boldsymbol{\theta}}}(\widetilde{A})$ can be demonstrated as follows:

$$\widetilde{P}_X^{\widetilde{\boldsymbol{\theta}}}(\widetilde{A}) = \int_0^1 \int P_X^{\widetilde{\boldsymbol{\theta}}_0}(\widetilde{A}[w])\pi_{\widetilde{\boldsymbol{\theta}}_0}(\boldsymbol{z})d\boldsymbol{z}dw.$$

Theorem 2.2 *Let* $(\Omega, \mathcal{B}(\mathbb{R}), P)$ *be a probability space. Then,* $\widetilde{P}_X^{\widetilde{\boldsymbol{\theta}}}$ *meets the requirements of Theorem 2.1.*

Proof *The verification of this claim is left as an exercise.*

Example 2.7 *Assume that X a normal random variable with parameters of* μ *and* σ^2 *where* μ *and* σ^2 *are* ***TFNs*** *as* $\widetilde{\mu} = (0; 1, 1)_T$ *and* $\widetilde{\sigma}^2 = (1; 0.5, 0.5)_T$. *Consider probability of a* ***FE*** *as 'X is about −1' with a membership function of* $\widetilde{A} = (-1; 0.3, 0.4)_T$. *For this purpose, first, note that* $\widetilde{\boldsymbol{\theta}} = (\widetilde{\mu}, \widetilde{\sigma}^2)^\top$ *and* $\boldsymbol{z} = (\mu, \sigma^2)^\top$ *and*

$$\pi_{\widetilde{\boldsymbol{\theta}}}(\boldsymbol{z}) = \frac{\widetilde{\mu}(\mu)\widetilde{\sigma}^2(\sigma^2)}{|\widetilde{\mu}||\widetilde{\sigma}^2|} = 2\widetilde{\mu}(\mu)\widetilde{\sigma}^2(\sigma^2).$$

Therefore, one has:

$$E_{\pi_{\widetilde{\boldsymbol{\theta}}}}\left[P_X^{\widetilde{\boldsymbol{\theta}}}(\widetilde{A}[W])\right] =$$

$$\int_{-\infty}^{\infty}\int_0^{\infty}\int_{0.3w-1.3}^{-0.6-0.4w} \frac{2}{\sqrt{2\pi\sigma^2}}\widetilde{\mu}(\mu)\widetilde{\sigma}^2(\sigma^2)\exp\left(\frac{-(x-\mu)^2}{2\sigma^2}\right)dxd\sigma^2d\mu.$$

This simply concludes to:

$$\begin{aligned}
\widetilde{P}_X^{\widetilde{\boldsymbol{\theta}}}(\widetilde{A}) &= E_W\left[E_{\pi_{\widetilde{\boldsymbol{\theta}}}}\left[P_X^{\widetilde{\boldsymbol{\theta}}}(\widetilde{A}[W])\right]\right] \\
&= \int_0^1 E_{\pi_{\widetilde{\boldsymbol{\theta}}}}\left[P_X^{\widetilde{\boldsymbol{\theta}}}(\widetilde{A}[w])\right]dw \\
&= \int_0^1\int_{-\infty}^{\infty}\int_0^{\infty}\int_{0.3w-1.3}^{-0.6-0.4w} \frac{2}{\sqrt{2\pi\sigma^2}}\widetilde{\mu}(\mu)\widetilde{\sigma}^2(\sigma^2)\exp\left(\frac{-(x-\mu)^2}{2\sigma^2}\right) \\
&\quad \times dxd\sigma^2d\mu dw \\
&= 0.048.
\end{aligned}$$

Example 2.8 *Let X be a binomial random variable with* $n = 7$ *and the fuzzy parameter of* $\widetilde{p} = (0.85; 0.05, 0.03)_{LR}$ *with* $L(x) = \sqrt{1-x^2}$ *and* $R(x) = (1-x)^2$. *Consider a* ***FE*** *to describe 'X is about 5' with membership function* $\widetilde{A} = \{\frac{0.6}{4}, \frac{1}{5}, \frac{0.6}{6}\}$. *To compute* $\widetilde{P}_X^{\widetilde{p}}(\widetilde{A})$, *first, note that:*

$$\widetilde{p}(p) = \begin{cases} \sqrt{1-(\frac{0.85-p}{0.05})^2}, & 0.80 \le p \le 0.85, \\ 1-(\frac{p-0.85}{0.03})^2, & 0.85 < p \le 0.88, \\ 0, & p \in [0,1] - (0.80, 0.88). \end{cases}$$

This concludes that:

$$\int_0^1 \widetilde{p}(p)dp = \int_{0.8}^{0.85} \sqrt{1-(\frac{0.85-p}{0.05})^2}dp + \int_{0.85}^{0.88} 1-(\frac{p-0.85}{0.03})^2 dp = 0.046,$$

and

$$\pi(p) = \begin{cases} \frac{\sqrt{1-(\frac{0.85-p}{0.05})^2}}{0.046}, & 0.80 \le p \le 0.85, \\ \frac{1-(\frac{p-0.85}{0.03})^2}{0.046}, & 0.85 < p \le 0.88, \\ 0, & p \in [0,1]-(0.80,0.88). \end{cases}$$

Thus, according to Definition 2.3,

$$\begin{aligned} &\widetilde{P}_X^{\widetilde{p}}(\widetilde{A}) \\ &= E_W\left[E_{\pi_{\widetilde{p}}}\left[P_X^{\widetilde{p}}(\widetilde{A}[W])\right]\right] \\ &= \int_0^1 \sum_{x=4}^{7} \binom{7}{x} \pi(p)\widetilde{A}(x)p^x(1-p)^x dp \\ &= \frac{1}{0.0467}\int_0^1 \widetilde{p}(p)(0.6\binom{7}{4}p^4(1-p)^6 + \binom{7}{5}p^5(1-p)^5 \\ &+ 0.6\binom{7}{6}p^6(1-p)^4)dp = 0.0014. \end{aligned}$$

2.2 Probability of a Fuzzy Product Event

Let $(\Omega^n = \Omega \times \Omega \times ... \times \Omega, \mathcal{A}^n = \mathcal{A} \times \mathcal{A} \times ...\mathcal{A})$ be a product space and $\boldsymbol{X} = (X_1, X_2, \ldots, X_n)$ be the vector function from this product space to $(\mathbb{R}^n, (\mathcal{B}(\mathbb{R}))^n)$. Here, inspired by Jershan and Yao [100], a notion of probability of a fuzzy product event can be rewritten according to our notion of a fuzzy event. The main results of this section are explored based on [100].

For a finite collection of **FEs** $\widetilde{A}_i$ in $\mathcal{B}(\mathbb{R})$, the fuzzy product event is considered as a Cartesian product of the **FPE** $\widetilde{\boldsymbol{A}} = \widetilde{A}_1 \times \widetilde{A}_2 \times ... \times \widetilde{A}_n$, say $\widetilde{\boldsymbol{A}} \in (\mathcal{B}(\mathbb{R}))^n$, with the following membership function:

$$\widetilde{\boldsymbol{A}}(\boldsymbol{x}) = \min_{i=1}^{n}\{\widetilde{A}_i(x_i)\}, \ \boldsymbol{x} = (x_1, x_2, ..., x_n), \ x_i \in \mathbb{R}.$$

Definition 2.4 *Let* $\widetilde{\boldsymbol{A}} = \widetilde{A}_1 \times \widetilde{A}_2 \times ... \times \widetilde{A}_n$ *be a* ***FPE***. *The probability of a* $\widetilde{\boldsymbol{A}}$ *is defined by:*

$$P_{\boldsymbol{X}}(\widetilde{\boldsymbol{A}}) = \int \widetilde{\boldsymbol{A}} dF_{\boldsymbol{X}}.$$

Example 2.9 *[100] Consider driving a small car and a truck. There are three kinds of velocities, around 40 km/h, around 50 km/h, and around 60 km/h. For the small car, the membership functions of the three kinds of velocities are:*

$$\widetilde{A}_{1j_1} = (k_1; 1, 4)_T, \ j_1 = \frac{k_1 - 30}{10}, \ k_1 = 40, 50, 60.$$

For the truck, the membership functions for the three kinds of velocities are:

$$\widetilde{A}_{2j_2} = (k_2; 2, 2)_T, \ j_2 = \frac{k_2 - 30}{10}, \ k_2 = 40, 50, 60.$$

In such cases, we can generate four fuzzy product events of $\boldsymbol{C}_{j_1 j_2} = \widetilde{A}_{j_1} \times \widetilde{B}_{j_2}$. *In this way, we have:*

$$\boldsymbol{C}_{j_1 j_2}(x_1, x_2) = \begin{cases} \frac{x_1 - k_1 + 4}{4}, & k_1 - 4 \le x_1 \le k_1, x_1 + k_2 - k_1 + 2 \le x_2 \le k_2 + 2, \\ \frac{x_2 - k_2 + 2}{4}, & k_1 - 4 \le x_1 \le k_1, k_2 - 2 \le x_2 \le x_1 + k_2 - k_1 + 2, \\ 0, & otherwise. \end{cases}$$

$$\boldsymbol{C}_{j_1 j_2}(x_1, x_2) = \begin{cases} \frac{x_1 - k_1 + 4}{4}, & k_1 - 4 \le x_1 \le k_1, k_2 + 2 \le x_2 \le k_1 + k_2 + 2 - x_1, \\ \frac{k_2 + 6 - x_2}{4}, & k_1 - 4 \le x_1 \le k_1, k_1 + k_2 + 2 - x_1 \le x_2 \\ & \le x_1 + k_2 + 6, \\ 0, & otherwise. \end{cases}$$

$$\boldsymbol{C}_{j_1 j_2}(x_1, x_2) = \begin{cases} \frac{k_1 + 4 - x_1}{4}, & k_1 \le x_1 \le k_1 + 4, k_1 + k_2 + 2 - x_1 \le k_2 + 2, \\ \frac{x_2 - k_2 + 2}{4}, & k_1 \le x_1 \le k_1 + 4, k_2 + 2 \le x_2 \le k_1 + k_2 + 2 - x_1, \\ 0, & otherwise. \end{cases}$$

$$\boldsymbol{C}_{j_1 j_2}(x_1, x_2) = \begin{cases} \frac{k_1 + 4 - x_1}{4}, & k_1 \le x_1 \le k_1 + 4, k_2 + 2 \le x_2 \le x_1 + k_2 - k_1 + 2, \\ \frac{k_2 + 6 - x_2}{4}, & k_1 \le x_1 \le k_1 + 4, x_1 + k_2 - k_1 + 2 \le x_2 \le k_2 + 6. \\ 0, & otherwise. \end{cases}$$

Let $\boldsymbol{X} = (X_1, X_2)$ *be the bivariate uniform distribution function on* $[36, 64] \times [38, 66]$. *Therefore, one can show that:*

$$P_{\boldsymbol{X}}(\boldsymbol{C}_{11}) = \frac{1}{784} \int_{36}^{64} \int_{38}^{68} \boldsymbol{C}_{11}(x_1, x_2) dx_2 dx_1 = \frac{1}{49}.$$

We can treat other cases in a similar way.

The following results summarized some of the main properties of $P_{\boldsymbol{X}}$.

Proposition 2.8 *For two* ***FPEs*** $\widetilde{\boldsymbol{A}}_1 = \widetilde{A}_{11} \times \widetilde{A}_2 \times \ldots \times \widetilde{A}_n$ *and* $\widetilde{\boldsymbol{A}}_2 = \widetilde{A}_{22} \times \widetilde{A}_2 \times \ldots \times \widetilde{A}_n$ *if* $\widetilde{A}_{11} \cap \widetilde{A}_{22} = \emptyset$ *then* $P_{\boldsymbol{X}}(\widetilde{\boldsymbol{A}}_1 \cup \widetilde{\boldsymbol{A}}_2) = P_{\boldsymbol{X}}(\widetilde{\boldsymbol{A}}_1) + P_{\boldsymbol{X}}(\widetilde{\boldsymbol{A}}_2)$.

Proof *First note that:*

$$P_{\boldsymbol{X}}(\widetilde{\boldsymbol{A}}_1 \cup \widetilde{\boldsymbol{A}}_2) = P_{\boldsymbol{X}}((\widetilde{A}_{11} \cup \widetilde{A}_{22}) \times \widetilde{A}_2 \times \ldots \times \widetilde{A}_n).$$

Now, setting $\widetilde{\boldsymbol{C}} = (\widetilde{A}_{11} \cup \widetilde{A}_{22}) \times \widetilde{A}_2 \times ... \times \widetilde{A}_n$, $\widetilde{A}_{11} \cap \widetilde{A}_{22} = \emptyset$ *yields* $\widetilde{\boldsymbol{A}}_1 \cap \widetilde{\boldsymbol{A}}_2 = \emptyset$ *and thus:*

$$\begin{aligned} P_{\boldsymbol{X}}(\widetilde{\boldsymbol{A}}_1 \cup \widetilde{\boldsymbol{A}}_2) &= P_{\boldsymbol{X}}(\widetilde{\boldsymbol{C}}) \\ &= \int_{\widetilde{A}_{11} \geq \widetilde{A}_{22}} \widetilde{\boldsymbol{C}} dF_{\boldsymbol{X}} + \int_{\widetilde{A}_{11} < \widetilde{A}_{22}} \widetilde{\boldsymbol{C}} dF_{\boldsymbol{X}} \\ &= P_{\boldsymbol{X}}(\widetilde{\boldsymbol{A}}_1) + P_{\boldsymbol{X}}(\widetilde{\boldsymbol{A}}_2). \end{aligned}$$

Proposition 2.9 *Let* $\{\widetilde{\boldsymbol{A}}_j = \widetilde{A}_{1j} \times \widetilde{A}_{2j} \times ... \times \widetilde{A}_{nj}\}_{j=1}^{m}$ *be a sequence of* ***FPEs*** *where* $\widetilde{A}_{kj} \cap \widetilde{A}_{2j'} = \emptyset$ *for every* $k = 1, 2, \ldots, n$ *and* $j \neq j' \in \{1, 2, \ldots, m\}$. *Then* $P_{\boldsymbol{X}}(\bigcup_{j=1}^{m} \widetilde{\boldsymbol{A}}_j) = \sum_{j=1}^{m} P_{\boldsymbol{X}}(\widetilde{\boldsymbol{A}}_j)$.

Proof *Note that* $P_{\boldsymbol{X}}(\bigcup_{j=1}^{m} \widetilde{\boldsymbol{A}}_j) = P_{\boldsymbol{X}}(\bigcup_{j=1}^{m} \widetilde{A}_{1j} \times \ldots \times \bigcup_{j=1}^{m} \widetilde{A}_{nj})$. *For every* $k = 1, 2, \ldots, n$, *since* $\widetilde{A}_{kj} \cap \widetilde{A}_{2j'} = \emptyset$, $j \neq j' \in \{1, 2, \ldots, m\}$, *the proof is a simple conclusion of Proposition 2.8 through a simple mathematical induction on* m.

Proposition 2.10 *Let* $\{\widetilde{\boldsymbol{A}}_j = \widetilde{A}_{1j} \times \widetilde{A}_{2j} \times ... \times \widetilde{A}_{nj}\}_{j=1}^{m}$ *be a finite collection of* ***FPEs*** *in which* $\widetilde{\boldsymbol{A}}_j \uparrow \boldsymbol{A} \in (\mathcal{B}(\mathbb{R}))^n$, *say* $\boldsymbol{A} = \bigcup_{j=1}^{\infty} \widetilde{\boldsymbol{A}}_j$. *If* $\widetilde{\boldsymbol{A}}_j \cap \widetilde{\boldsymbol{A}}_{j'} = \emptyset$, *for any* $j \neq j' \in \{1, 2, \ldots, m\}$, *then* $P_{\boldsymbol{X}}(\bigcup_{j=1}^{\infty} \widetilde{\boldsymbol{A}}_j) = \sum_{j=1}^{\infty} P_{\boldsymbol{X}}(\widetilde{\boldsymbol{A}}_j)$.

Proof *The proof is similar to that of Proposition 2.6.*

Proposition 2.11 *Let* $\{\widetilde{\boldsymbol{A}}_j = \widetilde{A}_{1j} \times \widetilde{A}_{2j} \times ... \times \widetilde{A}_{nj}\}_{j=1}^{\infty}$ *be a finite collection of* ***FPEs***. *Then,*

(1) *If* $\widetilde{\boldsymbol{A}}_j \uparrow \widetilde{\boldsymbol{A}} = \widetilde{A}_1 \times \widetilde{A}_2 \times ... \times \widetilde{A}_n \in (\mathcal{B}(\mathbb{R}))^n$. *then* $\lim_{j\to\infty} P_{\boldsymbol{X}}(\widetilde{\boldsymbol{A}}_j) \uparrow P_{\boldsymbol{X}}(\widetilde{\boldsymbol{A}})$.

(2) *If* $\widetilde{\boldsymbol{A}}_j \downarrow \widetilde{\boldsymbol{A}} = \widetilde{A}_1 \times \widetilde{A}_2 \times ... \times \widetilde{A}_n \in (\mathcal{B}(\mathbb{R}))^n$, *then* $\lim_{j\to\infty} P_{\boldsymbol{X}}(\widetilde{\boldsymbol{A}}_j) \downarrow P_{\boldsymbol{X}}(\widetilde{\boldsymbol{A}})$.

Proof *The proof can be verified similar to that of Proposition 2.3.*

Example 2.10 *[100] Let* $\boldsymbol{X} = (X_1, X_2)$ *be a bivariate uniform distribution function on* $[17/2, 23/2] \times [7/2, 13/2]$. *Consider a finite collection of* ***FPEs*** $\{\widetilde{\boldsymbol{A}}_j = \widetilde{A}_{1j} \times \widetilde{A}_{2j}\}_{j=1}^{\infty}$ *where* $\widetilde{A}_{1j} = (10; 1 + 2^{-j})_T$ *and* $\widetilde{A}_{2j} = (5; 1 + 2^{-j})_T$. *Set* $\widetilde{\boldsymbol{A}} = \widetilde{A}_1 \times \widetilde{A}_2$ *with* $\widetilde{A}_1 = (10; 1)_T$ *and* $\widetilde{A}_2 = (5; 1)_T$. *It is easy to check that* $\widetilde{\boldsymbol{A}}_j \downarrow \widetilde{\boldsymbol{A}} = \widetilde{A}_1 \times \widetilde{A}_2$. *Therefore, by Proposition 2.11, we have* $\lim_{j\to\infty} P_{\boldsymbol{X}}(\widetilde{\boldsymbol{A}}_j) \to P_{\boldsymbol{X}}(\widetilde{\boldsymbol{A}})$. *Note that, such result can be also checked by finding* $P_{\boldsymbol{X}}(\widetilde{\boldsymbol{A}}) = P_{\boldsymbol{X}}(\widetilde{A}_1 \times \widetilde{A}_2) = 4/27$ *and*

$$P_{\boldsymbol{X}}(\widetilde{\boldsymbol{A}}_j) = P_{\boldsymbol{X}}(\widetilde{A}_{1j} \times \widetilde{A}_{2j}) = \frac{4}{9(1 + 2^{-j})}[\frac{1}{3} + \frac{2^{-j}}{3} + 2^{-2j} + 2^{-j}] \to 4/27, \quad \text{as } j \to \infty.$$

Therefore, $\lim_{j\to\infty} P_{\boldsymbol{X}}(\widetilde{\boldsymbol{A}}_j) \to P_{\boldsymbol{X}}(\widetilde{\boldsymbol{A}})$.

2.3 Conditional Probability, Independence, and Bayes' Theorem

In this section, some notions of conditional probability and independence of **FE**s are defined and discussed. Then, the concept of Bayes' theorem is established based on **FE**s for a discrete random variable.

Definition 2.5 *Let $(\mathcal{B}(\mathbb{R}), \mathbb{F}_m(\mathbb{R}))$ be a fuzzy measurable space and $X : \mathbb{F}_m(\mathbb{R})) \to \mathbb{R}$ be a random variable. The conditional probability of $\widetilde{A} \in \mathbb{F}_m(\mathbb{R})$ given $\widetilde{B} \in \mathbb{F}_m(\mathbb{R})$ can be defined as follows:*

$$P_X(\widetilde{A}|\widetilde{B}) = \frac{P_X(\widetilde{A} \odot \widetilde{B})}{P_X(\widetilde{B})}, \quad P_X(\widetilde{B}) \neq 0.$$

*where $(\widetilde{A} \odot \widetilde{B})(x) = \widetilde{A}(x)\widetilde{B}(x)$, $x \in \mathbb{R}$. Therefore, the conditional probability of $\widetilde{A}$ is the probability that the event will occur given the knowledge that a **FE** of $\widetilde{B}$ has already occurred.*

Example 2.11 *Consider a bag containing six beads numbered from 1 to 6. A bead is randomly pulled out from the bag. Let X be the number of a randomly selected bead as uniformly distributed on $\{1, 2, 3, 4, 5, 6\}$. Now, consider two **FE**s of $\widetilde{A}$ to interpret 'the observed number is too small' and $\widetilde{B}$ as 'the observed number is about 2' with the following membership functions, respectively:*

$$\widetilde{A} = \{\frac{1}{1}, \frac{0.7}{2}, \frac{0.4}{3}, \frac{0.2}{4}, \frac{0.1}{5}\}, \quad \widetilde{B} = \{\frac{0.8}{1}, \frac{1}{2}, \frac{0.8}{3}, \frac{0.3}{4}\}.$$

In this way, it can be checked that $P_X(\widetilde{A}) = 2.4/6$ and $P_X(\widetilde{A} \odot \widetilde{B}) = 1.08/6$ and hence $P_X(\widetilde{B}|\widetilde{A}) = 1.08/2.4 = 0.45$.

Theorem 2.3 *Recall the assumptions in Definition 2.2. For any fixed **FE** $\widetilde{C}$, $P_X(.|\widetilde{C})$ provides a probability measure, i.e.,*

(1) $P_X(I(\mathbb{R})|\widetilde{C}) = 1$,

(2) *For two disjoint **FE**s of $\widetilde{A}$ and $\widetilde{B}$, $P_X(\widetilde{A} \cup \widetilde{B}|\widetilde{C}) = P_X(\widetilde{A}|\widetilde{C}) + P_X(\widetilde{B}|\widetilde{C}) - P_X(\widetilde{A} \cap \widetilde{B}|\widetilde{C})$.*

Proof *It is easy to verify that (1) is valid. For (2), it is observed that:*

$$\begin{aligned}
P_X(\widetilde{A} \cup \widetilde{B}|\widetilde{C}) &= \frac{P_X((\widetilde{A} \cup \widetilde{B}) \odot \widetilde{C})}{P_X(\widetilde{C})} \\
&= \frac{1}{P_X(\widetilde{C})}(P_X((\widetilde{A} \odot \widetilde{C}) \cup (\widetilde{B} \odot \widetilde{C}))) \\
&= \frac{1}{P_X(\widetilde{C})}(P_X(\widetilde{A} \odot \widetilde{C}) + P_X(\widetilde{B} \odot \widetilde{C}) - P_X((\widetilde{A} \odot \widetilde{C}) \cap (\widetilde{B} \odot \widetilde{C})) \\
&= \frac{1}{P_X(\widetilde{C})}(P_X(\widetilde{A} \odot \widetilde{C}) + P_X(\widetilde{B} \odot \widetilde{C}) - P_X(\widetilde{C} \odot (\widetilde{B} \cap \widetilde{C})) \\
&= P_X(\widetilde{A}|\widetilde{C}) + P_X(\widetilde{B}|\widetilde{C}) - P_X(\widetilde{A} \cap \widetilde{B}|\widetilde{C}).
\end{aligned}$$

It is easy to verify that the above conditional probability measure meets the following results.

Proposition 2.12 *For any fixed* ***FE*** *of* $\widetilde{C}$, $P_X(.|\widetilde{C})$ *satisfies the following conditions:*

(1) *If* $\widetilde{A} \subseteq \widetilde{B}$ *then* $P_X(\widetilde{A}|\widetilde{C}) \leq P_X(\widetilde{B}|\widetilde{C})$,

(2) $P_X(\widetilde{A}^c|\widetilde{C}) = 1 - P_X(\widetilde{A}|\widetilde{C})$,

(3) $P_X(\widetilde{A} - \widetilde{B}|\widetilde{C}) = P_X(\widetilde{A}|\widetilde{C}) - P_X(\widetilde{A} \cap \widetilde{B}|\widetilde{C})$, *and*

(4) $P_X(\widetilde{A} \triangle \widetilde{B}|\widetilde{C}) = P_X(\widetilde{A}|\widetilde{C}) + P_X(\widetilde{B}|\widetilde{C}) - 2P_X(\widetilde{A} \cap \widetilde{B}|\widetilde{C})$.

A concept of independence of two **FE**s is given next.

Definition 2.6 *Let* $(\mathcal{B}(\mathbb{R}), \mathbb{F}_m(\mathbb{R}))$ *be a fuzzy measurable space and* $X : \mathbb{F}_m(\mathbb{R})) \to \mathbb{R}$ *be a random variable. We say that* $\widetilde{A}$ *and* $\widetilde{B}$ *(with non-zero probability) are independent if* $P_X(\widetilde{A}|\widetilde{B}) = P_X(\widetilde{A})$ *and* $P_X(\widetilde{B}|\widetilde{A}) = P_X(\widetilde{B})$, *i.e.* $P_X(\widetilde{A} \odot \widetilde{B}) = P_X(\widetilde{A})P_X(\widetilde{B})$.

Proposition 2.13 *If* $\widetilde{A}$ *and* $\widetilde{B}$ *are two independent* ***FE****s then 1 –* $\widetilde{A}^c$ *and* $\widetilde{B}$, *2 –* $\widetilde{A}$ *and* $\widetilde{B}^c$, *and 3 –* $\widetilde{A}^c$ *and* $\widetilde{B}^c$ *are also independent* ***FE****s.*

Proof *Let* $\widetilde{A}$ *and* $\widetilde{B}$ *be two independent* ***FE****s. We only show that* $\widetilde{A}^c$ *and* $\widetilde{B}$ *are independent* ***FE****s. For this,*

$$P_X(\widetilde{A}^c \odot \widetilde{B}) = \int (1 - \widetilde{A})\widetilde{B} dF_X = P_X(\widetilde{B})(1 - P_X(\widetilde{A})) = P_X(\widetilde{B})P_X(\widetilde{A}^c).$$

The other claims are left as an exercise.

Example 2.12 *Let X be the number of heads when three coins are tossed. Assume that* $\widetilde{A}$ *and* $\widetilde{B}$ *are two* ***FE****s to describe 'high value of X' and 'moderate value of X' with the following respective membership functions:*

$$\widetilde{A} = \{\frac{0.3}{1}, \frac{0.7}{2}, \frac{1}{3}\}, \ \widetilde{B} = \{\frac{0.3}{0}, \frac{0.8}{1}, \frac{0.8}{2}, \frac{0.3}{3}\}.$$

First note that $P(X=x)=\binom{8}{x}/8$ *for* $x=0,1,2,3$ *and*

$$\widetilde{A}\cap\widetilde{B}=\{\frac{0.3}{1},\frac{0.7}{2},\frac{0.3}{3}\}.$$

It can be easily shown that $P_X(\widetilde{A})=4/8$, $P_X(\widetilde{B})=5.1/8$ *implies* $P_X(\widetilde{A}\cap\widetilde{B})=2.7/8$. *But it can be seen that* $P_X(\widetilde{A}\cap\widetilde{B})\neq P_X(\widetilde{A})P_X(\widetilde{B})$. *This concludes that* $\widetilde{A}$ *and* $\widetilde{B}$ *are not independent.*

Here, a notion of Bayes' theorem is introduced to describe how to update the probabilities of hypotheses when given evidence in the fuzzy domain.

Theorem 2.4 *Let* $\{\widetilde{B}_k\}_{k=1}^n$ *be a collection of the fuzzy partition of* $\mathbb{X}$, *i.e.,* $\sum_{k=1}^n \widetilde{B}_k(x)=1$ *for every* $x\in\mathbb{X}$. *Then, for every* ***FE*** *of* $\widetilde{A}$ *with non-zero probability, we have:*

$$P_X(\widetilde{B}_i|\widetilde{A})=\frac{P_X(\widetilde{A}|\widetilde{B}_i)P_X(\widetilde{B}_i)}{\sum_{i=1}^n P_X(\widetilde{A}|\widetilde{B}_i)P_X(\widetilde{B}_i)}.$$

Proof *By Definition 2.5, we have* $P_X(\widetilde{B}_i|\widetilde{A})=\frac{P_X(\widetilde{A}|\widetilde{B}_i)P_X(\widetilde{B}_i)}{P_X(\widetilde{A})}$. *We claim that* $P_X(\widetilde{A})=\sum_{i=1}^n P_X(\widetilde{B}_i|\widetilde{A})P_X(\widetilde{B}_i)$. *To do this, note that:*

$$\begin{aligned}\sum_{i=1}^n P_X(\widetilde{B}_i|\widetilde{A})P_X(\widetilde{B}_i) &= \sum_{i=1}^n P_X(\widetilde{A}\cap\widetilde{B}_i)\\ &= \sum_{i=1}^n E_X(\widetilde{B}_i\widetilde{A}) = E_X((\sum_{i=1}^n \widetilde{B}_i)\widetilde{A}) = P_X(\widetilde{A}).\end{aligned}$$

Example 2.13 *Assume that the educational level can be classified by some linguistic levels as non, elementary school, junior high school, bachelor, master science, and Doctor of Philosophy. Let* X *be the educational level of a person with discrete uniform distribution on* $\{N,ES,JHS,SHS,BA,MS,PhD\}$. *Fuzzy partitions of education can be assumed as follows:*

$$\begin{aligned}Low\ education: \widetilde{B}_1 &= \{\frac{1}{N},\frac{0.8}{ES},\frac{0.5}{JHS}\},\\ Medium\ education: \widetilde{B}_2 &= \{\frac{0.2}{ES},\frac{0.5}{JHS},\frac{0.9}{SHS},\frac{0.2}{BA}\},\\ High\ education: \widetilde{B}_3 &= \{\frac{0.1}{SHS},\frac{0.8}{BA},\frac{1}{MS},\frac{1}{PhD}\}.\end{aligned}$$

Therefore, one can find $P_X(\widetilde{B}_1)=23/70$, $P_X(\widetilde{B}_2)=29/70$, *and* $P_X(\widetilde{B}_3)=18/70$. *Now, assume one person is randomly selected for a survey involving salary. Let* $\widetilde{A}$ *denotes a* ***FE*** *to describe 'high salary'. If* 10% *of the people had 'Low education',* 12% *of them had 'Medium education', and* 25% *were classified* *as 'High education' and earned 'high salary' then, according to Theorem 2.4, the posterior probability $P_X(\widetilde{B}_1|\widetilde{A})$ can be evaluated as:*

$$P_X(\widetilde{B}_1|\widetilde{A}) = \frac{\frac{23}{70} \times \frac{10}{100}}{\frac{23}{70} \times \frac{10}{100} + \frac{29}{70} \times \frac{12}{100} + \frac{18}{70} \times \frac{25}{100}} = \frac{230}{794} \simeq 0.29.$$

Similarly, $P_X(\widetilde{B}_2|\widetilde{A})$ and $P_X(\widetilde{B}_3|\widetilde{A})$ can be obtained.

2.4 Fuzzy Probability of a FE and Its Properties

This section attempts to extend Zadeh's probability to the fuzzy sets.

Proposition 2.14 *Let X be a random variable with density function of $f_X^{\boldsymbol{\theta}}$, $\boldsymbol{\theta} = (\theta_1, \theta_2, \ldots, \theta_p)^\top \in \mathbb{R}^p$. For every **FE** of $\widetilde{A}$, $P_X^{\boldsymbol{\theta}}(\widetilde{A})$ introduced in Theorem 2.1, can be rewritten as follows:*

$$P_X^{\boldsymbol{\theta}}(\widetilde{A}) = E_W[P_X^{\boldsymbol{\theta}}(\widetilde{A}[W])], \tag{2.2}$$

where W is a uniform random variable on $(0,1)$ and $E_W(g(W))$ stands for the expectation of $g(.)$ with respect to W.

Proof *The proof is left as an exercise.*

This probability measure shows the expected value of the membership function of a **FE** over a classical distribution function, thereby providing a flexible criterion to combine the fuzziness with randomness. Such a probability measure is non-fuzzy quantity. However, one may interested to evaluate the chance of a fuzzy event as a fuzzy sets. For instance, consider a random experiment to find an interesting roman's book in a book shop. In such cases, the chance of the fuzzy event 'an interesting roman's book' can be interpreted as some fuzzy quantities such as 'rather probable' or 'low probable'. In this regard, the probability of a **FE** of $\widetilde{A}$ as 'about $P_X^{\boldsymbol{\theta}}(\widetilde{A})$' can be defined as follows.

Definition 2.7 *Let X be a random variable with the density function of $f_X^{\boldsymbol{\theta}}$, $\boldsymbol{\theta} = (\theta_1, \theta_2, \ldots, \theta_p)^\top \in \mathbb{R}^p$. A fuzzy probability of a **FE** $\widetilde{A}$ is defined to be a fuzzy set of $\widetilde{P}_X^{\boldsymbol{\theta}}(\widetilde{A})$ with the following membership function:*

$$\widetilde{P}_X^{\boldsymbol{\theta}}(\widetilde{A})(p) = \sup\left\{\alpha \in [0,1] : p \in \left[(\widetilde{P}_X^{\boldsymbol{\theta}}(\widetilde{A}))_\alpha^L, (\widetilde{P}_X^{\boldsymbol{\theta}}(\widetilde{A}))_\alpha^U\right]\right\}, \; p \in (0,1], \tag{2.3}$$

where

$$(\widetilde{P}_X^{\boldsymbol{\theta}}(\widetilde{A}))_\alpha^L = E_W\left[P_X^{\boldsymbol{\theta}}(\widetilde{A}[1-(1-W)\alpha])\right], \; (\widetilde{P}_X^{\boldsymbol{\theta}}(\widetilde{A}))_\alpha^U = E_W\left[P_X^{\boldsymbol{\theta}}(\widetilde{A}[W\alpha])\right]. \tag{2.4}$$

Lemma 2.2 *The fuzzy set $\widetilde{P}^{\boldsymbol{\theta}}_X(\widetilde{A})$ in Definition 2.7 is a **FN**.*

Proof *It is enough to show that $\widetilde{P}^{\boldsymbol{\theta}}_X(\widetilde{A})$ is a normal fuzzy set and $\{\widetilde{P}^{\boldsymbol{\theta}}_X(\widetilde{A})\}_{\alpha\in[0,1]}$ constructs a collection of nested closed intervals. To this end, for $\alpha = 1$, note that $\widetilde{P}^{\boldsymbol{\theta}}_X(\widetilde{A})[1] = I(E_W\left[P^{\boldsymbol{\theta}}_X(\widetilde{A}[W])\right])$. Therefore, $\widetilde{P}^{\boldsymbol{\theta}}_X(\widetilde{A})$ is a normal fuzzy set. Moreover, it is easy to show that $E_W\left[P^{\boldsymbol{\theta}}_X(\widetilde{A}[1-(1-W)\alpha])\right]$ and $E_W\left[P^{\boldsymbol{\theta}}_X(\widetilde{A}[W\alpha])\right]$ are increasing and decreasing function of α, respectively. Therefore, $\{\widetilde{P}^{\boldsymbol{\theta}}_X(\widetilde{A})[\alpha]\}_{\alpha\in[0,1]}$ is a collection of closed intervals and $\widetilde{P}^{\boldsymbol{\theta}}_X(\widetilde{A})$ is normal. These conclude that $\widetilde{P}^{\boldsymbol{\theta}}_X(\widetilde{A})$ is a **FN**.*

Example 2.14 *Recall Example 2.2. Here, we wish to evaluate the probability of $\widetilde{A} = \{\frac{0.5}{6}, \frac{0.9}{7}, \frac{1}{8}, \frac{0.6}{9}, \frac{0.4}{10}\}$ as a **FN**. For this, both lower and upper α-cuts $(\widetilde{P}^{\boldsymbol{\theta}}_X(\widetilde{A}))^L_\alpha = E_W\left[P^{\boldsymbol{\theta}}_X(\widetilde{A}[1-(1-W)\alpha])\right]$ and $(\widetilde{P}^{\boldsymbol{\theta}}_X(\widetilde{A}))^U_\alpha = E_W\left[P^{\boldsymbol{\theta}}_X(\widetilde{A}[W\alpha])\right]$ should be computed for any $\alpha \in (0, 1]$. To do this end, first, note that*

$$\widetilde{A}[\alpha] = \begin{cases} \{6,7,8,9,10\}, & 0.0 < \alpha \le 0.40, \\ \{6,7,8,9\}, & 0.40 < \alpha \le 0.50, \\ \{7,8,9\}, & 0.50 < \alpha \le 0.60, \\ \{8,9\}, & 0.60 < \alpha \le 0.90, \\ \{8\}. & 0.90 < \alpha \le 1.0. \end{cases}$$

Therefore, for a fixed $\alpha \in (0, 1]$, the upper α-cuts of $\widetilde{P}^{\boldsymbol{\theta}}_X(\widetilde{A})$ can be evaluated as follows:

$$\begin{aligned} E_W\left[\widetilde{P}^{\boldsymbol{\theta}}_X(\widetilde{A}[W\alpha])\right. &= \int_{\{w:0\le w\alpha\le 0.40\}} P(X \in \{6,7,8,9,10\})dw \\ &+ \int_{\{w:0.40< w\alpha\le 0.50\}} P(X \in \{6,7,8,9\})dw \\ &+ \int_{\{w:0.50< w\alpha\le 0.60\}} P(X \in \{7,8,9\})dw \\ &+ \int_{\{w:0.60< w\alpha\le 0.90\}} P(X \in \{8,9\})dw \\ &+ \int_{\{w:0.90< w\alpha\le 1\}} P(X = 8)dw \\ &= \frac{5}{12}\min\{1, \frac{0.4}{\alpha}\} + \frac{4}{12}(\min\{1, \frac{0.5}{\alpha}\} - \min\{1, \frac{0.4}{\alpha}\}) \\ &+ \frac{3}{12}(\min\{1, \frac{0.6}{\alpha}\} - \min\{1, \frac{0.5}{\alpha}\}) + \frac{2}{12}(\min\{1, \frac{0.9}{\alpha}\} \\ &- \min\{1, \frac{0.6}{\alpha}\}) + \frac{1}{12}(\min\{1, \frac{1}{\alpha}\} - \min\{1, \frac{0.9}{\alpha}\}). \end{aligned}$$

Furthermore, the lower α-cuts of $\widetilde{P}_X^{\boldsymbol{\theta}}(\widetilde{A})$ can be obtained as:

$$
\begin{aligned}
E_W\Big[\widetilde{P}_X^{\boldsymbol{\theta}}(\widetilde{A}[1-(1-W)\alpha])\Big] &= \int_{\{w:0\leq 1-(1-w)\alpha\leq 0.40\}} P(X\in\{6,7,8,9,10\})dw \\
&+ \int_{\{w:0.40<1-(1-w)\alpha\leq 0.50\}} P(X\in\{6,7,8,9\})dw \\
&+ \int_{\{w:0.50<1-(1-w)\alpha\leq 0.60\}} P(X\in\{7,8,9\})dw \\
&+ \int_{\{w:0.60<1-(1-w)\alpha\leq 0.90\}} P(X\in\{8,9\})dw \\
&+ \int_{\{w:0.90<1-(1-w)\alpha\leq 1\}} P(X\in\{8\})dw \\
&= \frac{5}{12}(\max\{0,1-\frac{0.6}{\alpha}\}-\max\{0,1-\frac{1}{\alpha}\}) \\
&+ \frac{4}{12}(\max\{0,1-\frac{0.5}{\alpha}\}-\max\{0,1-\frac{0.6}{\alpha}\}) \\
&+ \frac{3}{12}(\max\{0,1-\frac{0.4}{\alpha}\}-\max\{0,1-\frac{0.5}{\alpha}\}) \\
&+ \frac{2}{12}(\max\{0,1-\frac{0.1}{\alpha}\}-\max\{0,1-\frac{0.4}{\alpha}\}) \\
&+ \frac{1}{12}(1-\max\{0,1-\frac{0.1}{\alpha}\}).
\end{aligned}
$$

For plotting the membership function of $\widetilde{P}_X^{\boldsymbol{\theta}}(\widetilde{A})$, there is a need to evaluate the lower and upper bounds of its α-cuts for some values of $[0,1]$. For this purpose, some α-cuts of $\widetilde{P}_X^{\boldsymbol{\theta}}(\widetilde{A})$ are also listed in Table 2.1. Then the values of α are plotted in y-axis while the values of $(\widetilde{P}_X^{\boldsymbol{\theta}}(\widetilde{A}))_\alpha^L$ and $(\widetilde{P}_X^{\boldsymbol{\theta}}(\widetilde{A}))_\alpha^U$ are plotted in x-axis. Now, by computing all values of $(\widetilde{P}_X^{\boldsymbol{\theta}}(\widetilde{A}))_\alpha^L$ and $(\widetilde{P}_X^{\boldsymbol{\theta}}(\widetilde{A}))_\alpha^U$, the membership function of $\widetilde{P}_X^{\boldsymbol{\theta}}(\widetilde{A})$ can be plotted as shown in Fig. 2.1.

Example 2.15 *Recall Example 2.1. To evaluate $\widetilde{P}_X^{\boldsymbol{\theta}}(\widetilde{A})$, note that:*

$$
\begin{aligned}
(\widetilde{P}_X^{\boldsymbol{\theta}}(\widetilde{A}))_\alpha^L &= E_W\Big[P_X^{\boldsymbol{\theta}}(\widetilde{A}[1-\alpha(1-W)])\Big] \\
&= \int_0^1\int_{66+4(1-(1-w))\alpha}^{74-4(1-(1-w))\alpha}\frac{1}{80}dxdw = \frac{4\alpha}{80},
\end{aligned}
$$

and

$$
\begin{aligned}
(\widetilde{P}_X^{\boldsymbol{\theta}}(\widetilde{A}))_\alpha^U &= E_W\Big[P_X^{\boldsymbol{\theta}}(\widetilde{A}[W\alpha])\Big] \\
&= \int_0^1\int_{66+4w\alpha}^{74-4w\alpha}\frac{1}{80}dxdw = \frac{8-4\alpha}{80}.
\end{aligned}
$$

*This simply concludes that $\widetilde{P}_X^{\boldsymbol{\theta}}(\widetilde{A})$ is a **TFN** as $(0.05;0.05)_T$.*

TABLE 2.1
Some α-cuts of $\widetilde{P}^{\boldsymbol{\theta}}_X(\widetilde{A})$ in Example 2.14.

α	$(\widetilde{P}^{\boldsymbol{\theta}}_X(\widetilde{A}))^L_\alpha$	$(\widetilde{P}^{\boldsymbol{\theta}}_X(\widetilde{A}))^U_\alpha$
0.02	0.085	0.417
0.12	0.110	0.417
0.22	0.125	0.417
0.32	0.320	0.417
0.42	0.419	0.405
0.52	0.518	0.380
0.62	0.621	0.351
0.72	0.246	0.337
0.82	0.273	0.325
0.92	0.298	0.305
1	0.2833	0.2833

Next, some basic properties of $\widetilde{P}^{\boldsymbol{\theta}}_X$ are established and discussed by the following results.

Theorem 2.5 *Let X be a random variable with the density function $f^{\boldsymbol{\theta}}_X$, $\boldsymbol{\theta} = (\theta_1, \theta_2, \ldots, \theta_p)^\top \in \mathbb{R}^p$. Then the fuzzy probability of $\widetilde{P}^{\boldsymbol{\theta}}_X(\widetilde{A})$ introduced in Eq. (2.3) satisfies the following conditions:*

(1) *For two **FE**s of $\widetilde{A}$ and $\widetilde{B}$ if $\widetilde{A} \subseteq \widetilde{B}$ then $P_d(\widetilde{P}^{\boldsymbol{\theta}}_X(\widetilde{A}) \preceq \widetilde{P}^{\boldsymbol{\theta}}_X(\widetilde{B})) = 1$.*

(2) *For a **FE** of $\widetilde{A}$, $\widetilde{P}^{\boldsymbol{\theta}}_X(\widetilde{A}^c) = 1 \ominus \widetilde{P}^{\boldsymbol{\theta}}_X(\widetilde{A})$.*

(3) *For two **FE**s of $\widetilde{A}$ and $\widetilde{B}$, $P_d(\widetilde{P}^{\boldsymbol{\theta}}_X(\widetilde{A} \cup \widetilde{B}) \preceq \widetilde{P}^{\boldsymbol{\theta}}_X(\widetilde{A}) \oplus \widetilde{P}^{\boldsymbol{\theta}}_X(\widetilde{B})) = 1$.*

(4) *If $\widetilde{A}$ reduces to an ordinary event A then $\widetilde{P}^{\boldsymbol{\theta}}_X(\widetilde{A})$ will be reduced to an ordinary probability of an event $P^{\boldsymbol{\theta}}_X(A)$.*

Proof. Item (4) can be easily verified. For (2), assuming $\widetilde{A} \subseteq \widetilde{B}$ simply concludes that:

$$E_W\Big[P^{\boldsymbol{\theta}}_X(\widetilde{A}[1-(1-W)\alpha])\Big] \subseteq E_W\Big[P^{\boldsymbol{\theta}}_X(\widetilde{B}[1-(1-W)\alpha])\Big], \text{ and}$$
$$E_W\Big[P^{\boldsymbol{\theta}}_X(\widetilde{A}[W\alpha])\Big] \subseteq E_W\Big[P^{\boldsymbol{\theta}}_X(\widetilde{B}[W\alpha])\Big],$$

for all $\alpha \in [0,1]$. These show that:

$$(\widetilde{P}^{\boldsymbol{\theta}}_X(\widetilde{A}))^L_\alpha \leq (\widetilde{P}^{\boldsymbol{\theta}}_X(\widetilde{B}))^L_\alpha,\ (\widetilde{P}^{\boldsymbol{\theta}}_X(\widetilde{A}))^U_\alpha \leq (\widetilde{P}^{\boldsymbol{\theta}}_X(\widetilde{B}))^U_\alpha,$$

for all $\alpha \in [0,1]$. That is $P_d(\widetilde{P}^{\boldsymbol{\theta}}_X(\widetilde{A}) \preceq \widetilde{P}^{\boldsymbol{\theta}}_X(\widetilde{B})) = 1$ and thus (1) follows. To establish (2), note that:

$$\begin{aligned}(\widetilde{P}^{\boldsymbol{\theta}}_X(\widetilde{A}^c))^L_\alpha &= E_W\Big[P^{\boldsymbol{\theta}}_X((\widetilde{A}[(1-W)\alpha])^c)\Big] \\ &= E_W\Big[1 - P^{\boldsymbol{\theta}}_X(\widetilde{A}[W\alpha])\Big] = 1 - (\widetilde{P}^{\boldsymbol{\theta}}_X(\widetilde{A}^c))^U_\alpha,\end{aligned}$$

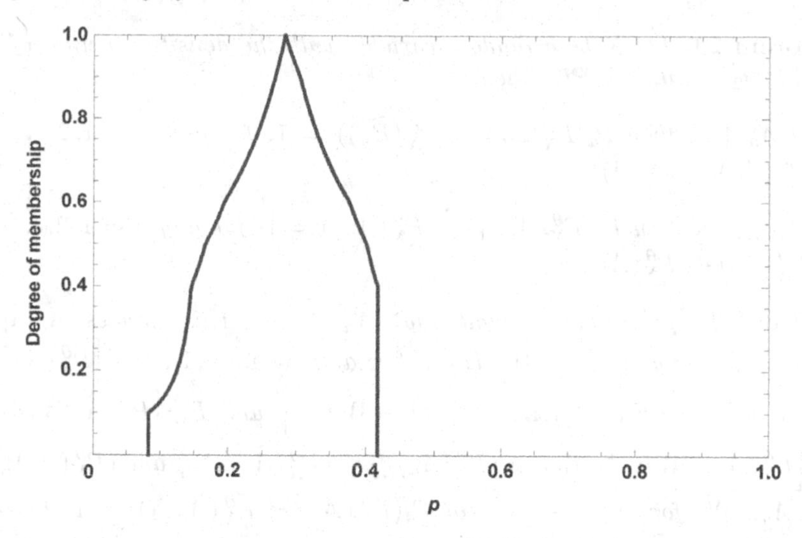

FIGURE 2.1
Membership function of $\widetilde{P}^{\boldsymbol{\theta}}_X(\widetilde{A})$ in Example 2.14.

and

$$\begin{aligned}(\widetilde{P}^{\boldsymbol{\theta}}_X(\widetilde{A}^c))^U_\alpha &= E_W\left[P^{\boldsymbol{\theta}}_X((\widetilde{A}[1-W\alpha])^c)\right]\\ &= E_W\left[1-P^{\boldsymbol{\theta}}_X(\widetilde{A}[1-(1-W)\alpha])\right] = 1-(\widetilde{P}^{\boldsymbol{\theta}}_X(\widetilde{A}^c))^L_\alpha.\end{aligned}$$

Therefore, $\widetilde{P}^{\boldsymbol{\theta}}_X(\widetilde{A}^c)[\alpha] = [1-\widetilde{P}^{\boldsymbol{\theta}}_X(\widetilde{A}^c))^U_\alpha, 1-\widetilde{P}^{\boldsymbol{\theta}}_X(\widetilde{A}^c))^L_\alpha]$ verifying (2). Next consider (3). Let $\widetilde{A}$ and $\widetilde{B}$ be two **FE**s. It can be shown that:

$$\begin{aligned}&(\widetilde{P}^{\boldsymbol{\theta}}_X(\widetilde{A}\cup\widetilde{B}))^L_\alpha\\ &\quad= E_W\left[P^{\boldsymbol{\theta}}_X((\widetilde{A}\cup\widetilde{B})[(1-(1-W))\alpha])\right]\\ &\quad= E_W\left[P^{\boldsymbol{\theta}}_X(\widetilde{A}[(1-(1-W))\alpha]\cup\widetilde{B}[(1-W)\alpha])\right]\\ &\quad= E_W\left[P^{\boldsymbol{\theta}}_X(\widetilde{A}[(1-(1-W))\alpha])\right]+E_W\left[P^{\boldsymbol{\theta}}_X(\widetilde{B}[(1-W)\alpha])\right]\\ &\quad- E_W\left[P^{\boldsymbol{\theta}}_X(\widetilde{A}[(1-(1-W))\alpha]\cap\widetilde{B}[(1-W)\alpha])\right]\\ &\quad\le E_W\left[P^{\boldsymbol{\theta}}_X(\widetilde{A}[(1-(1-W))\alpha])\right]+E_W\left[P^{\boldsymbol{\theta}}_X(\widetilde{B}[(1-W)\alpha])\right]\\ &\quad= (\widetilde{P}^{\boldsymbol{\theta}}_X(\widetilde{A}))^L_\alpha+(\widetilde{P}^{\boldsymbol{\theta}}_X(\widetilde{B}))^L_\alpha.\end{aligned}$$

These conclude that: $(\widetilde{P}^{\boldsymbol{\theta}}_X(\widetilde{A}\cup\widetilde{B}))^L_\alpha \le ((\widetilde{P}^{\boldsymbol{\theta}}_X(\widetilde{A}))^L_\alpha+(\widetilde{P}^{\boldsymbol{\theta}}_X(\widetilde{B}))^L_\alpha)$. Similarly, it can be shown that $(\widetilde{P}^{\boldsymbol{\theta}}_X(\widetilde{A}\cup\widetilde{B}))^U_\alpha \le ((\widetilde{P}^{\boldsymbol{\theta}}_X(\widetilde{A}))^U_\alpha+(\widetilde{P}^{\boldsymbol{\theta}}_X(\widetilde{B}))^U_\alpha)$. According to the ordering procedure mentioned in introductory chapter, we finally get $P_d(\widetilde{P}^{\boldsymbol{\theta}}_X(\widetilde{A}\cup\widetilde{B}) \preceq \widetilde{P}^{\boldsymbol{\theta}}_X(\widetilde{A})\oplus\widetilde{P}^{\boldsymbol{\theta}}_X(\widetilde{B}) = 1$. This completes the proof.

Theorem 2.6 *Let X be a random variable with the density function of $f_X^{\boldsymbol{\theta}}$, $\boldsymbol{\theta} = (\theta_1, \theta_2, \ldots, \theta_p)^\top \in \mathbb{R}^p$. Then:*

(1) *If $\widetilde{A}_n \uparrow \widetilde{A}$ then $P_d(\widetilde{P}_X^{\boldsymbol{\theta}}(\widetilde{A}_n) \preceq \widetilde{P}_X^{\boldsymbol{\theta}}(\widetilde{B}_n)) = 1$, for any $n = 1, 2, \ldots$ and $\widetilde{P}_X^{\boldsymbol{\theta}}(\widetilde{A}_n) \to \widetilde{P}_X^{\boldsymbol{\theta}}(\widetilde{A})$.*

(2) *If $\widetilde{A}_n \downarrow \widetilde{A}$ then $P_d(\widetilde{P}_X^{\boldsymbol{\theta}}(\widetilde{A}_{n+1}) \preceq \widetilde{P}_X^{\boldsymbol{\theta}}(\widetilde{A}_n)) = 1$, for any $n = 1, 2, \ldots$ and $\widetilde{P}_X^{\boldsymbol{\theta}}(\widetilde{A}_n) \to \widetilde{P}_X^{\boldsymbol{\theta}}(\widetilde{A})$.*

Proof *To prove (1), assume that $\widetilde{A}_n \uparrow \widetilde{A}$. This implies $\widetilde{A}_n[\alpha] \subseteq \widetilde{A}_{n+1}[\alpha]$ for any $\alpha \in [0,1]$. Thus, for any $\alpha \in [0,1]$, $E_W\Big[P_X^{\boldsymbol{\theta}}(\widetilde{A}_n[1-(1-W)\alpha])\Big] \leq E_W\Big[P_X^{\boldsymbol{\theta}}(\widetilde{A}_{n+1}[1-(1-W)\alpha])\Big]$ and $E_W\Big[P_X^{\boldsymbol{\theta}}(\widetilde{A}_n[W\alpha])\Big] \downarrow E_W\Big[P_X^{\boldsymbol{\theta}}(\widetilde{A}_{n+1}[W\alpha])\Big]$. That is $(\widetilde{P}_X^{\boldsymbol{\theta}}(\widetilde{A}_n))_\alpha^L \leq (\widetilde{P}_X^{\boldsymbol{\theta}}(\widetilde{A}_{n+1}))_\alpha^L$ and $(\widetilde{P}_X^{\boldsymbol{\theta}}(\widetilde{A}_n))_\alpha^U \leq (\widetilde{P}_X^{\boldsymbol{\theta}}(\widetilde{A}_{n+1}))_\alpha^U$ for any $\alpha \in (0,1]$ or $P_d(\widetilde{P}_X^{\boldsymbol{\theta}}(\widetilde{A}_n) \preceq \widetilde{P}_X^{\boldsymbol{\theta}}(\widetilde{A}_{n+1})) = 1$. Furthermore, the M.C.T. concludes that:*

$$E_W\Big[P_X^{\boldsymbol{\theta}}(\widetilde{A}_n[1-(1-W)\alpha])\Big] \downarrow E_W\Big[P_X^{\boldsymbol{\theta}}(\widetilde{A}[1-(1-W)\alpha])\Big],$$

and

$$E_W\Big[P_X^{\boldsymbol{\theta}}(\widetilde{A}_n[W\alpha])\Big] \downarrow E_W\Big[P_X^{\boldsymbol{\theta}}(\widetilde{A}[W\alpha])\Big],$$

for all $\alpha \in (0,1]$. This yields $(\widetilde{P}_X^{\boldsymbol{\theta}}(\widetilde{A}_n))_\alpha^L \to (\widetilde{P}_X^{\boldsymbol{\theta}}(\widetilde{A}))_\alpha^L$ and $(\widetilde{P}_X^{\boldsymbol{\theta}}(\widetilde{A}_n))U \to (\widetilde{P}_X^{\boldsymbol{\theta}}(\widetilde{A}))U$ for any $\alpha \in (0,1]$ or equivalently $\widetilde{P}_X^{\boldsymbol{\theta}}(\widetilde{A}_n) \to \widetilde{P}_X^{\boldsymbol{\theta}}(\widetilde{A})$. Part (2) can be verified similar to that of (1).

Lemma 2.3 *Let $\{\widetilde{A}_k\}_{k=1}^n$ be a finite collection of mutually disjoint* **FE***s, i.e., $\widetilde{A}_i \cap \widetilde{A}_j = \emptyset \ \forall i \neq j$. For any $n = 1, 2, \ldots$, if $\bigcup_{k=1}^n \widetilde{A}_k \subseteq \widetilde{A}$ and $\bigcup_{k=1}^n \widetilde{A}_k \to \widetilde{A}$, then $\widetilde{P}_X^{\boldsymbol{\theta}}(\bigcup_{k=1}^\infty \widetilde{A}_k) = \bigoplus_{k=1}^\infty \widetilde{P}_X^{\boldsymbol{\theta}}(\widetilde{A}_k)$.*

Proof *Let $\widetilde{Q}_n = \lim_{m\to\infty} \widetilde{S}_m(n)$ with $\widetilde{S}_m(n) = \bigcup_{k=n+1}^m \widetilde{A}_k$ where n is a fixed integer and $m = n+1, n+2, \ldots$. This concludes that $\widetilde{P}_X^{\boldsymbol{\theta}}(\bigcup_{k=1}^\infty \widetilde{A}_k) = \widetilde{P}_X^{\boldsymbol{\theta}}(\cup_{k=1}^n \widetilde{A}_k \cup \widetilde{Q}_n)$. Since $\widetilde{A}_i \cap \widetilde{A}_j = \emptyset \ \forall i \neq j$, we also have $\widetilde{P}_X^{\boldsymbol{\theta}}(\bigcup_{k=1}^\infty \widetilde{A}_k) = \widetilde{P}_X^{\boldsymbol{\theta}}(\cup_{k=1}^n \widetilde{A}_k) \oplus \widetilde{P}_X^{\boldsymbol{\theta}}(\widetilde{Q}_n) = \bigoplus_{k=1}^n \widetilde{P}_X^{\boldsymbol{\theta}}(\widetilde{A}_k) \oplus \widetilde{P}_X^{\boldsymbol{\theta}}(\widetilde{Q}_n)$. Therefore, the proof can be simply verified as $\widetilde{P}_X^{\boldsymbol{\theta}}(\widetilde{Q}_n) \to 0$ and $\bigoplus_{k=1}^n \widetilde{P}_X^{\boldsymbol{\theta}}(\widetilde{A}_k) \to \bigoplus_{k=1}^\infty \widetilde{P}_X^{\boldsymbol{\theta}}(\widetilde{A}_k)$ as $n \to \infty$.*

The following result extends a notion of Borel-Cantelli Lemma for the fuzzy probability of **FE**s.

Proposition 2.15 *Let $\{\widetilde{A}_k\}_{k=1}^\infty$ be a countable collection of* **FE***s. If $P_d(\bigoplus_{n=1}^\infty \widetilde{P}_X^{\boldsymbol{\theta}}(\widetilde{A}_n) \preceq I\{\infty\}) = 1$, then $\widetilde{P}_X^{\boldsymbol{\theta}}(\lim_{n\to\infty} \sup \widetilde{A}_n) = I\{0\}$.*

Proof *It is enough to show that* $(\widetilde{P}^{\boldsymbol{\theta}}_X(\limsup_{n\to\infty}\widetilde{A}_n))^L_\alpha = 0$ *and* $(\widetilde{P}^{\boldsymbol{\theta}}_X(\limsup_{n\to\infty}\widetilde{A}_n))^U_\alpha = 0$ *for any* $\alpha \in (0,1]$. *For this purpose, we have:*

$$
\begin{aligned}
(\widetilde{P}^{\boldsymbol{\theta}}_X(\limsup_{n\to\infty}\widetilde{A}_n))^U_\alpha &= E_W(\lim_{n\to\infty} P^{\boldsymbol{\theta}}_X(\cup_{k=n}^{\infty}\widetilde{A}_k[W\alpha])) \\
&\leq E_W(\lim_{n\to\infty}\sum_{k=n}^{\infty} P^{\boldsymbol{\theta}}_X(\widetilde{A}_k[W\alpha])) = 0,
\end{aligned}
$$

for every $\alpha \in (0,1]$ *due to* $\sum_{n=1}^{\infty} P^{\boldsymbol{\theta}}_X(\widetilde{A}_1[W\alpha]) < \infty$. *Similarly, we have* $(\widetilde{P}^{\boldsymbol{\theta}}_X(\limsup_{n\to\infty}\widetilde{A}_n))^L_\alpha = 0$, *for every* $\alpha \in (0,1]$, *and hence the proof is complete.*

Remark 2.2 *Here, some common existing fuzzy probabilities of a **FE** are recalled for a probability space adopted with a random variable* $X \sim f^{\boldsymbol{\theta}}_X$ *with* $\boldsymbol{\theta} = (\theta_1, \theta_2, \ldots, \theta_p)^\top \in \mathbb{R}^p$. *Yager [211] introduced a probability of a **FE** of* $\widetilde{A}$ *as a fuzzy set of* $\widetilde{P}_{Y1}(\widetilde{A})$ *with the following membership function:*

$$
\widetilde{P}_{Y1}(\widetilde{A})(x) = \sup\left\{\alpha \in [0,1] : x \in \left[P^{\boldsymbol{\theta}}_X(\widetilde{A}[1]), P^{\boldsymbol{\theta}}_X(\widetilde{A}[\alpha])\right]\right\}, \; x \in (0,1].
$$

It should be noted that $\widetilde{P}_{Y1}(\widetilde{A})$ *can be regarded as 'about* $P(\widetilde{A}[1])$*'. However, (1) it has a decreasing membership function and (2) it does not satisfy the condition (2) in Theorem 2.5. Klement [112] modified* $\widetilde{P}_{Y1}(\widetilde{A})$ *by introducing a fuzzy probability* $\widetilde{P}_K(\widetilde{A})$ *with the following membership function:*

$$
\widetilde{P}_K(\widetilde{A})(x) = \sup\left\{\alpha \in [0,1] : x \in \left[0, P^{\boldsymbol{\theta}}_X(\widetilde{A}[\alpha])\right]\right\}, \; x \in (0,1].
$$

It can be regarded as 'about $P(\widetilde{A}[1])$*'. However,* $\widetilde{P}_K(\widetilde{A})$ *satisfies neither of the conditions (2) and (4) in Theorem 2.5 Moreover, Yager [212] modified* $\widetilde{P}_K(\widetilde{A})$ *by defining a fuzzy probability of* $\widetilde{A}$ *with the following membership function:*

$$
\widetilde{P}_{Y2}(\widetilde{A})(x) = \sup\left\{\alpha \in [0,1] : x \in \left[[1 - P^{\boldsymbol{\theta}}_X(\widetilde{A}^c[\alpha]), 1] \cap [0, P^{\boldsymbol{\theta}}_X(\widetilde{A}[\alpha])]\right]\right\}, \; x \in (0,1].
$$

It should be mentioned that $\widetilde{P}_{Y2}(\widetilde{A})$ *is not a normal fuzzy set in general and its membership function follows a decreasing trend. It also does not satisfy the condition (2) and (4) in Theorem 2.5. Moreover, Hesamian and Shams [88] proposed a fuzzy probability* ($\widetilde{P}_{HS}(\widetilde{A})$) *with the following membership function:*

$$
\widetilde{P}_{HS}(\widetilde{A})[\alpha] = \left[P^{\boldsymbol{\theta}}_X(\widetilde{A}[1-\alpha/2]), P^{\boldsymbol{\theta}}_X(\widetilde{A}[\alpha/2])\right].
$$

Note that $\widetilde{P}_{HS}$ *can be interpreted as 'about* $P(\widetilde{A}[0.5])$*'. Therefore, it does not care about highly possible values of the **FE** of* $\widetilde{A}$. *However, unlike the previous methods, it satisfies all conditions of Theorem 2.5. This is while the proposed* *method not only provided a fuzzy probability as 'about Zadeh's probability' but also satisfied all conditions of Theorem 2.5. Unlike the other methods, the main advantage of the proposed method is that it can be applied to evaluate the probability of a* **FE** *in cases where the parameters of the population are also fuzzy quantities rather than exact ones.*

Next, the above probability of a **FE** is developed for the cases that some parameters are fuzzy quantities instead of exact values.

Definition 2.8 *Let X be a random variable with the density function of $f_X^{\boldsymbol{\theta}}$, $\boldsymbol{\theta} = (\theta_1, \theta_2, \ldots, \theta_p)^\top \in \mathbb{R}^p$. Assume that some of elements $\boldsymbol{\theta}$ can be reported by fuzzy sets. For this purpose, assume that $\widetilde{\boldsymbol{\theta}} = (\widetilde{\boldsymbol{\theta}}_0, \boldsymbol{\theta}_1)^\top$ where $\widetilde{\boldsymbol{\theta}}_0 = (\widetilde{\theta}_1^0, \widetilde{\theta}_2^0, \ldots, \widetilde{\theta}_k^0)^\top$ and $\boldsymbol{\theta}_1 = (\theta_{k+1}, \ldots, \theta_p)^\top$ with $k \leq p$. It is assumed that $\widetilde{\theta}_j^0$, $j = 1, 2, \ldots, k$ are some LR-***FN**s with membership degrees of $\widetilde{\theta}_j^0(x)$. The fuzzy probability of a* **FE** *$\widetilde{A}$ is defined to be a* ***FN*** *as $\widetilde{P}_X^{\widetilde{\boldsymbol{\theta}}}(\widetilde{A})$ with the following membership function:*

$$\widetilde{P}_X^{\widetilde{\boldsymbol{\theta}}}(\widetilde{A})(x) = \sup\left\{\alpha \in [0,1] : p \in \left[(\widetilde{P}_X^{\widetilde{\boldsymbol{\theta}}}(\widetilde{A}))_\alpha^L, (\widetilde{P}_X^{\widetilde{\boldsymbol{\theta}}}(\widetilde{A}))_\alpha^U\right]\right\}, \ p \in [0,1], \quad (2.5)$$

where

$$(\widetilde{P}_X^{\widetilde{\boldsymbol{\theta}}}(\widetilde{A}))_\alpha^L = E_W\left[E_{\pi_{\widetilde{\boldsymbol{\theta}}_0}}\left[P_X^{\widetilde{\boldsymbol{\theta}}_0}(\widetilde{A}[1-(1-W)\alpha])\right]\right]$$
$$= \int_0^1 \int P_X^{\widetilde{\boldsymbol{\theta}}_0}(\widetilde{A}[1-(1-w)\alpha])\pi_{\widetilde{\boldsymbol{\theta}}_0}(\boldsymbol{z})d\boldsymbol{z}dw,$$

and

$$(\widetilde{P}_X^{\widetilde{\boldsymbol{\theta}}}(\widetilde{A}))_\alpha^U = E_W\left[E_{\pi_{\widetilde{\boldsymbol{\theta}}_0}}\left[P_X^{\widetilde{\boldsymbol{\theta}}_0}(\widetilde{A}[W\alpha])\right]\right]$$
$$= \int_0^1 \int P_X^{\widetilde{\boldsymbol{\theta}}_0}(\widetilde{A}[w\alpha])\pi_{\widetilde{\boldsymbol{\theta}}_0}(\boldsymbol{z})d\boldsymbol{z}dw.$$

in which

$$\pi_{\widetilde{\boldsymbol{\theta}}_0}(\boldsymbol{z}) = \frac{\prod_{i=1}^k \widetilde{\theta}_i^0(z_i)}{\int \prod_{i=1}^k \widetilde{\theta}_i^0(z_i)d\boldsymbol{z}},$$

is the joint density function of $(\widetilde{\theta}_1^0, \widetilde{\theta}_2^0, \ldots, \widetilde{\theta}_k^0)^\top$ at $\boldsymbol{z} = (z_1, z_2, \ldots, z_k)^\top$.

Example 2.16 *Let X be a binomial random variable with $n = 3$ and fuzzy parameter of $\widetilde{p} = (0.35; 0.05, 0.03)_{LR}$ where $L(x) = \sqrt{1-x^2}$ and $R(x) = (1-x)^2$. Assume that the membership function a* **FE** *to describe 'X is about*

2' can be expressed as $\widetilde{A} = \{\frac{0.7}{1}, \frac{1}{2}, \frac{0.7}{3}\}$. *Note that:*

$$\pi_{\widetilde{p}}(p) = \frac{\widetilde{p}(p)}{\int_0^1 \widetilde{p}(p)dp}$$
$$= 20.2964 \begin{cases} \sqrt{1 - (\frac{p-0.35}{0.05})^2}, & 0.30 < p \leq 0.35, \\ (1 - \frac{p-0.35}{0.03})^2, & 0.35 < p \leq 0.38, \\ 0, & \mathbb{R} - [0.30, 38]. \end{cases}$$

Therefore, for any $\alpha \in [0, 1]$, *the upper and lower* α-*cuts of* $\widetilde{P}_X^{\widetilde{p}}(\widetilde{A})$ *can be obtained as:*

$$(\widetilde{P}_X^{\widetilde{p}}(\widetilde{A}))_\alpha^U = E_W\Big[E_{\pi_{\widetilde{p}}}(P_X^{\widetilde{p}}(\widetilde{A}[W\alpha]))\Big]$$
$$= \int_{\{w:0<w\alpha\leq 0.7\}} \int_{0.30}^{0.38} P_X^p(X \in \{1,2,3\})\pi_{\widetilde{p}}(p)dpdw$$
$$+ \int_{\{w:0.7<w\alpha\leq 1\}} \int_{0.30}^{0.38} P_X^p(X = 2)\pi_{\widetilde{p}}(p)dpdw$$
$$= \min\{1, \frac{0.7}{\alpha}\} \int_{0.30}^{0.38} (3p(1-p)^2 + 3p^2(1-p) + p^3)\pi_{\widetilde{p}}(p)dp$$
$$+ (\min\{1, \frac{1}{\alpha}\} - \min\{1, \frac{0.7}{\alpha}\}) \int_{0.30}^{0.38} 3p^2(1-p)\pi_{\widetilde{p}}(p)dp$$
$$= 0.704843 \min\{1, \frac{0.7}{\alpha}\} + 0.223495(\min\{1, \frac{1}{\alpha}\} - \min\{1, \frac{0.7}{\alpha}\}),$$

and

$$(\widetilde{P}_X^{\widetilde{p}}(\widetilde{A}))_\alpha^L = E_W\Big[E_p(P_X^p(\widetilde{A}[1-(1-W)\alpha]))\Big]$$
$$= \int_{\{w:0<1-(1-w\alpha)\leq 0.7\}} \int_{0.30}^{0.38} P_X^p(X \in \{1,2,3\})\pi_{\widetilde{p}}(p)dpdw$$
$$+ \int_{\{w:0.7<1-(1-w\alpha)\leq 1\}} \int_{0.30}^{0.38} P_X^p(X = 2)\pi_{\widetilde{p}}(p)dpdw$$
$$= 0.704843(\max\{0, 1 - \frac{0.3}{\alpha}\} - \max\{0, 1 - \frac{1}{\alpha}\})$$
$$+ 0.223495(1 - \max\{0, 1 - \frac{0.3}{\alpha}\}).$$

Some α-*cuts of* $\widetilde{P}_X^{\widetilde{p}}(\widetilde{A})$ *are listed in Table 2.2. The membership function of* $\widetilde{P}_X^{\widetilde{p}}(\widetilde{A})$ *is also plotted in Fig. 2.2. Thus, it can be said that* $\widetilde{P}_X^{\widetilde{p}}(\widetilde{A})$ *is a* ***FN*** *as 'about 0.56'.*

Example 2.17 *Suppose* $X \sim U(0, 1/\theta)$ *with the density function of* $f(x) = \theta$, $0 < x < 1/\theta$. *Assume that* θ *is reported as* $\widetilde{\theta} = (2; 0.1)_T$. *Consider a* ***FE*** *to*

TABLE 2.2
Some α-cuts of $\widetilde{P}_X^{\widetilde{p}}(\widetilde{A})$ in Example 2.16.

α	$(\widetilde{P}_X^{\widetilde{p}}(\widetilde{A}))_\alpha^L$	$(\widetilde{P}_X^{\widetilde{p}}(\widetilde{A}))_\alpha^U$
0.37	0.31456	0.704843
0.47	0.397599	0.704843
0.57	0.451502	0.704843
0.67	0.489314	0.704843
0.77	0.517304	0.661084
0.87	0.538861	0.610786
0.97	0.555972	0.570859
1	0.560438	0.560438

establish 'X is essentially greater than 0.4' with the following membership function:

$$\widetilde{A}(x) = \begin{cases} 0, & x < 0.3, \\ \frac{x-0.3}{0.1}, & 0.3 \le x \le 0.4, \\ 1, & x > 0.4. \end{cases}$$

First note that $|\widetilde{\theta}| = \int_{1.9}^{2.1} \widetilde{\theta}(x)dx = 0.1$. *Then the lower and upper* α*-cuts of* $\widetilde{P}_X^{\widetilde{\theta}}(\widetilde{A})$ *are:*

$$(\widetilde{P}_X^{\widetilde{\theta}}(\widetilde{A}))_\alpha^L = E_W\left[P_X^{\widetilde{\theta}}(\widetilde{A}[1-(1-W)\alpha])\right] = \frac{1}{|\widetilde{\theta}|}\int_0^1\int_{1.9}^{2.1}\int_{0.4-0.1(1-w)\alpha}^{1/z} zdx\widetilde{\theta}(z)dzdw$$

and

$$(\widetilde{P}_X^{\widetilde{\theta}}(\widetilde{A}))_\alpha^U = E_W\left[P_X^{\theta}(\widetilde{A}[W\alpha])\right] = \frac{1}{|\widetilde{\theta}|}\int_0^1\int_{1.9}^{2.1}\int_{0.3+0.1w\alpha}^{1/z} zdx\widetilde{\theta}(z)dzdw,$$

where

$$\widetilde{\theta}(z) = \begin{cases} 10\frac{z-1.9}{0.1}, & 1.9 \le z \le 2, \\ 10\frac{2.1-z}{0.1}, & 2 \le z \le 2.1, \\ 0, & z \in \mathbb{R} - [1.9, 2.1]. \end{cases}$$

Some α*-cuts of* $\widetilde{P}_X^{\widetilde{\theta}}(\widetilde{A})$ *are summarized in Table 2.3. The membership function of* $\widetilde{P}_X^{\widetilde{\theta}}(\widetilde{A})$ *is also plotted in Fig. 2.3.*

The following results summarize some of the main properties of $\widetilde{P}^{\widetilde{\theta}}$ whose verifications are left as exercises.

Theorem 2.7 *Let X be a random variable with the density function of* $f_X^{\widetilde{\theta}}$. *Then the fuzzy probability of* $\widetilde{P}_X^{\widetilde{\theta}}(\widetilde{A})$ *meets the following conditions:*

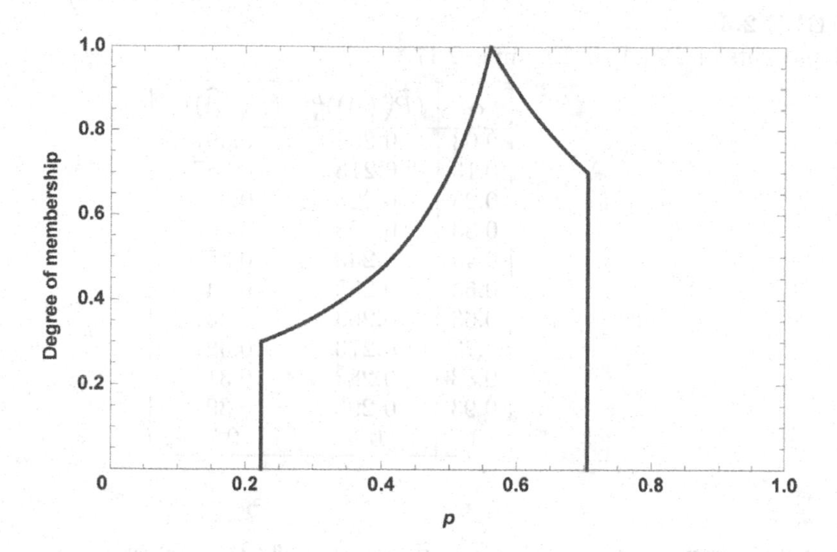

FIGURE 2.2
Membership function of $\widetilde{P}_X(\widetilde{A})$ in Example 2.16.

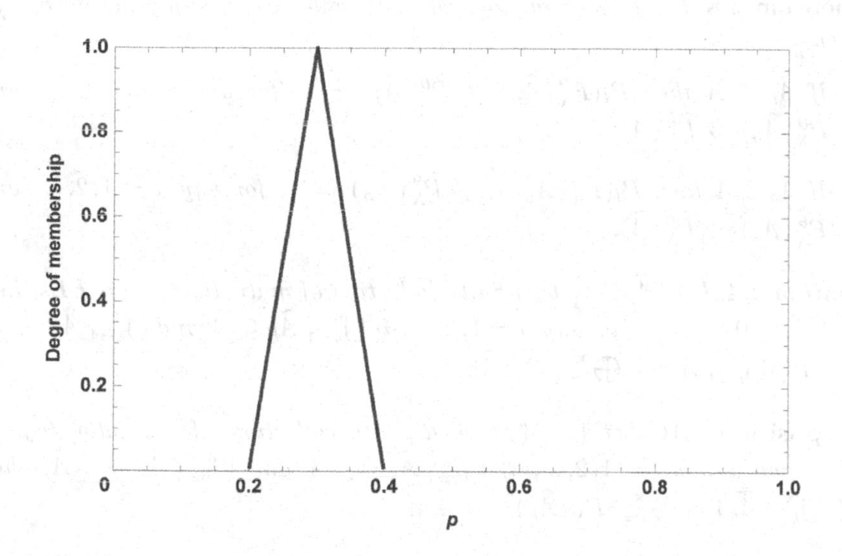

FIGURE 2.3
Membership function of $\widetilde{P}_X(\widetilde{A})$ in Example 2.17.

TABLE 2.3
Some α-cuts of $\widetilde{P}_X^{\widetilde{\theta}}(\widetilde{A})$ in Example 2.17.

α	$(\widetilde{P}_X^{\widetilde{\theta}}(\widetilde{A}))_\alpha^L$	$(\widetilde{P}_X^{\widetilde{\theta}}(\widetilde{A}))_\alpha^U$
0.03	0.203	0.397
0.13	0.213	0.387
0.23	0.223	0.377
0.33	0.233	0.367
0.43	0.243	0.357
0.53	0.253	0.347
0.63	0.263	0.337
0.73	0.273	0.327
0.83	0.283	0.317
0.93	0.293	0.307
1	0.3	0.3

(1) *For two* ***FE****s of* $\widetilde{A}$ *and* $\widetilde{B}$, *if* $\widetilde{A} \subseteq \widetilde{B}$ *then* $P_d(\widetilde{P}_X^{\widetilde{\theta}}(\widetilde{A}) \preceq \widetilde{P}_X^{\widetilde{\theta}}(\widetilde{B})) = 1$.

(2) *For a* ***FE*** *of* $\widetilde{A}$, $\widetilde{P}_X^{\widetilde{\theta}}(\widetilde{A}^c) = 1 \ominus \widetilde{P}_X^{\widetilde{\theta}}(\widetilde{A})$.

(3) *For two* ***FE****s of* $\widetilde{A}$ *and* $\widetilde{B}$, $P_d(\widetilde{P}_X^{\widetilde{\theta}}(\widetilde{A} \cup \widetilde{B}) \preceq \widetilde{P}_X^{\widetilde{\theta}}(\widetilde{A}) \oplus \widetilde{P}_X^{\widetilde{\theta}}(\widetilde{B})) = 1$.

Theorem 2.8 *Let* X *be a random variable with the density function of* $f_X^{\widetilde{\theta}}$. *Then:*

(1) *If* $\widetilde{A}_n \uparrow \widetilde{A}$ *then* $P_d(\widetilde{P}_X^{\widetilde{\theta}}(\widetilde{A}_n) \preceq \widetilde{P}_X^{\widetilde{\theta}}(\widetilde{A})) = 1$, *for any* $n = 1, 2, \ldots$ *and* $\widetilde{P}_X^{\widetilde{\theta}}(\widetilde{A}_n) \to \widetilde{P}_X^{\widetilde{\theta}}(\widetilde{A})$.

(2) *If* $\widetilde{A}_n \downarrow \widetilde{A}$ *then* $P_d(\widetilde{P}_X^{\widetilde{\theta}}(\widetilde{A}_{n+1}) \preceq \widetilde{P}_X^{\widetilde{\theta}}(\widetilde{A}_n)) = 1$, *for any* $n = 1, 2, \ldots$ *and* $\widetilde{P}_X^{\widetilde{\theta}}(\widetilde{A}_n) \to \widetilde{P}_X^{\widetilde{\theta}}(\widetilde{A})$.

Lemma 2.4 *Let* $\{\widetilde{A}_k\}_{k=1}^n$ *be a finite collection of mutually disjoint* ***FE****s, i.e.,* $\widetilde{A}_i \cap \widetilde{A}_j = \emptyset \; \forall i \neq j$. *For any* $n = 1, 2, ...$, *if* $\bigcup_{k=1}^n \widetilde{A}_k \subseteq \widetilde{A}$ *and* $\bigcup_{k=1}^n \widetilde{A}_k \to \widetilde{A}$, *then* $\widetilde{P}_X^{\widetilde{\theta}}(\bigcup_{k=1}^\infty \widetilde{A}_k) = \bigoplus_{k=1}^\infty \widetilde{P}_X^{\widetilde{\theta}}(\widetilde{A}_k)$.

Proposition 2.16 *Let* $\{\widetilde{A}_k\}_{k=1}^n$ *be a finite collection of mutually disjoint* ***FE****s. For any* $n = 1, 2, ...$, *if* $\bigcup_{k=1}^n \widetilde{A}_k \subseteq \widetilde{A}$ *and* $\bigcup_{k=1}^n \widetilde{A}_k \to \widetilde{A}$, *then* $\widetilde{P}_X^{\widetilde{\theta}}(\bigcup_{k=1}^\infty \widetilde{A}_k) = \bigoplus_{k=1}^\infty \widetilde{P}_X^{\widetilde{\theta}}(\widetilde{A}_k)$.

Proposition 2.17 *Let* $\{\widetilde{A}_k\}_{k=1}^\infty$ *be a countable collection of* ***FE****s. If* $P_d(\bigoplus_{n=1}^\infty \widetilde{P}_X^{\widetilde{\theta}}(\widetilde{A}_n) \preceq I\{\infty\}) = 1$, *then* $\widetilde{P}_X^{\widetilde{\theta}}(\lim_{n\to\infty} \sup \widetilde{A}_n) = I\{0\}$.

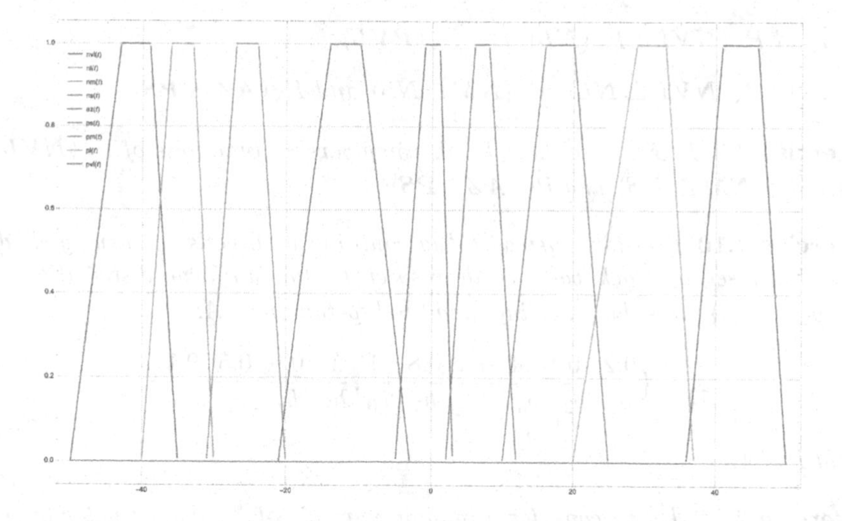

FIGURE 2.4
Membership functions of (**NVL**, **NL**, ...,**PVL**) in Exercise 2.8.

2.5 Exercise

Exercise 2.1 *Prove Theorem 1.20.*

Exercise 2.2 *Find the probability of* $\widetilde{C}_{12}$, $\widetilde{C}_{21}$, *and* $\widetilde{C}_{22}$ *in Example 2.9.*

Exercise 2.3 *Prove Proposition 1.27.*

Exercise 2.4 *Prove Proposition 1.28.*

Exercise 2.5 *Prove Proposition 1.33.*

Exercise 2.6 *Complete the proof of Proposition 1.35.*

Exercise 2.7 *Prove Proposition 1.39.*

Exercise 2.8 *Suppose that yields (in %) on stocks A is is a continuous random variable uniformly distributed on* $(-50, 50)$. *To describe the real variable 'stock yield', a fuzzy partition Negative Very Large (**NVL**), Negative Large (**NL**), Negative Medium (**NM**), Negative Small (**NS**), Approximately Zero (**AZ**), Positive Small (**PS**), Positive Medium (**PM**), Positive Large (**PL**), and Positive Very Large (**PVL**) of possible stock yields are used. The membership functions of such fuzzy sets are shown in Fig. 2.4.*

(1) *Find* $P_X(\boldsymbol{NVL})$, $P_X(\boldsymbol{NL})$, ..., $P_X(\boldsymbol{PVL})$.

(2) *Find* $P_X(\boldsymbol{NVL} \triangle \boldsymbol{NL})$, $P_X(\boldsymbol{NM} \cap \boldsymbol{NS})$ *and* $P_X(\boldsymbol{AZ} \triangle \boldsymbol{PS})$.

Exercise 2.9 *In Exercise 2.8, plot the membership functions of* $\widetilde{P}_X(\boldsymbol{NVL} \cup \boldsymbol{NL})$, $\widetilde{P}_X(\boldsymbol{NM} \triangle \boldsymbol{NS})$ *and* $\widetilde{P}_X(\boldsymbol{AZ} \cap \boldsymbol{PS})$.

Exercise 2.10 *Consider instead a bag containing 10 balls of varying shade. The fuzzy set of black balls (with respect to the uniform distribution on* $\{b_1, b_2, ..., b_{10}\}$*) has the following membership function,* $\widetilde{A}$*:*

$$\widetilde{A} = \{\frac{0.2}{b_1}, \frac{0}{b_2}, \frac{0.4}{b_3}, \frac{0.6}{b_4}, \frac{0.8}{b_5}, \frac{0}{b_6}, \frac{1}{b_7}, \frac{0.9}{b_8}, \frac{0.5}{b_9}, \frac{0.1}{b_{10}}\}.$$

Find $P_X(\widetilde{A})$.

Exercise 2.11 *Let us consider a random variable of* X*, the diameter of particles, which is uniformly distributed on* $(0,1)$ *and three* ***FEs*** *of* $\widetilde{A}$*: 'fine',* $\widetilde{B}$*: 'middle size', and* $\widetilde{C}$*: 'coarse'. The membership functions of* $\widetilde{A}$*,* $\widetilde{B}$*, and* $\widetilde{C}$ *are given as:*

$$\widetilde{A}(x) = \begin{cases} 1, & 0 \le x \le 0.2, \\ \frac{0.35-x}{0.15}, & 0.2 < x \le 0.35, \end{cases}$$

$$\widetilde{B}(x) = \begin{cases} \frac{x-0.2}{0.15}, & 0.2 \le x \le 0.35, \\ 1, & 0.35 <\le x \le 0.65, \\ \frac{0.80-x}{0.15}, & 0.65 < x \le 0.80, \end{cases}$$

$$\widetilde{C}(x) = \begin{cases} \frac{0.65-x}{0.15}, & 0.65 < x \le 0.80, \\ 1, & 0.8 \le x \le 1. \end{cases}$$

(1) *Find the probability of* $\widetilde{A}$*,* $\widetilde{B}$*, and* $\widetilde{C}$.

(2) *Find the probability of* $\widetilde{B} - \widetilde{A}$ *and* $\widetilde{B} \cup \widetilde{C}$.

Exercise 2.12 *The performance evaluation in an aerospace firm (on* $[0, 100]$*) is made by using a system based on the fuzzy sets. According to this system, the evaluator assigns a fuzzy value of performance to each worker by taking into account some certain criteria such as experience, responsibility, mental, and physical efforts. Let* X*: performance evaluation be a normal random variable with respective mean and variance of 60 and 25. If* $\widetilde{A} = (80; 10)_T$ *denotes 'more or less high performance', then identify and plot the membership functions of* $\widetilde{P}_{Y1}(\widetilde{A})$, $\widetilde{P}_K(\widetilde{A})$, $\widetilde{P}_{Y2}(\widetilde{A})$ *and* $\widetilde{P}(\widetilde{A})$.

Exercise 2.13 *In Exercise 12, if* $X \sim N(\widetilde{\mu}, 25)$ *where* $\widetilde{\mu} = (60; 10)_T$ *then plot the membership function of* $\widetilde{P}_X(\widetilde{A})$.

Exercise 2.14 *Let $\{\widetilde{A}_k\}_{k=1}^{\infty}$ be a finite sequence of* **FE***s. Prove that*

(1) $P_X(\bigcup_{k=1}^{n}\widetilde{A}_k) \leq \sum_{k=1}^{n} P_X(\widetilde{A}_k)$.

(2) *If* $\bigcup_{k=1}^{n}\widetilde{A}_k \uparrow \widetilde{A}$, *then* $P_X(\bigcup_{k=1}^{\infty}\widetilde{A}_k) \leq \sum_{k=1}^{\infty} P_X(\widetilde{A}_k)$.

Exercise 2.15 *Let $\{\widetilde{A}_k\}_{k=1}^{\infty}$ be a finite sequence of* **FE***s.*

(1) *Prove that* $P_d(\widetilde{P}_X(\bigcup_{k=1}^{n}\widetilde{A}_k) \preceq \bigoplus_{k=1}^{n}\widetilde{P}_X(\widetilde{A}_k)) = 1$:

(2) *If* $\bigcup_{k=1}^{n}\widetilde{A}_k \uparrow \widetilde{A}$ *then* $P_d(\widetilde{P}_X(\bigcup_{k=1}^{\infty}\widetilde{A}_k) \preceq \bigoplus_{k=1}^{\infty}\widetilde{P}_X(\widetilde{A}_k)) = 1$.

Exercise 2.16 *Let $\{\widetilde{A}_k\}_{k=1}^{\infty}$ be a finite sequence of* **FE***s in which $\widetilde{A}_i \cap \widetilde{A}_j = \emptyset$ for all $i \neq j$ and $\bigcup_{k=1}^{n}\widetilde{A}_k \uparrow \widetilde{A}$. Show that $P_X(\bigcup_{k=1}^{\infty}\widetilde{A}_k \leq P_X(\widetilde{B})$.*

Exercise 2.17 *Let $\{\widetilde{A}_k\}_{k=1}^{\infty}$ be a finite sequence of* **FE***s in which $\widetilde{A}_i \cap \widetilde{A}_j = \emptyset$ for all $i \neq j$ and $\bigcup_{k=1}^{n}\widetilde{A}_k \uparrow \widetilde{A}$. Prove that $P_d(\widetilde{P}_X(\bigcup_{k=1}^{\infty}\widetilde{A}_k) \preceq \widetilde{P}_X(\widetilde{B})) = 1$.*

Exercise 2.18 *Assume that $\widetilde{A}_{kj_k}$ is a set of mutually disjoints of* **FE***s for $k = 1, 2, \ldots, n$ and $j_k = 1, 2, \ldots, m_k$. Prove that:*

$$P_X\Big((\bigcup_{j_1=1}^{m_1}\widetilde{A}_{1j_1}) \times (\bigcup_{j_2=1}^{m_2}\widetilde{A}_{2j_2}) \times \ldots \times (\bigcup_{j_n=1}^{m_n}\widetilde{A}_{nj_n})\Big) =$$

$$\sum_{j_1=1}^{m_1}\sum_{j_2=1}^{m_2}\cdots\sum_{j_n=1}^{m_n} P_X(\widetilde{A}_{1j_1} \times \widetilde{A}_{2j_2} \times \ldots \times \widetilde{A}_{nj_n}).$$

Exercise 2.19 *Prove that $\widetilde{P}_{Y1}$ does not satisfy condition (2) in Theorem 2.5.*

Exercise 2.20 *Prove that $\widetilde{P}_K$ satisfies neither of conditions (2) and (4) in Theorem 2.5.*

Exercise 2.21 *Prove that $\widetilde{P}_{Y2}$ is not a normal fuzzy set in general and it does not satisfy condition (2) and (4) in Theorem 2.5.*

Exercise 2.22 *Prove that $\widetilde{P}_{HS}$ satisfied all conditions of Theorem 2.5.*

Exercise 2.23 *In Exercise 10, assume that $\widetilde{B}$ indicates the 'about 5' with the following membership function:*

$$\widetilde{B} = \{\frac{0.2}{1}, \frac{0.4}{2}, \frac{0.6}{3}, \frac{0.8}{4}, \frac{1}{5}, \frac{0.8}{6}, \frac{0.6}{7}, \frac{0.4}{8}, \frac{0.2}{9}\}.$$

(1) *Find* $P_X(\widetilde{A}|\widetilde{B})$.

(2) *Are $\widetilde{A}$ and $\widetilde{B}$ independent?*

Exercise 2.24 *23- Let X be a random variable with probability density function of* $P_X(X=3)=0.25$, $P_X(X=5)=0.40$, *and* $P_X(X=7)=0.35$. *Assuming* $\widetilde{A}=\{\frac{0.5}{3},\frac{1}{5},\frac{0.6}{7}\}$ *and* $\widetilde{B}=\{\frac{1}{3},\frac{0.6}{5}\}$ *find* $P_X(\widetilde{A}\times\widetilde{B})$.

Exercise 2.25 *For a countable collection of* **FEs** $\{\widetilde{A}_n\}_{n\geq 1}$, *prove that* $\lim_{n\to\infty} P_X(\liminf_{k\to\infty}(\widetilde{A}_n\cap\widetilde{A}_k^c))=0$.

Exercise 2.26 *For a countable collection of* **FEs** $\{\widetilde{E}_n\}_{n\geq 1}$, *prove that:*

$$P_X(\cup_{i=1}^N\widetilde{E}_i)=\sum_{i=1}^N P_X(\widetilde{E}_i)+\sum_{i_1<i_2} P_X(\widetilde{E}_{i_1}\cap\widetilde{E}_{i_2})+...$$

$$(-1)^{n+1}\sum_{i_1<i_2<...<i_n} P_X(\widetilde{E}_{i_1}\cap\widetilde{E}_{i_2}\cap...\widetilde{E}_{i_n})+...+(-1)^{N+1}P_X(\widetilde{E}_1\cap\widetilde{E}_2\cap...\widetilde{E}_N).$$

Exercise 2.27 *To numerically compare the proposed fuzzy probability of* $\widetilde{P}_X$ *with respect to* $\widetilde{P}_{Y1}$, $\widetilde{P}_K$, $\widetilde{P}_{HS}$, *and* $\widetilde{P}_{Y2}$, *assume* $\mathbb{X}=\{x_1,x_2,x_3,x_4,x_5\}$ *with* $P_X(\{x_1\})=0.05$, $P_X(\{x_2\})=0.2$, $P_X(\{x_3\})=0.15$, $P_X(\{x_4\})=0.4$, $P_X(\{x_5\})=0.2$ *and* $\widetilde{A}=\{\frac{x_1}{0.2},\frac{x_2}{0.7},\frac{x_3}{0.8},\frac{x_4}{1},\frac{x_5}{0.6}\}$. *Show that the membership functions of* $\widetilde{P}_{Y1}(\widetilde{A})$, $\widetilde{P}_K(\widetilde{A})$ *and* $\widetilde{P}_{Y2}(\widetilde{A})$ *are as follows:*

$$\widetilde{P}_{Y1}(\widetilde{A})=\left\{\frac{0.2}{1},\frac{0.6}{0.95},\frac{0.7}{0.75},\frac{0.8}{0.55},\frac{1}{0.4}\right\},$$

$$\widetilde{P}_K(\widetilde{A})(x)=\begin{cases}1, & 0.00<x\leq 0.40,\\ 0.8, & 0.4\leq x\leq 0.55,\\ 0.7, & 0.55<x\leq 0.75,\\ 0.6, & 0.75<x\leq 0.95,\\ 0.2, & 0.95<x\leq 1.00,\end{cases}$$

$$\widetilde{P}_{Y2}(\widetilde{A})(x)=\begin{cases}0, & x=0.00,\\ 0.2, & 0.0<x\leq 0.05,\\ 0.6, & 0.05<x\leq 0.20,\\ 0.7, & 0.20<x\leq 0.45,\\ 0.8, & 0.45<x\leq 0.55,\\ 0.7, & 0.55<x\leq 0.75,\\ 0.6, & 0.75<x\leq 0.95,\\ 0.2, & 0.95<x\leq 1.00,\end{cases}$$

$$\widetilde{P}_{HS}(\widetilde{A})=\left\{\frac{0.4}{0.4},\frac{0.6}{0.55},\frac{0.8}{0.75},\frac{1}{0.95},\frac{0.4}{1}\right\}.$$

Further, find the lower and upper α*-cuts of* $\widetilde{P}(\widetilde{A})$ *and compare them with other methods.*

Exercise 2.28 *Prove Theorem 1.52.*

Exercise 2.29 *Prove Theorem 1.53.*

Exercise 2.30 *Prove Lemma 1.54.*

Exercise 2.31 *Prove Proposition 1.55.*

Exercise 2.32 *Prove Proposition 1.56.*

2.6 Glossary

Conditional probability: Rule for the calculation of conditional probability of fuzzy events.

Fuzzy event: A fuzzy-valued event.

Fuzzy probability: Generalization of a probability measure whose values are fuzzy sets.

Fuzzy Bayes' formula: Rule for the calculation of fuzzy probabilities of fuzzy hypotheses conditional on observed fuzzy events.

Fuzzy independence: A generalization of independence of fuzzy events.

Fuzzy parameter: Generalization of parameters as fuzzy numbers.

Fuzzy product event: Generalization of product event to the situation of fuzzy event.

Non-fuzzy probability: Synonym for exact probability of a fuzzy event.

3

Descriptive Statistics Based on Fuzzy Data

The descriptive statistics is aimed to present a mass of data in a more understandable form. The data may be summarized in numbers in the form of (a) average, or in some cases a proportion, (b) measure of variability or spread, and (c) quantities such as quartiles or percentiles. This type of classification divides the data so that certain values or percentages of the data will be above or below these marks. Furthermore, we may choose to describe the data by various graphical representations or by the bar graphs called histograms to show the data distribution over various intervals of the varying quantity. It is also necessary or desirable to consider the data in groups and determine their frequency. One shortcoming of classical descriptive statistics is that the values of relevant attribute in the sample and/or population should be exact real numbers. However, in many practical situations, it is non-exact quantities and can be recorded as fuzzy quantities. For instance, consider a frequency table for life of tire produced by a tire and rubber company. Since, under some unexpected situations, we cannot measure the tire life precisely, we can just obtain the tire life around a number. For such cases, fuzzy set theory can be used to extend the conventional descriptive statistics based on fuzzy data. For this purpose, consider a set of fuzzy data $\widetilde{x}_1, \cdots, \widetilde{x}_n$ as **TFN**s throughout this chapter. It should be noted that Fruhwirth-Schnatter [48] also suggested a procedure for descriptive statistics in the case when the values of relevant attribute in a sample are set in the form of fuzzy categories. Viertle [198] also proposed a notion of fuzzy histogram based on fuzzy data.

3.1 Central Tendency

In this section, the concept of 'average' is extended to indicate a fuzzy central value of a set of fuzzy data.

Definition 3.1 ***(Arithmetic mean based on fuzzy data)*** *Let $\widetilde{x}_1, \cdots, \widetilde{x}_n$ be a set of fuzzy data. The fuzzy arithmetic mean can be defined by:*

$$\widetilde{\overline{x}} = \frac{1}{n} \otimes \bigoplus_{i=1}^{n} \widetilde{x}_i = (\frac{1}{n}\sum_{i=1}^{n} x_i^L, \frac{1}{n}\sum_{i=1}^{n} x_i, \frac{1}{n}\sum_{i=1}^{n} x_i^U)_T. \quad (3.1)$$

DOI: 10.1201/9781003248644-3

Lemma 3.1 *Let $\widetilde{x}_1, \cdots, \widetilde{x}_n$ be a set of fuzzy data and $\widetilde{y}_i = a \otimes \widetilde{x}_i \oplus b$ for $i = 1, 2, \ldots, n$ where $a \neq 0$ and $b \in \mathbb{R}$. Then $\overline{\widetilde{y}} = a \otimes \overline{\widetilde{x}} \oplus b$.*

Proof *By definition of a fuzzy arithmetic mean,*

$$\overline{\widetilde{y}} = \frac{1}{n} \otimes \bigoplus_{i=1}^{n} \widetilde{y}_i = \frac{1}{n} \otimes \bigoplus_{i=1}^{n} (a \otimes \widetilde{x}_i \oplus b).$$

According to the arithmetic operation of ***TFNs****, we easily have:*

$$\begin{aligned}\overline{\widetilde{y}} =& \Big(\frac{1}{n}\sum_{i=1}^{n}(I(a>0)x_i^L + I(a<0)x_i^U + b),\\ & \frac{1}{n}\sum_{i=1}^{n}(ax_i + b),\\ & \frac{1}{n}\sum_{i=1}^{n}(I(a>0)x_i^U + I(a<0)x_i^L + b)\Big)_T\\ =& a \otimes \overline{\widetilde{x}} \oplus \widetilde{b}.\end{aligned}$$

Definition 3.2 ***(Geometric mean)*** *Let $\widetilde{x}_1, \cdots, \widetilde{x}_n$ be a set of fuzzy data. The fuzzy geometric mean is defined as the fuzzy n^{th} root of the product of the fuzzy data:*

$$\overline{\widetilde{x}}_g = \sqrt[n]{\otimes_{i=1}^{n}\widetilde{x}_i}. \tag{3.2}$$

Remark 3.1 *In special cases where all fuzzy data $\widetilde{x}_1, \cdots, \widetilde{x}_n$ are positive fuzzy quantities, we get:*

$$\overline{\widetilde{x}}_g = (\sqrt[n]{\prod_{i=1}^{n} x_i^L}, \sqrt[n]{\prod_{i=1}^{n} x_i}, \sqrt[n]{\prod_{i=1}^{n} x_i^U})_T. \tag{3.3}$$

Example 3.1 *The data set below include the consumption rate of eight cars (liter per 100 kilometers) from an automobile company:*

$$\begin{aligned} &\widetilde{x}_1 = (4, 5, 6)_T, && \widetilde{x}_2 = (3.5, 4, 4.5)_T,\\ &\widetilde{x}_3 = (4.5, 5, 6)_T, && \widetilde{x}_4 = (6, 6.5, 7)_T,\\ &\widetilde{x}_5 = (5, 5.5, 6)_T, && \widetilde{x}_6 = (5.5, 6, 6.5)_T,\\ &\widetilde{x}_7 = (3.5, 4.5, 5.5)_T, && \widetilde{x}_8 = (5, 6, 6.5)_T.\end{aligned}$$

Some partial calculations of $\overline{\widetilde{x}}_g$ are given in Table 3.1 showing that $\overline{\widetilde{x}}_g =$ $(4.54447, 5.25229, 5.95483)_T$.

TABLE 3.1
Calculation of the elements of fuzzy geometric mean in Example 3.1.

sample	x_i^L	x_i	x_i^U
1	4	5	6
2	3.5	4	4.5
3	4.5	5	6
4	6	6.5	7
5	5	5.5	6
6	5.5	6	6.5
7	3.5	4.5	5.5
8	5	6	6.5
	$\sqrt[8]{\prod_{i=1}^{8} x_i^L} = 4.54447$	$\sqrt[8]{\prod_{i=1}^{8} x_i} = 5.25229$	$\sqrt[8]{\prod_{i=1}^{8} x_i^U} = 5.95483$

TABLE 3.2
Calculation of the elements of fuzzy harmonic mean in Example 3.2.

sample	$1/x_i^L$	$1/x_i$	$1/x_i^U$
1	0.25	0.2	0.16667
2	0.28571	0.25	0.22222
3	0.22222	0.2	0.16667
4	0.16667	0.15385	0.14286
5	0.2	0.18182	0.16667
6	0.18182	0.16667	0.15385
7	0.28571	0.22222	0.18182
8	0.2	0.16667	0.15385
	$\frac{1}{\sum_{i=1}^{8}(\frac{8}{x_i^L})} = 4.46395$	$\frac{1}{\sum_{i=1}^{8}(\frac{8}{x_i})} = 5.19069$	$\frac{1}{\sum_{i=1}^{8}(\frac{8}{x_i^U})} = 5.90585$

Definition 3.3 ***(Harmonic mean)*** *Let $\widetilde{x}_1, \cdots, \widetilde{x}_n$ be a set of fuzzy data with $x_i^l, x_i, x_i^U \neq 0$ for all $i = 1, 2, ..., n$. The fuzzy harmonic mean can be defined as:*

$$\widetilde{\overline{x}}_h = 1 \oslash (\bigoplus_{i=1}^{n} (n \oslash \widetilde{x}_i)).$$

Example 3.2 *Recall Example 3.1. Therefore, according to Table 3.2, the fuzzy harmonic mean is $\widetilde{\overline{x}}_H = (4.46395, 5.19069, 5.90585)_T$.*

Lemma 3.2 *If $\widetilde{x}_1, \cdots, \widetilde{x}_n$ is a set of positive fuzzy data then the following relationship is valid between $\widetilde{\overline{x}}_h$, $\widetilde{\overline{x}}_g$, and $\widetilde{\overline{x}}$:*

$$\widetilde{\overline{x}}_h \preceq_{P_d} \widetilde{\overline{x}}_g \preceq_{P_d} \widetilde{\overline{x}}. \tag{3.4}$$

Proof *Since $\widetilde{x}_1, \cdots, \widetilde{x}_n$ are positive* ***FNs****,*

$$\widetilde{\overline{x}}_h = \left(\frac{1}{\sum_{i=1}^{n}(\frac{n}{x_i^L})}, \frac{1}{\sum_{i=1}^{n}(\frac{n}{x_i})}, \frac{1}{\sum_{i=1}^{n}(\frac{n}{x_i^U})}\right)_T. \tag{3.5}$$

According to the conventional inequality between arithmetic, geometric, and harmonic means, Note that:

$$\frac{1}{\sum_{i=1}^{n}(\frac{n}{x_i^L})} \leq \sqrt[n]{\prod_{i=1}^{n} x_i^L} \leq \sum_{i=1}^{n} x_i^L, \tag{3.6}$$

$$\frac{1}{\sum_{i=1}^{n}(\frac{n}{x_i})} \leq \sqrt[n]{\prod_{i=1}^{n} x_i} \leq \sum_{i=1}^{n} x_i, \tag{3.7}$$

and

$$\frac{1}{\sum_{i=1}^{n}(\frac{1}{x_i^U})} \leq \sqrt[n]{\prod_{i=1}^{n} x_i^U} \leq \sum_{i=1}^{n} x_i^U. \tag{3.8}$$

Therefore, $P_d(\widetilde{\overline{x}}_h \preceq_{P_d} \widetilde{\overline{x}}_g) = 1$ *and* $P_d(\widetilde{\overline{x}}_g \preceq_{P_d} \widetilde{\overline{x}}) = 1$ *which also concludes that* $P_d(\widetilde{\overline{x}}_h \preceq_{P_d} \widetilde{\overline{x}}) = 1$*. This completes the proof.*

Definition 3.4 ***(Median)*** *Assume that all the fuzzy data* $\widetilde{x}_1, \cdots, \widetilde{x}_n$ *are sorted in an ascending order that is* $\widetilde{x}_{(1)} \leq_{P_d} \widetilde{x}_{(2)} \preceq_{P_d} \cdots \preceq_{P_d} \widetilde{x}_{(n)}$*. The fuzzy median is defined to be the middle item of* $\widetilde{x}_{(1)}, \widetilde{x}_{(2)}, \ldots, \widetilde{x}_{(n)}$ *as:*

$$\widetilde{m} = \begin{cases} \widetilde{x}_{(\frac{n+1}{2})}, & n \text{ is odd}, \\ \frac{1}{2} \otimes (\widetilde{x}_{(\frac{n}{2})} \oplus \widetilde{x}_{(\frac{n}{2}+1)}), & n \text{ is even}. \end{cases} \tag{3.9}$$

Example 3.3 *Suppose that a portfolio with these annual returns would lose about 5% of value during five years as* ***TFNs*** *when* $\widetilde{x}_1 = (0.76, 0.78, 0.85)_T$, $\widetilde{x}_2 = (0.78, 0.80, 0.82)_T$, $\widetilde{x}_3 = (0.75, 0.83, 0.95)_T$, $\widetilde{x}_4 = (0.83, 0.87, 0.93)_T$, *and* $\widetilde{x}_5 = (0.88, 0.91, 0.92)_T$*. The preference degrees of* $P_d(\widetilde{x}_i \succeq \widetilde{x}_j)$, $i, j = 1, 2, 3, 4, 5$ *are listed in Table 3.3. It can be checked that the fuzzy data can be ordered as:*

$$\widetilde{x}_1 \preceq_{P_d} \widetilde{x}_2 \preceq_{P_d} \widetilde{x}_3 \preceq_{P_d} \widetilde{x}_5 \preceq_{P_d} \widetilde{x}_4.$$

This concludes that the fuzzy median is:

$$\widetilde{m} = \frac{1}{2} \otimes (\widetilde{x}_{(3)} \oplus \widetilde{x}_{(4)}) = \frac{1}{2} \otimes (\widetilde{x}_3 \oplus \widetilde{x}_5) = (0.875, 0.925, 0.975)_T.$$

Definition 3.5 ***(Quantile based on fuzzy data)*** *Let* $\widetilde{x}_{(1)} \preceq_{P_d} \widetilde{x}_{(2)} \preceq_{P_d} \ldots \preceq_{P_d} \widetilde{x}_{(n)}$ *be the ordered set of fuzzy data* $\widetilde{x}_1, \cdots, \widetilde{x}_n$*. For* $i = (p/100)(n+1)$*, the* p^{th} *fuzzy quantile of fuzzy data* $\widetilde{x}_1, \widetilde{x}_2, ..., \widetilde{x}_n$ *is defined as follows:*

$$\widetilde{H}_p = \begin{cases} \widetilde{x}_{(r)}, & i = r \in \mathbb{Z}^+, \\ \big((1-w) \otimes \widetilde{x}_{(r)}\big) \oplus \big(w \otimes \widetilde{x}_{(r+1)}\big), & i = r.w. \end{cases} \tag{3.10}$$

We say that $\widetilde{H}_p$ *is the fuzzy* p^{th} *quantile of fuzzy data if* $p\%$ *of fuzzy data is smaller than* $\widetilde{H}_p$*. Specifically, 1) for* $p = 1, 2, \ldots, 99$, $\widetilde{H}_p$ *is called the* p^{th} *fuzzy percentile, 2)* $\widetilde{H}_p$ *is called the* p^{th} *fuzzy decade for* $p = 10, 20, \ldots, 90$*, and 3)* $\widetilde{H}_p$ *is called* p^{th} *fuzzy quartile for* $p = 25, 50, 75$*.*

TABLE 3.3
Preference degrees in Example 3.3.

j	1	2	3	4	5
$P_d(\widetilde{x}_1 \succeq \widetilde{x}_j)$	0.5	0.2727	0.0086	0	0
$P_d(\widetilde{x}_2 \succeq \widetilde{x}_j)$	0.7273	0.5	0.0789	0	0
$P_d(\widetilde{x}_3 \succeq \widetilde{x}_j)$	0.9914	0.9211	0.5	0.0435	0.0296
$P_d(\widetilde{x}_4 \succeq \widetilde{x}_j)$	1	1	0.9565	0.5	0.0161
$P_d(\widetilde{x}_5 \succeq \widetilde{x}_j)$	1	1	0.9704	0.9839	0.5

For simplicity, the first, second, and third quantiles are denoted by $\widetilde{Q}_1 = \widetilde{H}_{25}$, $\widetilde{Q}_2 = \widetilde{H}_{50}$, and $\widetilde{Q}_3 = \widetilde{H}_{75}$.

Example 3.4 *The daily emissions of sulfur dioxide from an industrial plant in tonnes/day are reported as:*

$$\widetilde{x}_1 = (4.1, 4.2, 4.7)_T, \quad \widetilde{x}_2 = (4.4, 4.8, 4.9)_T,$$
$$\widetilde{x}_3 = (5.2, 5.4, 5.5)_T, \quad \widetilde{x}_4 = (5, 5.8, 6.3)_T,$$
$$\widetilde{x}_5 = (4.7, 5, 5.9)_T, \quad \widetilde{x}_6 = (5.4, 6, 6.1)_T,$$
$$\widetilde{x}_7 = (3.9, 4.5, 5)_T, \quad \widetilde{x}_8 = (4.5, 5.5, 5.7)_T,$$
$$\widetilde{x}_9 = (4.4, 4.7, 4.9)_T, \quad \widetilde{x}_{10} = (4.3, 5.1, 5.6)_T.$$

The preference degrees $P_d(\widetilde{x}_i \succeq \widetilde{x}_j)$ are evaluated in Table 3.4. It can be checked that the fuzzy data can be ordered as:

$$\widetilde{x}_1 \preceq_{P_d} \widetilde{x}_7 \preceq_{P_d} \widetilde{x}_9 \preceq_{P_d} \widetilde{x}_2 \preceq_{P_d} \widetilde{x}_5 \preceq_{P_d} \widetilde{x}_{10} \preceq_{P_d} \widetilde{x}_3 \preceq_{P_d} \widetilde{x}_8 \preceq_{P_d} \widetilde{x}_4 \preceq_{P_d} \widetilde{x}_6.$$

Therefore, the 60th fuzzy percentile and third fuzzy quantile can be evaluated as follows:

$$\widetilde{H}_{60} = (0.4 \otimes \widetilde{x}_{(6)}) \oplus (0.6 \otimes \widetilde{x}_{(7)}) = (0.4 \otimes \widetilde{x}_{10}) \oplus (0.6 \otimes \widetilde{x}_3) = (4.84, 5.28, 5.54)_T.$$

$$\widetilde{Q}_3 = (0.75 \otimes \widetilde{x}_{(8)}) \oplus (0.25 \otimes \widetilde{x}_{(9)}) = (0.75 \otimes \widetilde{x}_8) \oplus (0.25 \otimes \widetilde{x}_4) = (4.625, 5.575, 5.850)_T.$$

3.2 Fuzzy Deviation Criteria

To describe the quantitative difference of fuzzy data as $\widetilde{x}_i = (x_i; l_{x_i}, r_{x_i})_T$, some descriptive measures should indicate the variation or spread. Such descriptive measures are referred to as measures of variation or measures of

TABLE 3.4
Preference degrees in Example 3.4.

j	1	2	3	4	5	6	7	8	9	10
$P_d(\widetilde{x}_1 \succeq \widetilde{x}_j)$	0.5	0	0	0	0	0	0	0	0	0
$P_d(\widetilde{x}_2 \succeq \widetilde{x}_j)$	1	0.5	0.0217	0	0.0213	0	0.8947	0	1	0
$P_d(\widetilde{x}_3 \succeq \widetilde{x}_j)$	1	0.9783	0.5	0	0.973	0	0.9596	0	0.9952	0.6
$P_d(\widetilde{x}_4 \succeq \widetilde{x}_j)$	1	1	1	0.5	1	0.25	1	0.9452	1	1
$P_d(\widetilde{x}_5 \succeq \widetilde{x}_j)$	1	0.9787	0.027	0	0.5	0	0.9516	0	1	0.1739
$P_d(\widetilde{x}_6 \succeq \widetilde{x}_j)$	1	1	1	0.75	1	0.5	1	0.9042	1	0.9882
$P_d(\widetilde{x}_7 \succeq \widetilde{x}_j)$	1	0.1053	0.0404	0	0.0484	0	0.5	0	0.3214	0
$P_d(\widetilde{x}_8 \succeq \widetilde{x}_j)$	1	1	1	0.0548	1	0.0958	1	0.5	1	1
$P_d(\widetilde{x}_9 \succeq \widetilde{x}_j)$	1	0	0.0048	0	0	0	0.6786	0	0.5	0
$P_d(\widetilde{x}_{10} \succeq \widetilde{x}_j)$	1	1	0.4	0	0.8261	0.0118	1	0	1	0.5

spread. Here, we introduce some most frequently used measures of variation for fuzzy data.

Definition 3.6 ***(Sample range)*** *The generalized fuzzy difference between the smallest and the largest fuzzy data is called the fuzzy sample range and can be defined as:*

$$\widetilde{R} = \widetilde{x}_{(n)} \ominus_G \widetilde{x}_{(1)} = (x_{(n)} - x_{(1)}; |l_{x_{(n)}} - l_{x_{(1)}}|)_T. \tag{3.11}$$

Example 3.5 *Recall Example 3.4. Therefore,* $\widetilde{R} = \widetilde{x}_{(10)} \ominus_G \widetilde{x}_{(1)} = \widetilde{x}_6 \ominus_G \widetilde{x}_1 = (1.3, 1.8, 2.2)_T$.

Definition 3.7 ***(Interquartile range)*** *The fuzzy interquartile range refers to the difference between the third and first fuzzy quartiles:*

$$\widetilde{IQR} = \widetilde{Q}_3 \ominus_G \widetilde{Q}_1.$$

TABLE 3.5
Data set in Example 3.6.

No.	$\widetilde{x}_i$	No.	$\widetilde{x}_i$
1	$(83.6; 0.9, 3.9)_T$	6	$(78.7; 5.1, 2.4)_T$
2	$(86.8; 4.2, 3.4)_T$	7	$(84.7; 7, 1.2)_T$
3	$(86.4; 5.5, 7)_T$	8	$(91.1; 3.5, 0.9)_T$
4	$(87.0; 4.6, 5.7)_T$	9	$(80.0; 3.4, 6.7)_T$
5	$(90.5; 4, 3.7)_T$	10	$(81.76; 3.41, 5.89)_T$

Example 3.6 *The data set in Table 3.5 lists the values of five perceptive attributes related to eight objects, recorded during a sensory analysis experiment. The assessor was asked to evaluate his perception of each attribute for each*

TABLE 3.6
Preference degrees in Example 3.6.

j	1	2	3	4	5	6	7	8	9	10
$P_d(\widetilde{x}_1 \succeq \widetilde{x}_j)$	0.5	0.0003	0.0634	0.0021	0	1	0.8864	0	1	0.9986
$P_d(\widetilde{x}_2 \succeq \widetilde{x}_j)$	0.9997	0.5	0.4298	0.0345	0	1	1	0	1	1
$P_d(\widetilde{x}_3 \succeq \widetilde{x}_j)$	0.9366	0.5702	0.5	0.1369	0	1	1	0.0209	1	1
$P_d(\widetilde{x}_4 \succeq \widetilde{x}_j)$	0.9979	0.9655	0.8631	0.5	0	1	1	0.0079	1	1
$P_d(\widetilde{x}_5 \succeq \widetilde{x}_j)$	1	1	1	1	0.5	1	1	0.4859	1	1
$P_d(\widetilde{x}_6 \succeq \widetilde{x}_j)$	0	0	0	0	0	0.5	0	0	0	0
$P_d(\widetilde{x}_7 \succeq \widetilde{x}_j)$	0.1136	0	0	0	0	1	0.5	0	0.9883	0.8465
$P_d(\widetilde{x}_8 \succeq \widetilde{x}_j)$	1	1	0.9791	0.9921	0.5141	1	1	0.5	1	1
$P_d(\widetilde{x}_9 \succeq \widetilde{x}_j)$	0	0	0	0	0	1	0.0117	0	0.5	0
$P_d(\widetilde{x}_{10} \succeq \widetilde{x}_j)$	0.0014	0	0	0	0	1	0.1535	0	1	0.5

object on a scale between 0 and 100 as the form of lower and upper bounds, respectively. The preference degrees $P_d(\widetilde{x}_i \succeq \widetilde{x}_j)$ are evaluated in Table 3.6. Therefore, the fuzzy data can be ordered from the smallest to largest fuzzy values as:

$$\widetilde{x}_6, \widetilde{x}_9, \widetilde{x}_{10}, \widetilde{x}_7, \widetilde{x}_1, \widetilde{x}_2, \widetilde{x}_3, \widetilde{x}_4, \widetilde{x}_5, \widetilde{x}_8.$$

Thus, one can find $\widetilde{Q}_3 = (83.425, 87.875, 93.075)_T$ and $\widetilde{Q}_1 = (77.9125, 81.3200, 87.4125)_T$ and hence:

$$IQR = \widetilde{Q}_3 \ominus_G \widetilde{Q}_1 = (6.5550; 1.0425, 0.8925)_T.$$

Definition 3.8 ***(Mean absolute deviation from the mean)*** *The mean absolute deviation from the fuzzy mean is defined by:*

$$s_D = \frac{1}{n}\sum_{i=1}^{n} d_1(\widetilde{x}_i, \widetilde{\overline{x}}), \tag{3.12}$$

where $\widetilde{x}_i = (x_i^L, x_i, x_i^U)_T$ and d_1 is the absolute error distance between two ***TFNs*** *$\widetilde{A} = (a^L, a, a^U)_T$ and $\widetilde{B} = (b^L, b, b^U)_T$ given in Chapter 1 as:*

$$d_1(\widetilde{A}, \widetilde{B}) = \frac{|a^L - b^L| + |a - b| + |a^U - b^U|}{3}. \tag{3.13}$$

Example 3.7 *Consider a study on the lifetime of particular batteries in an industrial application. In this regard, 12 batteries were recorded to the nearest tenth of a year. The observed data were reported as* ***TFNs*** *$\widetilde{x}_i = (0.95x_i, x_i, 1.05x_i)_T$ where*

$$x_i:\ 4.1, 5.2, 2.8, 4.9, 5.6, 4.0, 4.1, 4.3, 5.4, 6.3, 3.3, 5.8, 3.7.$$

According to Table 3.7, we get

$$S_D = \frac{1}{13 \times 3}(10.9031 + 11.4769 + 12.0508) = 0.882840.$$

TABLE 3.7
Computing the elements of S_D in Example 3.7

i	$\|x_i^L - \overline{x}^L\|$	$\|x_i - \overline{x}\|$	$\|x_i^U - \overline{x}^U\|$
1	0.453077	0.476923	0.500769
2	0.591923	0.623077	0.654231
3	1.68808	1.776920	1.865770
4	0.306923	0.323077	0.339231
5	0.971923	1.023080	1.074230
6	0.548077	0.576923	0.605769
7	0.453077	0.476923	0.500769
8	0.263077	0.276923	0.290769
9	0.781923	0.823077	0.864231
10	1.63692	1.723080	1.809230
11	1.21308	1.276920	1.340770
12	1.16192	1.223080	1.284230
13	0.833077	0.876923	0.920769
Sum	10.9031	11.4769	12.0508

Definition 3.9 ***(Sample standard deviation)*** *The standard deviation of a set of fuzzy data can be expressed as:*

$$s^2 = \frac{1}{n-1}\sum_{i=1}^{n} d_2^2(\widetilde{x}_i, \widetilde{\overline{x}}), \tag{3.14}$$

*where $\widetilde{x}_i = (x_i^L, x_i, x_i^U)_T$ and d_2 is the square error distance between two **TFNs** of $\widetilde{A} = (a^L, a, a^U)_T$ and $\widetilde{B} = (b^L, b, b^U)_T$ defined by:*

$$d_2^2(\widetilde{A}, \widetilde{B}) = \frac{|a^L - b^L|^2 + |a - b|^2 + |a^U - b^U|^2}{3}. \tag{3.15}$$

Roughly speaking, s indicates how far, on average, the fuzzy data are from the fuzzy sample mean.

Lemma 3.3 *Assume that the fuzzy data of $\widetilde{x}_1, \widetilde{x}_2, ..., \widetilde{x}_n$ are converted to $\widetilde{y}_i = (a \otimes \widetilde{x}_i) \oplus \widetilde{b}$ for $i = 1, 2, \ldots, n$ where $a \neq 0$ and $b \in \mathbb{R}$. Then $s_{\widetilde{y}}^2 = a^2 s_{\widetilde{x}}^2$. where $s_{\widetilde{y}}^2$ and $s_{\widetilde{x}}^2$ denotes the mean absolute deviation from the fuzzy mean and sample standard deviation relevant to the fuzzy data of $\widetilde{x}_1, \widetilde{x}_2, \ldots, \widetilde{x}_n$ and $\widetilde{y}_1, \widetilde{y}_2, \ldots, \widetilde{y}_n$, respectively.*

Proof *The proof is immediately followed since $d_2^2(\widetilde{y}_i, \widetilde{\overline{y}}) = a^2 d_2^2(\widetilde{x}_i, \widetilde{\overline{x}})$ for any $a \neq 0$.*

TABLE 3.8
Computing the elements of s^2 in Example 3.8.

i	$\|x_i^L - \overline{x}^L\|^2$	$\|x_i - \overline{x}\|^2$	$\|x_i^U - \overline{x}^U\|^2$
1	12.62980	25.77510	37.11620
2	8.14444	16.62130	23.93470
3	2.11367	4.31361	6.21160
4	0.00290	0.00592	0.00853
5	1.81213	3.69822	5.32544
6	7.54136	15.39050	22.16240
7	17.19060	35.08280	50.51930
8	4.18675	8.54438	12.30390
9	0.41752	0.85207	1.22698
10	0.56828	1.15976	1.67006
11	4.63905	9.46746	13.63310
12	24.54060	50.08280	72.11930
13	23.48520	47.92900	69.01780
Sum	107.272	218.923	315.249

Example 3.8 *Consider a data set generated during the development of a sophisticated software system for running a mall on the internet on an HTTP server. The observed outputs are 13 monthly global scores for assessing the performance of the system under development and the score in each month was expressed as* $\widetilde{x}_i = (0.7x_i, x_i, 1.2x_i)_T$ *where*

$$x_i: \ 12, 13, 15, 17, 19, 21, 23, 20, 18, 16, 14, 10, 24.$$

The elements of sample variance are evaluated as shown in Table 3.8. These show that

$$s^2 = \frac{1}{3 \times 12}(107.272 + 218.923 + 315.249) = 17.8179.$$

3.3 Box Plot Based on Fuzzy Data

A boxplot can be utilized as a graphical tool to display certain characteristics of the frequency distribution of fuzzy data. A narrow box extends from the first quartile to the third one. Thus the interquartile range shows the measure of variability. The median is marked by a line extending across the box. The smallest and the largest fuzzy data of the distribution are marked, each of which is joined to the box by a straight line.

Now, consider a set of fuzzy data $\widetilde{x}_1, \widetilde{x}_2, ..., \widetilde{x}_n$ where $\widetilde{x}_i = (x_i^L, x_i, x_i^U)_T$. Here, a notion of a boxplot is introduced for fuzzy data. First assume that $\widetilde{x}_{(1)} \preceq_{P_d} \widetilde{x}_{(2)} \preceq_{P_d} \cdots \preceq_{P_d} \widetilde{x}_{(n)}$ denotes the sorted values of fuzzy data. The fuzzy lower and upper extremes of a boxplot are defined as:

$$\widetilde{x}_L = \widetilde{\max}\{\widetilde{x}_{(1)}, \widetilde{Q}_1 \ominus_G (1.5 \otimes \widetilde{IQR})\}, \tag{3.16}$$

$$\widetilde{x}_U = \widetilde{\min}\{\widetilde{x}_{(n)}, \widetilde{Q}_3 \oplus (1.5 \otimes \widetilde{IQR})\}, \tag{3.17}$$

where

1. $$\widetilde{m} = \begin{cases} \widetilde{x}_{(\frac{n+1}{2})}, & \text{n is odd,} \\ \frac{1}{2} \otimes (\widetilde{x}_{(\frac{n}{2})} \oplus \widetilde{x}_{(\frac{n}{2}+1)}), & \text{n is even,} \end{cases} \tag{3.18}$$

   is the fuzzy median.

2. $\widetilde{Q}_1 = (Q_1^L, Q_1, Q_1^U)_T$ is the first quantile that is the fuzzy median of $\widetilde{x}_{(1)} \preceq_{P_d} \widetilde{x}_{(2)} \preceq_{P_d} \cdots \preceq_{P_d} \widetilde{m}$,

3. $\widetilde{Q}_3 = (Q_3^L, Q_3, Q_3^U)_T$ the third quantile is the fuzzy median of $\widetilde{m} \preceq_{P_d} \ldots \preceq_{P_d} \widetilde{x}_{(n)}$, and

4. $\widetilde{IQR} = \widetilde{Q}_3 \ominus_G \widetilde{Q}_1 = (Q_3 - Q_1 - |(Q_3 - Q_3^L) - (Q_1 - Q_1^L)|, Q_3 - Q_1, Q_3 - Q_1 + |(Q_3^U - Q_3) - (Q_1^U - Q_1)|)_T$, is the fuzzy inter quantile range.

In this regard, we say that $\widetilde{x}^*$ is a fuzzy extreme (fuzzy outlier) if:

$$D_* = \max\{P_d(\widetilde{x}^* \succ \widetilde{x}_U), P_d(\widetilde{x}^* \prec \widetilde{x}_L)\} > 0.5. \tag{3.19}$$

A typical fuzzy boxplot is shown in Fig. 3.1.

TABLE 3.9
Data set in Example 3.9.

No.	$\widetilde{x}_i$	No.	$\widetilde{x}_i$
1	$(1.15, 1.25, 1.45)_T$	9	$(1.10, 1.35, 1.40)_T$
2	$(1.57, 1.85, 2.20)_T$	10	$(1.45, 1.70, 2.30)_T$
3	$(1.85, 2.05, 2.17)_T$	11	$(1.80, 2.10, 2.13)_T$
4	$(1.45, 1.60, 1.85)_T$	12	$(1.32, 1.40, 1.42)_T$
5	$(2.50, 2.68, 2.84)_T$	13	$(1.35, 1.65, 1.75)_T$
6	$(2.43, 2.75, 2.81)_T$	14	$(1.70, 2, 2.45)_T$
7	$(2.56, 2.65, 2.87)_T$	15	$(2.25, 2.35, 2.85)_T$
8	$(2.33, 2.40, 2.85)_T$	16	$(2.08, 2.15, 2.35)_T$

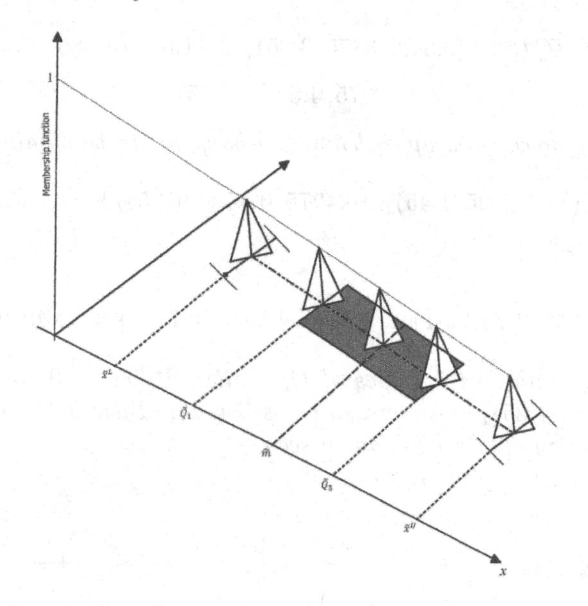

FIGURE 3.1
A typical Boxplot for fuzzy data.

Example 3.9 *The data set listed in Table 3.9 includes 16 daily atmospheric concentration of carbon monoxide. The preference degrees of $P_d(\widetilde{x}_i \succeq \widetilde{x}_j)$ are listed in Table 3.9. It can be checked that the fuzzy data can be ordered as:*

$$\begin{aligned} &\widetilde{x}_1 \preceq_{P_d} \widetilde{x}_9 \preceq_{P_d} \widetilde{x}_{12} \preceq_{P_d} \widetilde{x}_{13} \\ &\preceq_{P_d} \widetilde{x}_4 \preceq_{P_d} \widetilde{x}_{10} \preceq_{P_d} \widetilde{x}_2 \preceq_{P_d} \widetilde{x}_3 \\ &\preceq_{P_d} \widetilde{x}_{11} \preceq_{P_d} \widetilde{x}_{14} \preceq_{P_d} \widetilde{x}_{16} \preceq_{P_d} \widetilde{x}_{15} \\ &\preceq_{P_d} \widetilde{x}_8 \preceq_{P_d} \widetilde{x}_5 \preceq_{P_d} \widetilde{x}_7 \preceq_{P_d} \widetilde{x}_6. \end{aligned}$$

This simply concludes that the fuzzy median is:

$$\widetilde{m} = (1.825, 2.075, 2.150)_T.$$

Further, the first and third fuzzy quantiles are:

$$\widetilde{Q}_1 = (1.425, 1.6125, 1.825)_T, \widetilde{Q}_3 = (2.31, 3.3875, 3.85)_T.$$

Therefore, we have the following results:

$$\widetilde{IQR} = (2.31, 3.3875, 3.85)_T \ominus_G (1.425, 1.6125, 1.825)_T = (0.885, 1.775, 2.05)_T,$$

$$\begin{aligned} \widetilde{Q}_1 \ominus_G (1.5 \otimes \widetilde{IQR}) &= (1.425, 1.6125, 1.825)_T \ominus_G (1.5 \otimes (0.885, 1.775, 2.05)_T) \\ &= (0.4275, 0.45, 0.6125)_T, \end{aligned}$$

$$\widetilde{Q}_3 \oplus (1.5 \otimes \widetilde{IQR}) = (2.31, 3.3875, 3.85)_T \oplus (1.5 \otimes (0.885, 1.775, 2.05)_T)$$
$$= (3.3075, 4.55, 5.3875)_T.$$

Thus, the fuzzy lower and upper bounds of boxplot can be obtained as:

$$\widetilde{x}_L = \widetilde{\max}\{(1.15, 1.25, 1.45)_T, (0.4275, 0.45, 0.6125)_T\} = (1, 1.15, 1.25)_T,$$

and

$$\widetilde{x}_U = \widetilde{\min}\{(2.43, 2.75, 2.81)_T, (3.3075, 4.55, 5.3875)_T\} = (2.43, 2.75, 2.81)_T.$$

For each fuzzy data $\widetilde{x}_i$ *the values of* $D_i = \max\{P_d(\widetilde{x}_i \succ (2.43, 2.75, 2.81)_T), P_d(\widetilde{x}_i \prec (1, 1.15, 1.25)_T)\}$ *are evaluated as listed in Table 3.11. No outlier was found in the fuzzy data set as can be seen.*

3.4 Descriptive Statistics Based on Grouped Fuzzy Data

In many situations, one may face an overwhelming amount of fuzzy data. In such cases, the big or complicated set of fuzzy data can be compacted for easier understanding by summarizing in a table or chart. In this section, the common approaches fr organizing a set of fuzzy data will be described by summary **TFN**s of $\widetilde{x}_1, \cdots, \widetilde{x}_n$ where $\widetilde{x}_i = (x_i; l_{x_i}, r_{x_i})_T$. For this purpose, the fuzzy data were organized in a table that gives the frequency of occurrence of each distinct value. This is called the fuzzy grouped frequency approach. The width of each class is selected such that the difference between its lower and upper boundaries remains constant from one class to another. The number of classes often depends on the size of the represented sample. The Sturges' Rule (as $k \simeq 1 + 3.3 \log n$) is an empirical relation to approximate value of the appropriate number of classes. For a specified k, following steps should be taken to construct k fuzzy intervals $[\widetilde{a}_i, \widetilde{a}_{i+1}]$, $i = 1, 2, \ldots, k$ with $P_d(\widetilde{a}_{i+1} \succeq \widetilde{a}_i) \geq 0.5$ where $\widetilde{a}_i = (a_i; l_{a_i}, r_{a_i})_T$.

1) Assume that $\widetilde{x}_{(1)}$ and $\widetilde{x}_{(n)}$ shows the minimum and maximum values of fuzzy data in which $l_{\widetilde{x}_{(1)}} < l_{\widetilde{x}_{(n)}}$ and $r_{\widetilde{x}_{(1)}} < r_{\widetilde{x}_n}$.

2) Let $\widetilde{a}_i = \widetilde{x}_{(1)} \oplus ((i-1) \otimes \widetilde{l})$ where $\widetilde{l} = \widetilde{R} \oslash k$ for $i = 1, 2, ..., k+1$. According to the arithmetic operations of **TFN**s, the fuzzy boundaries of the ith class can be shown by **TFN**s as:

$$\widetilde{a}_i = (x_{(1)} + (i-1)\frac{(x_{(n)} - x_{(1)})}{k};$$
$$l_{x_{(1)}} + (i-1)\frac{l_{x_{(n)}} - l_{x_{(1)}}}{k}, r_{x_{(1)}} + (i-1)\frac{r_{x_{(n)}} - r_{x_{(1)}}}{k})_T.$$

TABLE 3.10
Preference degrees in Example 3.9.

j	1	2	3	4	5	6	7	8	9	10	11	12	13	14	15	16
$P_d(\widetilde{x}_1 \succeq \widetilde{x}_j)$	0.5	0	0	0	0	0	0	0	0.2	0	0	0.0111	0	0	0	0
$P_d(\widetilde{x}_2 \succeq \widetilde{x}_j)$	1	0.5	0.006	1	0	0	0	0	1	0.9	0.0222	1	1	0	0	0
$P_d(\widetilde{x}_3 \succeq \widetilde{x}_j)$	1	0.994	0.5	1	0	0	0	0	1	0.966	0.447	1	1	0.466	0	0
$P_d(\widetilde{x}_4 \succeq \widetilde{x}_j)$	1	0	0	0.5	0	0	0	0	1	0	0	1	0.8	0	0	0
$P_d(\widetilde{x}_5 \succeq \widetilde{x}_j)$	1	1	1	1	0.5	0.0315	0.312	0.9995	1	1	1	1	1	1	0.9997	1
$P_d(\widetilde{x}_6 \succeq \widetilde{x}_j)$	1	1	1	1	0.968	0.5	0.936	1	1	1	1	1	1	1	1	1
$P_d(\widetilde{x}_7 \succeq \widetilde{x}_j)$	1	1	1	1	0.687	0.063	0.5	1	1	1	1	1	1	1	1	1
$P_d(\widetilde{x}_8 \succeq \widetilde{x}_j)$	1	1	1	1	0.0005	0	0	0.5	1	1	1	1	1	1	1	1
$P_d(\widetilde{x}_9 \succeq \widetilde{x}_j)$	0.8	0	0	0	0	0	0	0	0.5	0	0	0	0	0	0	0
$P_d(\widetilde{x}_{10} \succeq \widetilde{x}_j)$	1	0.1	0.0338	1	0	0	0	0	1	0.5	0.0469	1	1	0	0	0
$P_d(\widetilde{x}_{11} \succeq \widetilde{x}_j)$	1	0.9778	0.552	1	0	0	0	0	1	0.953	0.5	1	1	0.478	0	0
$P_d(\widetilde{x}_{12} \succeq \widetilde{x}_j)$	0.989	0	0	0	0	0	0	0	1	0	0	0.5	0	0	0	0
$P_d(\widetilde{x}_{13} \succeq \widetilde{x}_j)$	1	0	0	0.2	0	0	0	0	1	0	0	1	0.5	0	0	0
$P_d(\widetilde{x}_{14} \succeq \widetilde{x}_j)$	1	1	0.533	1	0	0	0	0	1	1	0.521	1	1	0.5	0	0.061
$P_d(\widetilde{x}_{15} \succeq \widetilde{x}_j)$	1	1	1	1	0.0003	0	0	0	1	1	1	1	1	1	0.5	1
$P_d(\widetilde{x}_{16} \succeq \widetilde{x}_j)$	1	1	1	1	0	0	0	0	1	1	1	1	1	0.939	0	0.5

TABLE 3.11
Outlier detecting in Example 3.9.

No.	$P_d(\widetilde{x}_i \succ \widetilde{x}_U)$	$P_d(\widetilde{x}^* \prec \widetilde{x}^L)$	D_i
1	0	0	0
2	0	0	0
3	0	0	0
4	0	0	0
5	0.3438	0	0.3438
6	1	0	1
7	0.4752	0	0.4752
8	0.0053	0	0.0053
9	0	0	0
10	0	0	0
11	0	0	0
12	0	0	0
13	0	0	0
14	0	0	0
15	0.003839	0	0.003839
16	0	0	0

3) We say $\widetilde{x}_j$ belongs to $[\widetilde{a}_i, \widetilde{a}_{i+1}]$, $i = 1, 2, ...k-1$ ($\widetilde{x}_j \in_{P_d} [\widetilde{a}_i, \widetilde{a}_{i+1}]$) if the following conditions are simultaneously satisfied:

   1) $P_d(\widetilde{x}_j \succeq \widetilde{a}_i) \geq 0.5$ and $P_d(\widetilde{x}_j \preceq \widetilde{a}_{i+1})\} > 0.5$ for $i = 1, 2, ..., k-2$,
   2) $\min\{P_d(\widetilde{x}_j \succeq \widetilde{a}_{k-1}), P_d(\widetilde{x}_j \preceq \widetilde{a}_k)\} \geq 0.5$.

The frequency of fuzzy observations belong to the ith class is defined by:

$$f_i = \#\{i \in \{1, 2, ..., n\} : \widetilde{x}_i \epsilon_{P_d} [\widetilde{a}_i, \widetilde{a}_{i+1}]\}, \tag{3.20}$$

where # denotes the number of elements in a set. Therefore, we can construct a frequency table based on fuzzy data (Table 3.12). In this table, F_i, and r_i stand for cumulative frequency and relative frequency, respectively. Therefore, a histogram can be constructed as a visual display of a frequency distribution table with k classes as depicted in Fig. 3.2. This chart can show frequencies for k classes in this frequency table.

According to the frequency Table 3.12, some common measures of central tendency and deviations can be defined as follows:

1) Fuzzy mean:

$$\widetilde{\overline{x}}_m = \frac{1}{k} \bigoplus_{l=1}^{k} (f_l \otimes \widetilde{m}_l). \tag{3.21}$$

2) Standard deviation:

$$s_m^2 = \frac{1}{k-1} \sum_{l=1}^{k} f_l d_2^2(\widetilde{m}_l, \widetilde{\overline{x}}_m). \tag{3.22}$$

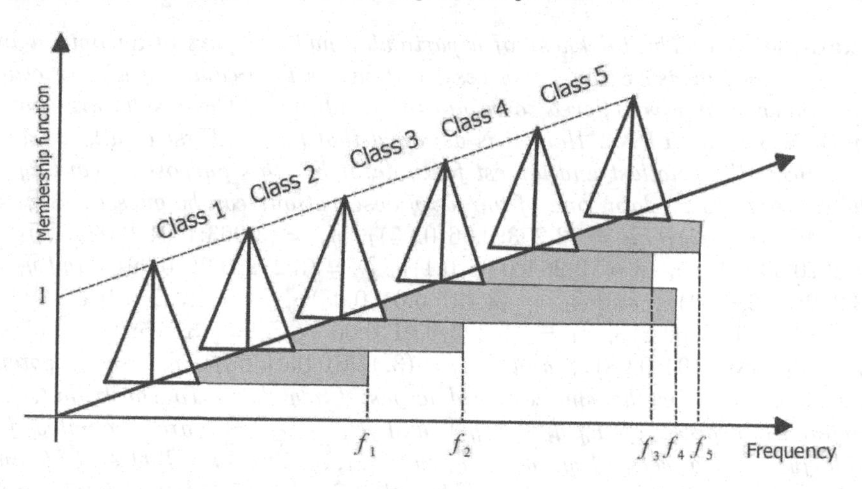

FIGURE 3.2
A typical fuzzy frequency distribution.

TABLE 3.12
A typical frequency table based on fuzzy data.

i	Class	$\widetilde{m}_i = \frac{\widetilde{a}_i \oplus \widetilde{a}_{i+1}}{2}$	f_i	$F_i = \sum_{j=1}^{i} f_i$	$r_i = \frac{f_i}{n}$
1	$[\widetilde{a}_1, \widetilde{a}_2]$	$\widetilde{m}_1$	f_1	F_1	r_1
2	$[\widetilde{a}_2, \widetilde{a}_3]$	$\widetilde{m}_2$	f_2	F_2	r_2
⋮	⋮	⋮	⋮	⋮	⋮
k	$[\widetilde{a}_{k-1}, \widetilde{a}_k]$	$\widetilde{m}_k$	f_k	F_k	r_k

3) Fuzzy median:

$$\widetilde{m} = \widetilde{a}_m \oplus ((\frac{n/2 - F_{m-1}}{f_m}) \otimes \widetilde{l}), \tag{3.23}$$

where $\widetilde{a}_m$ is the fuzzy lower bound of the the first class $i \in \{1, 2, ..., k\}$ such that $F_m \geq n/2$, F_{m-1} is the cumulative frequency of the class m and f_m is the frequency of the class m.

4) Fuzzy mod $\widetilde{M}$ is the fuzzy midpoint of a class with maximum frequency.

Remark 3.2 *1 It should be pointed out that the above technique to provide a frequency table based on fuzzy data performs well for cases with some intersections between fuzzy data otherwise the obtained frequency table coincides with an ordinary one. Moreover, Viertl [198] also proposed a histogram based on fuzzy data. For this purpose, he proposed a fuzzy cumulative sum via α-cuts approach while we employed a different approach with more discussions in this chapter.*

Example 3.10 *The thickness of a particular metallic part of an optical instrument was measured on 50 successive items as they came off a production line under what was believed to be normal conditions. The results are shown in Table 3.13 as* ***TFNs****. Here it is assumed that $k = 5$. First of all, we need to indicate the smallest and largest fuzzy data. For this purpose, according to the center of fuzzy data, one of the fuzzy observations can be guessed as* $\widetilde{y}_1 = (3.214; 0.08, 0.07)_T$, $\widetilde{y}_2 = (3.253; 0.06, 0.05)_T$, $\widetilde{y}_3 = (3.263; 0.01, 0.02)_T$, $\widetilde{y}_4 = (3.26; 0.09, 0.12)_T$, $\widetilde{y}_5 = (3.264; 0.08, 0.1)_T$, $\widetilde{y}_6 = (3.272; 0.08, 0.09)_T$ *and* $\widetilde{y}_7 = (3.282; 0.06, 0.04)_T$, *and* $\widetilde{z}_1 = (3.427; 0.01, 0.03)_T$, $\widetilde{z}_2 = (3.579; 0.1, 0.11)_T$, $\widetilde{z}_3 = (3.3; 0.03, 0.02)_T$, $\widetilde{z}_4 = (3.323; 0.01, 0.06)_T$, $\widetilde{z}_5 = (3.375; 0.07, 0.05)_T$, $\widetilde{z}_6 = (3.387; 0.04, 0.08)_T$, *and* $\widetilde{z}_7 = (3.435; 0.05, 0.05)_T$ *are some potential candidates for the smallest and largest fuzzy data. To check that, the preference degrees of* $P_d(\widetilde{y}_i \succeq \widetilde{y}_j)$ *and* $P_d(\widetilde{z}_i \succeq \widetilde{z}_j)$ *are evaluated for both fuzzy data sets of* $\widetilde{y}_1, \widetilde{y}_2, ..., \widetilde{y}_7$ *and* $\widetilde{z}_1, \widetilde{z}_2, ..., \widetilde{z}_7$ *in Tables 3.14 and 3.15. In this regards, it can be checked that* $\widetilde{x}_{(1)} = (3.214; 0.08, 0.07)_T$ *and* $\widetilde{x}_{(n)} = (3.435; 0.05, 0.05)_T$. *Therefore, the range is* $\widetilde{R} = \widetilde{x}_{(77)} \ominus_G \widetilde{x}_{(1)} = (3.579; 0.1, 0.11)_T \ominus_G (3.214; 0.08, 0.07)_T = (0.365; 0.02, 0.04)_T$. *Thus, the fuzzy lower and upper bounds of each five classes can be evaluated as follows:*

$$\widetilde{a}_1 = \widetilde{x}_{(1)} = (3.214; 0.08, 0.07)_T,$$

$$\begin{aligned}\widetilde{a}_2 &= (x_{(1)} + \frac{(x_{(n)} - x_{(1)})}{5}; l_{x_{(1)}} + \frac{(l_{(x_{(n)}} - l_{x_{(1)}})}{5}, r_{x_{(1)}} + \frac{(r_{x_{(n)}} - r_{x_{(1)}})}{5})_T \\ &= (3.287; 0.084, 0.078)_T.\end{aligned}$$

$$\begin{aligned}\widetilde{a}_3 &= (x_{(1)} + 2\frac{(x_{(n)} - x_{(1)})}{5}; l_{x_{(1)}} + 2\frac{(l_{x_{(n)}} - l_{x_{(1)}})}{5}, r_{x_{(1)}} + 2\frac{(r_{x_{(n)}} - r_{x_{(1)}})}{5})_T \\ &= (3.36; 0.088, 0.086)_T.\end{aligned}$$

$$\begin{aligned}\widetilde{a}_4 &= (x_{(1)} + 3\frac{(x_{(n)} - x_{(1)})}{5}; l_{x_{(1)}} + 3\frac{(l_{x_{(n)}} - l_{x_{(1)}})}{5}, r_{x_{(1)}} + 3\frac{(r_{x_{(n)}} - r_{x_{(1)}})}{5})_T \\ &= (3.433; 0.092, 0.094)_T.\end{aligned}$$

$$\begin{aligned}\widetilde{a}_5 &= (x_{(1)} + 4\frac{(x_{(n)} - x_{(1)})}{5}; l_{x_{(1)}} + 4\frac{(l_{x_{(n)}} - l_{x_{(1)}})}{5}, r_{x_{(1)}} + 4\frac{(r_{x_{(n)}} - r_{x_{(1)}})}{5})_T \\ &= (3.506; 0.096, 0.102)_T, \\ \widetilde{a}_6 &= \widetilde{x}_{(n)} = (3.579; 0.1, 0.11)_T.\end{aligned}$$

The values of $P_d(\widetilde{x}_j \succeq \widetilde{a}_i)$ *and* $P_d(\widetilde{x}_j \preceq \widetilde{a}_{i+1})$ *for* $i = 1, 2, ..., 5$ *and* $j = 1, 2, ..., 77$ *are summarized in Table 3.13. The values of* $P_d(\widetilde{x}_j \succeq \widetilde{a}_i)$ *and* $P_d(\widetilde{x}_j \preceq \widetilde{a}_{i+1})\}$ *for* $i = 1, 2, ..., 5$ *and* $j = 1, 2, ..., 77$ *are evaluated in Table*

TABLE 3.13
Data set in Example 3.10.

$(3.407, 0.1, 0.1)_T$	$(3.214; 0.1, 0.11)_T$	$(3.260; 0.09, 0.12)_T$	$(3.375; 0.07, 0.05)_T$	$(3.417; 0.08, 0.08)_T$
$(3.367; 0.1, 0.11)_T$	$(3.403; 0.01, 0.02)_T$	$(3.482; 0.07, 0.01)_T$	$(3.30; 0.03, 0.02)_T$	$(3.387; 0.04, 0.08)_T$
$(3.362; 0.07, 0.06)_T$	$(3.282; 0.06, 0.04)_T$	$(3.394; 0.1, 0.1)_T$	$(3.446; 0.09, 0.09)_T$	$(3.290, 0.02, 0.08)_T$
$(3.385; 0.5; 0.05)_T$	$(3.396; 0.1, 0.1)_T$	$(3.407; 0.02, 0.01)_T$	$(3.303; 0.02, 0.01)_T$	$(3.444; 0.04, 0.11)_T$
$(3.378; 0.04, 0.03)_T$	$(3.412; 0.1, 0.1)_T$	$(3.457; 0.09, 0.07)_T$	$(3.468; 0.02, 0.02)_T$	$(3.352; 0.08; 0.1)_T$
$(3.355; 0.03, 0.04)_T$	$(3.461; 0.08, 0.04)_T$	$(3.314; 0.05, 0.12)_T$	$(3.333; 0.08, 0.06)_T$	$(3.330; 0.05, 0.04)_T$
$(3.373; 0.08, 0.07)_T$	$(3.315; 0.07, 0.08)_T$	$(3.512; 0.02; 0.02)_T$	$(3.368; 0.1, 0.08)_T$	$(3.329; 0.1, 0.1)_T$
$(3.333; 0.04, 0.08)_T$	$(3.435; 0.05, 0.05)_T$	$(3.396; 0.04, 0.02)_T$	$(3.393; 0.07, 0.08)_T$	$(3.281; 0.08, 0.07)_T$
$(3.253; 0.06, 0.05)_T$	$(3.284; 0.04; 0.02)_T$	$(3.418; 0.02, 0.02)_T$	$(3.393; 0.07, 0.07)_T$	$(3.338; 0.03, 0.02)_T$
$(3.272; 0.08, 0.09)_T$	$(3.345; 0.01, 0.01)_T$	$(3.335; 0.08, 0.07)_T$	$(3.428; 0.06, 0.07)_T$	$(3.356; 0.06, 0.03)_T$
$(3.342, 0.1, 0.1)_T$	$(3.324; 0.02, 0.03)_T$	$(3.426, 0.02, 0.03)_T$	$(3.317; 0.1, 0.1)_T$	$(3.448; 0.07, 0.08)_T$
$(3.374; 0.07, 0.08)_T$	$(3.358; 0.02, 0.03)_T$	$(3.579; 0.08, 0.07)_T$	$(3.411, 0.1, 0.05)_T$	$(3.281; 0.04, 0.03)_T$
$(3.493; 0.1, 0.05)_T$	$(3.264; 0.08, 0.1)_T$	$(3.427; 0.01, 0.03)_T$	$(3.456; 0.03, 0.04)_T$	$(3.323; 0.01, 0.06)_T$
$(3.363; 0.02, 0.07)_T$	$(3.414; 0.03, 0.01)_T$	$(3.397; 0.02, 0.03)_T$	$(3.385; 0.08, 0.09)_T$	$(3.263; 0.01, 0.02)_T$
$(3.279; 0.03, 0.05)_T$	$(3.295; 0.08, 0.08)_T$	$(3.476; 0.02; 0.06)_T$	$(3.337; 0.01, 0.02)_T$	$(3.306; 0.02, 0.03)_T$
$(3.388; 0.1; 0.06)_T$	$(3.375; 0.03, 0.02)_T$			

TABLE 3.14
Preference degrees for the first seven fuzzy data in Example 3.10.

j	1	2	3	4	5	6	7
$P_d(\widetilde{y}_1 \succeq \widetilde{y}_j)$	0.5	1	1	0.86863	0.89865	0.97767	1
$P_d(\widetilde{y}_2 \succeq \widetilde{y}_j)$	0	0.5	0.74490	0.24338	0.4161	0.66653	0.46764
$P_d(\widetilde{y}_3 \succeq \widetilde{y}_j)$	0	0.25510	0.5	0	0.08478	0.56405	0
$P_d(\widetilde{y}_4 \succeq \widetilde{y}_j)$	0.13137	0.75662	1	0.5	0.97739	1	0.95031
$P_d(\widetilde{y}_5 \succeq \widetilde{y}_j)$	0.10135	0.5839	0.91523	0.02261	0.5	1	0.62308
$P_d(\widetilde{y}_6 \succeq \widetilde{y}_j)$	0.02233	0.33347	0.43595	0	0	0.5	0.22031
$P_d(\widetilde{y}_7 \succeq \widetilde{y}_j)$	0	0.53236	1	0.04969	0.37692	0.77969	0.5

TABLE 3.15
Preference degrees for the last seven fuzzy data in Example 3.10.

j	1	2	3	4	5	6	7
$P_d(\widetilde{z}_1 \succeq \widetilde{z}_j)$	0.00005	0.13049	0.05882	0	0	0	0.03020
$P_d(\widetilde{z}_2 \succeq \widetilde{z}_j)$	0	0.1668	0.03390	0	0	0.00368	0.00506
$P_d(\widetilde{z}_3 \succeq \widetilde{z}_j)$	1	1	1	1	1	1	1
$P_d(\widetilde{z}_4 \succeq \widetilde{z}_j)$	0	0	0	0	0	0	0
$P_d(\widetilde{z}_5 \succeq \widetilde{z}_j)$	0.00139	0.12846	0.0664	0	0	0	0.03774
$P_d(\widetilde{z}_6 \succeq \widetilde{z}_j)$	0	0	0	0	0	0	0
$P_d(\widetilde{z}_7 \succeq \widetilde{z}_j)$	0	0.01783	0	0	0	0	0

3.16. Therefore, the frequency of each fuzzy observation of $\widetilde{x}_i$ belong to the ith class which can be evaluated by:

$$f_i = \#\{l \in \{1, 2, ..., 77\} : \widetilde{x}_l \in_{P_d} [\widetilde{a}_i, \widetilde{a}_{i+1}]\}.$$

The frequency of each class is listed in Table 3.17. Thus a frequency table can be provided by calculating F_i and r_i as presented in Table 3.17.

3.5 Measure of Spread Based on Fuzzy Data

To investigate the symmetry of data distribution, the skewness and kurtosis measures were evaluated. In this section, some common measures of spread are extended based on fuzzy data $x_i = (x_i^L, x_i, x_i^U)_T$, $i = 1, 2, ..., n$.

Definition 3.10 ***(Fuzzy measure of skewness)*** *Based on the fuzzy observation of $\widetilde{x}_1, \widetilde{x}_2, ..., \widetilde{x}_n$, the fuzzy (Pearson) measure of skewness can be defined by:*

$$\widetilde{s}_k = \frac{3}{s} \otimes (\widetilde{\overline{x}} \ominus_G \widetilde{m}), \tag{3.24}$$

TABLE 3.16
The values of $D_{ij} = \{P_d(\widetilde{x}_j \succeq \widetilde{a}_i), P_d(\widetilde{x}_j \preceq \widetilde{a}_{i+1})\}$ in Example 3.10.

No.	D_{1j}	No.	D_{2j}	No.	D_{3j}	No.	D_{4j}	No.	D_{5j}
1	{1, 0}	1	{1, 0}	1	{1, 1}	1	{0, 1}	1	{0, 0}
2	{1, 0}	2	{1, 0.0472}	2	{0.9528, 1}	2	{0, 1}	2	{0, 0}
3	{1, 0}	3	{1, 0.5}	3	{0.5, 1}	3	{0, 1}	3	{0, 0}
4	{1, 0}	4	{1, 0.0309}	4	{0.9691, 1}	4	{0, 1}	4	{0, 0}
5	{1, 0}	5	{1, 0.2231}	5	{0.7769, 1}	5	{0, 1}	5	{0, 0}
6	{1, 0}	6	{1, 0.5381}	6	{0.4619, 1}	6	{0, 1}	6	{0, 0}
7	{1, 0}	7	{1, 0.0125}	7	{0.9875, 1}	7	{0, 1}	7	{0, 0}
8	{1, 0}	8	{1, 0.8911}	8	{0.1089, 1}	8	{0, 1}	8	{0, 0}
9	{1, 1}	9	{0, 1}	9	{0, 1}	9	{0, 1}	9	{0, 0}
10	{1, 1}	10	{0, 1}	10	{0, 1}	10	{0, 1}	10	{0, 0}
11	{1, 0}	11	{1, 1}	11	{0, 1}	11	{0, 1}	11	{0, 0}
12	{1, 0}	12	{1, 0}	12	{1, 1}	12	{0, 1}	12	{0, 0}
13	{1, 0}	13	{1, 0}	13	{1, 0}	13	{1, 1}	13	{0, 0}
14	{1, 0}	14	{1, 0.1241}	14	{0.8759, 0.9998}	14	{0.0002, 1}	14	{0, 0}
15	{1, 0.5356}	15	{0.4644, 1}	15	{0, 1}	15	{0, 1}	15	{0, 0}
16	{1, 0}	16	{1, 0}	16	{1, 1}	16	{0, 1}	16	{0, 0}
17	{0.5, 1}	17	{0, 1}	17	{0, 1}	17	{0, 1}	17	{0, 0}
18	{1, 0}	18	{1, 0.0401}	18	{0.9599, 0.8147}	18	{0.1853, 1}	18	{0, 0}
19	{1, 0.7653}	19	{0.2347, 1}	19	{0, 1}	19	{0, 1}	19	{0, 0}
20	{1, 0}	20	{1, 0}	20	{1, 1}	20	{0, 1}	20	{0, 0}
21	{1, 0}	21	{1, 0}	21	{1, 1}	21	{0, 1}	21	{0, 0}
22	{1, 0}	22	{1, 0}	22	{1, 0.1317}	22	{0.8683, 1}	22	{0, 0}
23	{1, 0}	23	{1, 1}	23	{0, 1}	23	{0, 1}	23	{0, 0}
24	{1, 0}	24	{1, 0}	24	{1, 0.4652}	24	{0.5348, 1}	24	{0, 0}
25	{1, 0.6269}	25	{0.3731, 1}	25	{0, 1}	25	{0, 1}	25	{0, 0}
26	{1, 0.0061}	26	{0.9939, 0.6815}	26	{0.3185, 1}	26	{0, 1}	26	{0, 0}
27	{1, 0.0149}	27	{0.9851, 0.9071}	27	{0.0929, 1}	27	{0, 1}	27	{0, 0}
28	{1, 0}	28	{1, 0.4839}	28	{0.5161, 1}	28	{0, 1}	28	{0, 0}
29	{1, 1}	29	{0, 1}	29	{0, 1}	29	{0, 1}	29	{0, 0}
30	{1, 0}	30	{1, 0.0302}	30	{0.9698, 0.8108}	30	{0.1892, 1}	30	{0, 0}
31	{1, 0}	31	{1, 1}	31	{0, 1}	31	{0, 1}	31	{0, 0}
32	{1, 0}	32	{1, 0.3013}	32	{0.6987, 0.9989}	32	{0.0011, 1}	32	{0, 0}
33	{1, 0.9352}	33	{0.0648, 1}	33	{0, 1}	33	{0, 1}	33	{0, 0}
34	{1, 0}	34	{1, 0}	34	{1, 0.0894}	34	{0.9106, 0.9991}	34	{0.0009, 0}
35	{1, 0}	35	{1, 0}	35	{1, 1}	35	{0, 1}	35	{0, 0}
36	{1, 0}	36	{1, 0.0547}	36	{0.9453, 0.8319}	36	{0.1682, 1}	36	{0, 0}
37	{1, 0}	37	{1, 0}	37	{1, 0}	37	{1, 1}	37	{0, 0}
38	{1, 0}	38	{1, 1}	38	{0, 1}	38	{0, 1}	38	{0, 0}
39	{1, 0}	39	{1, 0}	39	{1, 0}	39	{1, 0.4434}	39	{0.5566, 0.0075}
40	{1, 0}	40	{1, 0.089}	40	{0.911, 0.9758}	40	{0.0242, 1}	40	{0, 0}
41	{1, 0}	41	{1, 0.0041}	41	{0.9959, 0.7036}	41	{0.2964, 1}	41	{0, 0}
42	{1, 0}	42	{1, 1}	42	{0, 1}	42	{0, 1}	42	{0, 0}
43	{1, 0}	43	{1, 0}	43	{1, 0.5728}	43	{0.4272, 1}	43	{0, 0}
44	{1, 0}	44	{1, 0}	44	{1, 0}	44	{1, 0}	44	{1, 0.5}
45	{1, 0}	45	{1, 0}	45	{1, 0.5204}	45	{0.4796, 0.9981}	45	{0.0019, 0}
46	{1, 0}	46	{1, 0.0373}	46	{0.9627, 0.8953}	46	{0.1047, 1}	46	{0, 0}
47	{1, 0}	47	{1, 0}	47	{1, 0}	47	{1, 0.8035}	47	{0.1965, 0}
48	{1, 0}	48	{1, 0.1842}	48	{0.8158, 1}	48	{0, 1}	48	{0, 0}
49	{1, 0.2963}	49	{0.7037, 1}	49	{0, 1}	49	{0, 1}	49	{0, 0}
50	{1, 0}	50	{1, 0}	50	{1, 0}	50	{1, 1}	50	{0, 0}
51	{1, 0.285}	51	{0.715, 0.9926}	51	{0.0074, 1}	51	{0, 1}	51	{0, 0}
52	{1, 0}	52	{1, 0}	52	{1, 0.1148}	52	{0.8852, 0.9031}	52	{0.0969, 0}
53	{1, 0}	53	{1, 1}	53	{0, 1}	53	{0, 1}	53	{0, 0}
54	{1, 0}	54	{1, 0.08}	54	{0.92, 1}	54	{0, 1}	54	{0, 0}
55	{1, 0}	55	{1, 0}	55	{1, 1}	55	{0, 1}	55	{0, 0}
56	{1, 0}	56	{1, 0}	56	{1, 1}	56	{0, 1}	56	{0, 0}
57	{1, 0}	57	{1, 0}	57	{1, 0.6042}	57	{0.3958, 1}	57	{0, 0}
58	{1, 0}	58	{1, 1}	58	{0, 1}	58	{0, 1}	58	{0, 0}
59	{1, 0}	59	{1, 0}	59	{1, 1}	59	{0, 1}	59	{0, 0}
60	{1, 0}	60	{1, 0}	60	{1, 0.1312}	60	{0.8688, 0.981}	60	{0.019, 0}
61	{1, 0}	61	{1, 0}	61	{1, 1}	61	{0, 1}	61	{0, 0}
62	{1, 0.0051}	62	{0.9949, 0.7539}	62	{0.2461, 1}	62	{0, 1}	62	{0, 0}
63	{1, 0}	63	{1, 0}	63	{1, 1}	63	{0, 1}	63	{0, 0}
64	{1, 0}	64	{1, 0}	64	{1, 0.9953}	64	{0.0047, 1}	64	{0, 0}
65	{1, 0}	65	{1, 1}	65	{0, 1}	65	{0, 1}	65	{0, 0}
66	{1, 0}	66	{1, 0}	66	{1, 0}	66	{1, 1}	66	{0, 0}
67	{1, 0}	67	{1, 0.8302}	67	{0.1698, 1}	67	{0, 1}	67	{0, 0}
68	{1, 0}	68	{1, 0.9872}	68	{0.0128, 1}	68	{0, 1}	68	{0, 0}
69	{1, 0}	69	{1, 1}	69	{0, 1}	69	{0, 1}	69	{0, 0}
70	{1, 1}	70	{0, 1}	70	{0, 1}	70	{0, 1}	70	{0, 0}
71	{1, 0.0042}	71	{0.9958, 0.8412}	71	{0.1588, 1}	71	{0, 1}	71	{0, 0}
72	{1, 0}	72	{1, 0.7584}	72	{0.2416, 1}	72	{0, 1}	72	{0, 0}
73	{1, 0}	73	{1, 0}	73	{1, 0}	73	{1, 1}	73	{0, 0}
74	{1, 0.6495}	74	{0.3505, 1}0.3505	74	{0, 1}	74	{0, 1}	74	{0, 0}
75	{1, 0}	75	{1, 0.8451}0.8451	75	{0.1549, 1}	75	{0, 1}	75	{0, 0}
76	{0.9999, 0.7711}	76	{0.2289, 1}0.2289	76	{0, 1}	76	{0, 1}	76	{0, 0}
77	{1, 0.1379}	77	{0.8621, 0.9863}	77	{0.0137, 1}	77	{0, 1}	77	{0, 0}

TABLE 3.17
Frequency table in Example 3.10.

i	Class	$\widetilde{m}_i = \frac{\widetilde{a}_i \oplus \widetilde{a}_{i+1}}{2}$	f_i	$F_i = \sum_{j=1}^{i} f_i$	$r_i = \frac{f_i}{n}$
1	$[(3.214; 0.08, 0.07)_T, (3.287; 0.084, 0.078)_T]$	$(3.2505; 0.082, 0.074)_T$	11	11	0.1429
2	$[(3.287; 0.084, 0.078)_T, (3.36; 0.088, 0.086)_T]$	$(3.3235; 0.086, 0.082)_T$	22	33	0.2857
3	$[(3.36; 0.088, 0.086)_T, (3.433; 0.092, 0.094)_T]$	$(3.3965; 0.090, 0.090)_T$	31	64	0.4026
4	$[(3.433; 0.092, 0.094)_T, (3.506; 0.096, 0.102)_T]$	$(3.4695; 0.094, 0.098)_T$	11	75	0.1429
5	$[(3.506; 0.096, 0.102)_T, (3.579; 0.1, 0.11)_T]$	$(3.5425; 0.098, 0.106)_T$	2	77	0.0260

where $\tilde{\bar{x}}$, $\tilde{m}$, and s are fuzzy mean, fuzzy median, and sample deviation of fuzzy data. We say that the distribution of observations is right-skewed if $d(\tilde{s}_k \in (0,3]) > 0.5$, it is left-skewed if $d(\tilde{s}_k \in [-3,0)) > 0.5$ and it is symmetric if $d(\tilde{s}_k \in (0,3]) = d(\tilde{s}_k \in [-3,0)) = 0.5$ in which

$$d(\tilde{s}_k \in (0,3]) = \frac{\int_{\{x:x\in(0,3]\}} \tilde{s}_k(x)dx}{\int \tilde{s}_k(x)dx}, \tag{3.25}$$

represents the degree to which $\tilde{s}_k$ belongs to $(0,3]$.

Next, a measure of skewness is defined based on the tth central moment of fuzzy data.

Definition 3.11 *(**Exact measure of skewness**) Let $\tilde{x}_1, \tilde{x}_2, ..., \tilde{x}_n$ be a set of fuzzy data. The measure of skewness is defined by:*

$$m_t = \frac{1}{n}\sum_{l=1}^{n} \frac{(x_l^L - \bar{x}^L)^t + (x_l - \bar{x})^t + (x_l^U - \bar{x}^U)^t}{3}. \tag{3.26}$$

Fisher's skewness is also defined by:

$$s_F = (\frac{n^2}{(n-1)(n-2)})\frac{m_3}{s^3}. \tag{3.27}$$

In this regard, we say that the distribution of fuzzy data is symmetric if $s_F = 0$, right-skewed if $s_F > 0$, and left-skewed if $s_F < 0$.

To the extent to which a histogram has a sharp peak based on fuzzy data, the conventional measure of steepness called kurtosis can be introduced.

Definition 3.12 *(**Kurtosis**) Let $\tilde{x}_1, \tilde{x}_2, ..., \tilde{x}_n$ be a set of fuzzy data. A measure of kurtosis can be defined as:*

$$g = (\frac{n(n+1)}{(n-1)(n-2)(n-3)})\frac{m_4}{s^4} - \frac{3(n-1)^2}{(n-2)(n-3)}. \tag{3.28}$$

If kurtosis is zero, then the distribution of fuzzy data is normal. A positive value for the kurtosis indicates a harper peak while a negative value reveals a flatter one.

Example 3.11 *Recall Example 3.10. According to the frequency table (Table 3.17), it is easy to calculate that the fuzzy mean is:*

$$\tilde{\bar{x}}_m = \frac{1}{5}\bigoplus_{l=1}^{5}(f_l \otimes \tilde{m}_l) = (50.5199, 51.8827, 53.2223)_T,$$

and hence the standard deviation is

$$s_m^2 = \frac{1}{4}\sum_{l=1}^{5} f_l d_2^2(\tilde{m}_l, \tilde{\bar{x}}_m) = 45327.3.$$

Moreover, since $\widetilde{m}_m = \widetilde{x}_{(39)} = (3.305, 3.375, 3.425)_T$*, one can find the fuzzy (Pearson) measure of skewness as:*

$$\widetilde{s}_k = \frac{3}{S_m} \otimes (\widetilde{\overline{x}}_m \ominus_G \widetilde{m}_m) = (0.665304, 0.683521, 0.701693)_T.$$

Since $\widetilde{s}_k[0] \subseteq [0,3)$*, we have:*

$$d(\widetilde{s}_k \in (0,3]) = \frac{\int_{\{x:x\in(0,3]\}} \widetilde{s}_k(x)dx}{\int \widetilde{s}_k(x)dx} = 1.$$

Therefore, it can be concluded that the distribution of observations is right-skewed. Further, to compute Fisher's measure of skewness, we first need to compute m_3*:*

$$m_3 = \frac{1}{5}\sum_{l=1}^{5} f_l\Big(\frac{(m_l^L - 50.5199)^3 + (m_l - 51.8827)^3 + (m_l^U - 53.2223)^3}{3}\Big)$$
$$= 1.76082 \times 10^6.$$

The above calculations reveal that $s_F = (\frac{(77)^2}{76\times 75})\frac{1.76082\times 10^6}{(45327.3)^{3/2}} = 0.189794$*. Thus, the distribution of observations is right-skewed according to Fisher's measure of skewness, as well. Moreover:*

$$m_4 = \frac{1}{5}\sum_{l=1}^{5} f_l\Big(\frac{(m_l^L - 50.5199)^4 + (m_l - 51.8827)^4 + (m_l^U - 53.2223)^4}{3}\Big)$$
$$= 8.55423 \times 10^7.$$

This implies that:

$$g = (\frac{77 \times 78}{76 \times 75 \times 74})\frac{8.55423 \times 10^7}{(45327.3)^2} - \frac{3(76)^2}{75 \times 74} = -3.12157.$$

Since $g < 0$*, it can be concluded that the distribution of fuzzy data has a flatter peak compared to the symmetric case.*

3.6 Exercise

Exercise 3.1 *In Example 3.1,*

1) Find $\widetilde{H}_{20}$, $\widetilde{H}_{60}$, $\widetilde{H}_{95}$*.*

2) Find s^2 *and* s_D*.*

3) Detect fuzzy extreme values.

TABLE 3.18
Frequency table in Exercise 3.2.

i	Class	f_i
1	$[(-3,1)_T,(-1;1.25)_T]$	6
2	$[(-1;1.25)_T,(1;1.5)_T]$	6
3	$[(1;1.5)_T,(3;1.75)_T]$	7
4	$[(3;1.75)_T,(5;2)_T]$	6
5	$[(5;2)_T,(7;2.25)_T]$	5
6	$[(7;2.25)_T,(9;2.5)_T]$	5
7	$[(9;2.5)_T,(11;2.75)_T]$	8

Exercise 3.2 *Consider the frequency table given in Table 3.18.*

1) Find $\widetilde{s}_k$ *and check either the distribution of fuzzy data is right or left skewed .*

2) Find the skewness criterion of S_F *and interpret it.*

3) Find the fuzzy median of $\widetilde{m}_m$.

4) Find the kurtosis criterion of g and interpret it.

Exercise 3.3 *Let* $\widetilde{y}_i = a \otimes \widetilde{x}_i \oplus \widetilde{b}$ *for* $i = 1, 2, \ldots, n$ *where* $a \neq 0$ *and* $b \in \mathbb{R}$. *Prove that* $s_D^{\widetilde{y}} = |a| s_D^{\widetilde{x}}$.

TABLE 3.19
The center values of fuzzy life of laptop batteries in Exercise 3.4.

130	145	126	146
164	130	132	152
145	129	133	155
140	127	139	137
131	126	145	148
125	132	126	126
126	135	131	129
147	136	129	136
156	146	130	146
132	142	132	132

Exercise 3.4 *A manufacturer is investigating the operating life of laptop computer batteries. The data set was reported as* $\widetilde{x}_i = (x_i; 0.1x_i)_T$ *where* x_i *are summarized in Table 3.19.*

1) Construct a frequency table with $k = 5$.

2) Find $\widetilde{H}_{25}$, $\widetilde{H}_{55}$, *and* $\widetilde{H}_{85}$.

3) Find s^2 and s_D.

4) Detect the potential fuzzy outliers.

5) Find $\widetilde{R}$.

6) Find $\widetilde{s}_k$. Whether the distribution of fuzzy data is right-skewed or left-skewed.

7) Find the skewness criterion of s_F and interpret it.

8) Find the fuzzy median of $\widetilde{m}_m$.

9) Find the kurtosis criterion of g.

3.7 Glossary

Fuzzy data: Data which are not precise numbers.

Fuzzy descriptive statistics: A generalized descriptive statistics based on fuzzy data.

4

*Probability Reasoning Based on Fuzzy Random Variable (**FRV**)*

A random variable is a numerical description of the outcome of a statistical experiment. In this chapter, introducing a notion of fuzzy random variables, some common statistical properties such as expectation, variance, and cumulative distribution function are developed based on the fuzzy random variables. The notion of fuzzy random variables is also developed for a common family of random variables called random variables. Furthermore, some probability reasonings including several limit theorems and some common inequalities will be addressed in the fuzzy domain.

4.1 Fuzzy Random Variables

Let $(\Omega, \mathcal{A}, P)$ be a probability space and $X : \Omega \to \mathcal{A}$ is a random variable. The outcomes of a random experiment in the classical statistical inference are exact values. However, there are many situations where the outcomes of X are not real numbers but rather fuzzy quantities. So let us briefly describe how this may be accomplished.

Example 4.1 *Let*

$$X = \{\textit{perception about the registration fee of a conference for editors,}\}$$

be a random variable on a probability space $(\Omega, \mathcal{A}, P)$ *with*

$$\Omega = \{\textit{conferences to be held 2021 to 2022 and involving fuzzy statistics among their covered topics}\}.$$

However, the perception of outcomes (the registration fee for editorial members) can be viewed as some fuzzy quantities as $\widetilde{x}_1$ = *'very cheap',* $\widetilde{x}_2$ = *'cheap',* $\widetilde{x}_3$ = *'rather cheap',* $\widetilde{x}_4$ = *'moderate',* $\widetilde{x}_5$ = *'rather expensive',* $\widetilde{x}_6$ = *'expensive',* $\widetilde{x}_7$ = *'very expensive' which can be described, for instance, by means of* ***TFNs*** *[55]. Such membership functions can be reported as shown in Fig. 4.1. It is noticeable that such situations involve two important sources of*

DOI: 10.1201/9781003248644-4

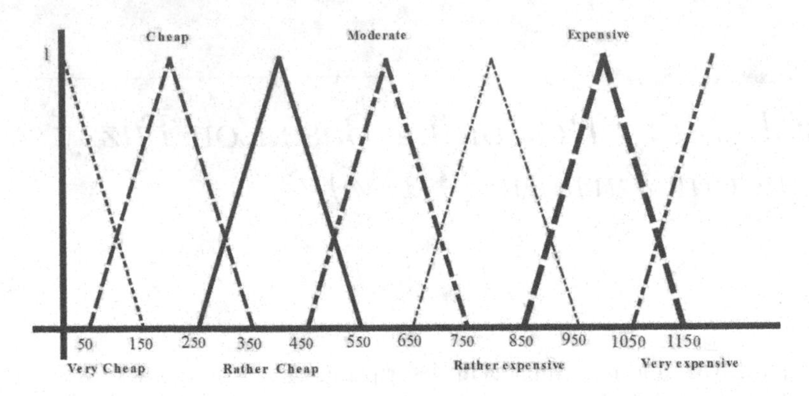

FIGURE 4.1
Fuzzy quantities of the 'perception of the registration fee'.

imprecision: randomness and fuzziness. Randomness relates to the stochastic variability of all possible outcomes. Fuzziness, on the other hand, can be traced to incomplete knowledge regarding the amount of registration fee. In such a situation, randomness and fuzziness can be merged to formulate a fuzzy random variable, that is, a function that assigns a fuzzy subset to each possible.

Roughly speaking, a fuzzy random variable is a random variable taking fuzzy quantities. Fuzzy random variables represent a well-formalized concept underlying many recent probabilistic and statistical studies involving data obtained from a random experiment when these data are assumed to be fuzzy quantities. In this regard, different approaches to this concept have been developed in the literature [32, 55, 117, 118, 121, 131, 179]. To see the relationship between such methods see Gill et al. [55].

A common notion of fuzzy random variables was introduced by Kwakernaak [121] as a mathematical model based on α-cuts of **FN**s. Here, a revised notion is given according to α-values of **FN**s.

Definition 4.1 *[72, 74] Let $(\Omega, \mathcal{A}, P)$ be a probability space. A fuzzy-valued mapping of $\widetilde{X} : \Omega \to \mathcal{F}(\mathbb{R})$ is said to be a fuzzy random variable (**FRV**) if for any $\alpha \in [0,1]$, the real-valued mapping of $\widetilde{X}_\alpha : \Omega \to \mathbb{R}$ is a random variable on $(\Omega, \mathcal{A}, P)$.*

Recall Example 4.1. For any $\alpha \in [0,1]$, since $(\widetilde{x}_i)_\alpha$, $i = 1, 2, ..., 7$ are some outcomes of a ordinary random variable 'X: true registration fee for editorial members in a conference', therefore

'perception about the registration fee of a conference for editors',

can be viewed as a **FRV** denoting by $\widetilde{X}$.

By the above definition, **FRV**s are interpreted as fuzzy perceptions of classical random variables which are referred to as the origin of their fuzzy counterparts [121, 118, 179].

Definition 4.2 *Two **FRVs** of $\widetilde{X}$ and $\widetilde{Y}$ are independent and identically distributed if $\widetilde{X}_\alpha$ and $\widetilde{Y}_\alpha$ are independent and identically distributed random variables for each $\alpha \in [0,1]$. Moreover, $\widetilde{X}_1, ..., \widetilde{X}_n$ is said to be a fuzzy random sample, say **FRS**, if $\widetilde{X}_i$'s are independent and identically distributed **FRVs**. Furthermore, we say that $\widetilde{X}$ is a continuous fuzzy random variable if any $\widetilde{X}_\alpha$ are continuous random variables. The observed **FRS** is denoted by $\widetilde{x}_1, \widetilde{x}_2, \ldots, \widetilde{x}_n$.*

4.1.1 Fuzzy expectation and exact variance

The expected value of a random variable is the weighted average of all possible values of the variable. Here, a notion of fuzzy expected value and exact value of a **FRV** is provided and discussed.

Definition 4.3 *Let $\widetilde{X}$ be a **FRV** on the probability space of $(\Omega, \mathcal{A}, P)$. The expectation value of $\widetilde{X}$ is defined to be a **FN** $\widetilde{E}(\widetilde{X})$ with α-values of $(\widetilde{E}(\widetilde{X}))_\alpha = E(\widetilde{X}_\alpha)$.*

The variance of random variable X is the measure of spread for a distribution of a random variable that determines the degree to which the values of a random variable differ from the expected value. Now, if the outcomes of a random experiment are **FN**s then the concept of variance of a **FRV** can be defined by an average of variances of its α-values.

Definition 4.4 *The variance of a **FRV** $\widetilde{X}$ is defined by $\boldsymbol{var}(\widetilde{X}) = E(d_2^2(\widetilde{X}, \widetilde{E}(\widetilde{X})))$ where $d_2(\widetilde{A}, \widetilde{B}) = (\int_0^1 (\widetilde{A}_\alpha - \widetilde{B}_\alpha)^2 d\alpha)^{0.5}$ is the square error distance between two **FNs** of $\widetilde{A}$ and $\widetilde{B}$..*

Lemma 4.1 *If $\widetilde{X}$ is a **FRV** with the mean $\widetilde{E}(\widetilde{X})$, then $\boldsymbol{var}(\widetilde{X}) = \int_0^1 var(\widetilde{X}_\alpha) d\alpha$.*

Proof *According to Definition 4.4, it can be simply concluded that:*

$$\begin{aligned}
\boldsymbol{var}(\widetilde{X}) &= \int_0^1 E(\widetilde{X}_\alpha - (\widetilde{E}(\widetilde{X}))_\alpha)^2 d\alpha \\
&= \int_0^1 E(\widetilde{X}_\alpha - E(\widetilde{X}_\alpha))^2 d\alpha \\
&= \int_0^1 var(\widetilde{X}_\alpha) d\alpha.
\end{aligned}$$

Example 4.2 *Assume that* $\widetilde{X} = (X; U_1, U_2)$ *where* $X \sim N(5,4)$, $U_1 \sim U(0,2)$, $U_2 \sim U(1,3)$ *and* X *is independent of* (U_1, U_2). *Note that:*

$$\widetilde{X}_\alpha = \begin{cases} X - (1-2\alpha)U_1, & 0 \le \alpha \le 0.5, \\ X - (1-2\alpha)U_2, & 0.5 < \alpha \le 1, \end{cases} \tag{4.1}$$

and hence:

$$E(\widetilde{X}_\alpha) = \begin{cases} 5 - (1-2\alpha), & 0 \le \alpha \le 0.5, \\ 5 - 2(1-2\alpha), & 0.5 < \alpha \le 1. \end{cases} \tag{4.2}$$

This shows $\widetilde{E}(\widetilde{X}) = (5,1,2)_T$. *Additionally,*

$$Var(\widetilde{X}_\alpha) = \begin{cases} 4 + \frac{1}{12}(1-2\alpha)^2, & 0 \le \alpha \le 0.5, \\ 4 + \frac{1}{3}(1-2\alpha)^2, & 0.5 < \alpha \le 1. \end{cases} \tag{4.3}$$

Therefore:

$$\begin{aligned} \boldsymbol{var}(\widetilde{X}) &= \int_0^1 var(\widetilde{X}_\alpha) d\alpha \\ &= \int_0^{0.5} (4 + \frac{1}{12}(1-2\alpha)^2) d\alpha + \int_{0.5}^1 (4 + \frac{1}{12}(1-2\alpha)^2) d\alpha = 4.06944. \end{aligned}$$

Example 4.3 *Consider a* ***FRV*** *as* $\widetilde{X} = (X; UX)_T$ *with* $X \sim exp(\lambda)$ *and* $U \sim U(0, 0.04)$ *where* U *and* X *are independent random variables. Since* $\widetilde{X}_\alpha = X(1 - (U(1-2\alpha))$, *we have* $E(\widetilde{X}_\alpha) = \frac{1}{\lambda}(1 - 0.02(1-2\alpha))$ *which concludes that* $\widetilde{E}(\widetilde{X}) = (\frac{1}{\lambda}; \frac{0.02}{\lambda})_T$. *In addition,*

$$\begin{aligned} var(\widetilde{X}_\alpha) &= E(var(X(1-(U(1-2\alpha))|U)) + var(E(X(1-(U(1-2\alpha)))|U) \\ &= E(1-(1-2\alpha)U)^2 var(X) + (E(X))^2 var((1-(1-2\alpha)U)). \end{aligned}$$

Since $E(U^2) = \frac{(0.04)^2}{12} + (0.02)^2$, $var(U) = \frac{(0.04)^2}{12}$ *and* $E(X^2) = 2/\lambda^2$ $var(X) = 1/\lambda^2$, *we have:*

$$var(\widetilde{X}_\alpha) = \frac{1}{\lambda^2} E(1 + (1-2\alpha)^2 U^2 - 2(1-2\alpha)U) + \frac{2}{\lambda^2}(1-2\alpha)^2 var(U).$$

This simply concludes that:

$$\begin{aligned} \boldsymbol{var}(\widetilde{X}) &= \int_0^1 var(\widetilde{X}_\alpha) d\alpha \\ &= \int_0^{0.5} (\frac{1}{\lambda^2}(1 + (1-2\alpha)^2(\frac{(0.04)^2}{12} + (0.02)^2)) - 0.04(1-2\alpha)) d\alpha \\ &+ \int_0^1 (\frac{2}{\lambda^2}(\frac{(0.04)^2}{12})(1-2\alpha)^2) d\alpha = \frac{1.00027}{\lambda^2}. \end{aligned}$$

Lemma 4.2 *If $\widetilde{X}$ is a **FRV** with a fuzzy mean of $\widetilde{E}(\widetilde{X})$, then $\widetilde{E}(a \otimes \widetilde{X} \oplus b) = a \otimes \widetilde{E}(\widetilde{X}) \oplus b$.*

Proof *For every $\alpha \in [0,1]$, using the arithmetic operations of **FNs** (based on α-values), we have:*

$$E((a \otimes \widetilde{X} \oplus b)_\alpha) = \begin{cases} aE(\widetilde{X}_\alpha) + b, & a \geq 0, \\ aE(\widetilde{X}_{1-\alpha}) + b, & a < 0, \end{cases} \tag{4.4}$$

which immediately verifies the claim.

Lemma 4.3 *Let $\{\widetilde{X}_i\}_{i=1}^n$ be a finite sequence of **FRVs** each with fuzzy mean of $\widetilde{E}(\widetilde{X}_i)$. Then:*

$$\widetilde{E}(\bigoplus_{i=1}^{n}(a_i \otimes \widetilde{X}_i)) = \bigoplus_{i=1}^{n}(a_i \otimes \widetilde{E}(\widetilde{X}_i)). \tag{4.5}$$

Proof *The proof is left for the reader.*

Lemma 4.4 *Let $\widetilde{X}$ be a **FRV** with $\widetilde{\mu} = \widetilde{E}(\widetilde{X})$ and $\widetilde{X}_1, \cdots, \widetilde{X}_n$ be a **FRS**. Then, $\widetilde{E}(\widetilde{\overline{X}}) = \widetilde{E}(\widetilde{X})$ where $\widetilde{\overline{X}} = (\frac{1}{n}) \otimes (\bigoplus_{i=1}^n \widetilde{X}_i)$.*

Proof *Since $\widetilde{X}_1, \cdots, \widetilde{X}_n$ are independent and identically distributed **FRVs**, for every $\alpha \in [0,1]$, we have:*

$$(\widetilde{E}(\widetilde{\overline{X}}))_\alpha = E(\widetilde{\overline{X}}_\alpha) = E(\frac{1}{n}\sum_{i=1}^{n}(\widetilde{X}_i)_\alpha) = E(\widetilde{X}_\alpha). \tag{4.6}$$

This means $\widetilde{E}(\widetilde{\overline{X}}) = \widetilde{E}(\widetilde{X})$.

Remark 4.1 *Let $\widetilde{X}$ be a **FRV** (with fuzzy mean $\widetilde{\mu} = \widetilde{E}(\widetilde{X})$) and $\widetilde{X}_1, \cdots, \widetilde{X}_n$ be a **FRS**. Roughly speaking, according to Lemma 4.4, $\widetilde{\overline{X}} = (1/n) \bigoplus_{i=1}^n \widetilde{X}_i$ is a (fuzzy) unbiased estimator for $\widetilde{\mu}$ denoted by $\widehat{\widetilde{\mu}} = \widetilde{\overline{X}}$.*

Definition 4.5 *Let $\widetilde{X}_1, \cdots, \widetilde{X}_n$ be a **FRS**. The (non-fuzzy) sample variance of $\widetilde{X}_1, \cdots, \widetilde{X}_n$ is defined as $S_n^2 = \frac{1}{n-1}\sum_{i=1}^n d_2^2(\widetilde{X}_i, \widetilde{\overline{X}})$ in which $\widetilde{\overline{X}} = (\frac{1}{n}) \bigoplus_{i=1}^n \widetilde{X}_i$ is the fuzzy (sample) mean.*

Theorem 4.1 *Let $\widetilde{X}$ be a **FRV** and $\widetilde{X}_1, \cdots, \widetilde{X}_n$ be a **FRS**. Then, $E(S_n^2) = \boldsymbol{var}(\widetilde{X})$ and S_n^2 converges to $\boldsymbol{var}(\widetilde{X})$ almost surely.*

Proof *Note that:*

$$E(S_n^2) = E(\frac{1}{n-1}\sum_{i=1}^{n} d_2^2(\widetilde{X}_i, \widetilde{\overline{X}})) = \frac{1}{n-1}\sum_{i=1}^{n} E(d_2^2(\widetilde{X}_i, \widetilde{\overline{X}})), \tag{4.7}$$

where $E(d_2^2(\widetilde{X}_i, \overline{\widetilde{X}})) = \int_0^1 E((\widetilde{X}_i)_\alpha - \overline{\widetilde{X}}_\alpha)^2 d\alpha = (n-1/n)\boldsymbol{var}(\widetilde{X})$. *Therefore,* $E(S_n^2) = \frac{1}{n-1} n \frac{n-1}{n} \boldsymbol{var}(\widetilde{X}) = \boldsymbol{var}(\widetilde{X})$. *Furthermore, it can be be checked that*

$$S_n^2 = \frac{1}{n}\sum_{i=1}^{n} d_2^2(\widetilde{X}_i, \widetilde{\mu}) - d_2^2(\overline{\widetilde{X}}, \widetilde{\mu}).$$

However, based on the strong law of large numbers (***SLLN***) *[132],* $\frac{1}{n}\sum_{i=1}^{n} d_2^2(\widetilde{X}_i, \widetilde{\mu}) \to E(d_2^2(\widetilde{X}_1, \widetilde{\mu})) = \boldsymbol{var}(\widetilde{X})$. *In addition,*

$$d_2^2(\overline{\widetilde{X}}, \widetilde{\mu}) = \int_0^1 (\overline{\widetilde{X}}_\alpha - \widetilde{\mu}_\alpha)^2 d\alpha \leq sup_{\alpha\in[0,1]}(\overline{\widetilde{X}}_\alpha - \widetilde{\mu}_\alpha)^2. \quad (4.8)$$

Since $\overline{\widetilde{X}}_\alpha \to \widetilde{\mu}_\alpha$ *almost surely for any* $\alpha \in [0,1]$, *it follows that* $d_2^2(\overline{\widetilde{X}}, \widetilde{\mu}) \to 0$ *almost surely. These conclude that* S_n^2 *converges to* $\boldsymbol{var}(\widetilde{X})$ *almost surely.*

4.1.2 Correlation between two FRVs

Correlation coefficients are used to measure how strong a relationship is between two variables. The most popular is Pearson's coefficient which is a correlation coefficient commonly used in statistical linear models. Here, the concept of covariance and correlation are extended for **FRV**s.

Definition 4.6 *Let* $\widetilde{X}$ *and* $\widetilde{Y}$ *be two* ***FRVs****. The covariance of two* ***FRVs*** *of* $\widetilde{X}$ *and* $\widetilde{Y}$ *is defined by:*

$$\boldsymbol{Cov}(\widetilde{X}, \widetilde{Y}) = \int_0^1 cov(\widetilde{X}_\alpha, \widetilde{Y}_\alpha) d\alpha. \quad (4.9)$$

Theorem 4.2 *Let* $\widetilde{X}$ *and* $\widetilde{Y}$ *be two* ***FRVs****. Then:*

1. $\boldsymbol{cov}(\widetilde{X}, \widetilde{Y}) = \boldsymbol{cov}(\widetilde{Y}, \widetilde{X})$.
2. $\boldsymbol{cov}(\widetilde{X}, \widetilde{X}) = \boldsymbol{var}(\widetilde{X})$.
3. $\boldsymbol{var}(\widetilde{X} \oplus \widetilde{Y}) = \boldsymbol{var}(\widetilde{X}) + \boldsymbol{var}(\widetilde{Y}) + 2\boldsymbol{cov}(\widetilde{X}, \widetilde{Y})$.
4. $\boldsymbol{cov}((a \otimes \widetilde{X}) \oplus \widetilde{b}, (c \otimes \widetilde{Y}) \oplus \widetilde{d}) = ac\boldsymbol{cov}(\widetilde{X}, \widetilde{Y})$ *provided that* $ac > 0$.

Proof *(1) and (2) are immediately verified. By (2), we have:*

$$\begin{aligned}\boldsymbol{var}(\widetilde{X} \oplus \widetilde{Y}) &= \boldsymbol{cov}(\widetilde{X} \oplus \widetilde{Y}, \widetilde{X} \oplus \widetilde{Y}) \\ &= \int_0^1 cov(\widetilde{X}_\alpha + \widetilde{Y}_\alpha, \widetilde{X}_\alpha + \widetilde{Y}_\alpha) d\alpha \\ &= \int_0^1 (var(\widetilde{X}_\alpha) + var(\widetilde{Y}_\alpha) + 2cov(\widetilde{X}_\alpha, \widetilde{Y}_\alpha)) d\alpha,\end{aligned}$$

which verifies (3). To prove (4), with out losing generality, assume that a and c are negative constants. Therefore:

$$\begin{aligned}
&\boldsymbol{cov}(((a\otimes\widetilde{X})\oplus\widetilde{b},(c\otimes\widetilde{Y})\oplus\widetilde{d})\\
&\qquad=\int_0^1 cov((a\otimes\widetilde{X})\oplus\widetilde{b})_\alpha,((c\otimes\widetilde{Y})\oplus\widetilde{d})_\alpha)d\alpha\\
&\qquad=\int_0^1 cov(a\widetilde{X}_{1-\alpha}+\widetilde{b}_\alpha,c\widetilde{Y}_{1-\alpha}+\widetilde{d}_\alpha)d\alpha\\
&\qquad=ac\int_0^1 cov(\widetilde{X}_{1-\alpha},\widetilde{Y}_{1-\alpha})d\alpha=ac\,\boldsymbol{cov}(\widetilde{X},\widetilde{Y}).
\end{aligned}$$

Similar reasoning is valid for the cases that $a>0, c>0$.

Example 4.4 *Consider two* ***FRVs*** *of* $\widetilde{X}=(X;U)$ *and* $\widetilde{Y}=(Y;U^2)$ *where* $(X,Y)\sim N_2(\mu_X=0,\mu_Y=0,\sigma_X^2=4,\sigma_Y^2=9,\rho=0.80)$ *and* $U\sim U(0,2)$. *Then:*

$$\begin{aligned}
\boldsymbol{cov}(\widetilde{X},\widetilde{Y})&=\int_0^1 cov(X-(1-2\alpha)U,Y-(1-2\alpha)U^2)d\alpha\\
&=cov(X,Y)+(\int_0^1(1-2\alpha)^2d\alpha)cov(U,U^2)\\
&=2\times3\times0.8+(\frac{1}{3})\times(\frac{2}{3})=\frac{226}{45}.
\end{aligned}$$

Definition 4.7 *Let* $\widetilde{X}$ *and* $\widetilde{Y}$ *be two* ***FRVs***. *The correlation of* $\widetilde{X}$ *and* $\widetilde{Y}$ *can be defined by:*

$$\boldsymbol{\rho}(\widetilde{X},\widetilde{Y})=\frac{\boldsymbol{Cov}(\widetilde{X},\widetilde{Y})}{\sqrt{\boldsymbol{var}(\widetilde{X})\boldsymbol{var}(\widetilde{Y})}}. \tag{4.10}$$

Theorem 4.3 *Let* $\widetilde{X}$ *and* $\widetilde{Y}$ *be two* ***FRVs***. *Then:*

1) $|\boldsymbol{\rho}(\widetilde{X},\widetilde{Y})|\leq 1$.

2) $\boldsymbol{\rho}(\widetilde{X},\widetilde{Y})=1$ *if and only if* $P(\widetilde{Y}\oplus(a_0\otimes\widetilde{E}(\widetilde{X}))=\widetilde{E}(\widetilde{Y})\oplus(a_0\otimes\widetilde{X}))=1$.

3) $\boldsymbol{\rho}(\widetilde{X},\widetilde{Y})=-1$ *if and only if* $P(\widetilde{Y}\oplus(a_0\otimes\widetilde{X})=\widetilde{E}(\widetilde{Y})\oplus(a_0\otimes\widetilde{E}(\widetilde{X})))=1$.

Proof *For any* $a\in\mathbb{R}$, *let:*

$$h(a)=\boldsymbol{var}(\widetilde{Y})-2a\,\boldsymbol{Cov}(\widetilde{X},\widetilde{Y})+\boldsymbol{var}(\widetilde{X}). \tag{4.11}$$

It is completely straight forward to show that:

$$h(a)=\begin{cases} E\big(d_2^2(\widetilde{Y}\oplus(a\otimes\widetilde{E}(\widetilde{X})),\widetilde{E}(\widetilde{Y})\oplus(a\otimes\widetilde{X}))\big), & a\geq 0,\\ E\big(d_2^2(\widetilde{Y}\oplus((-a)\otimes\widetilde{X}),\widetilde{E}(\widetilde{Y})\oplus((-a)\otimes\widetilde{E}(\widetilde{X})))\big), & a<0.\end{cases} \tag{4.12}$$

This concludes that $h(a) \geq 0$ *for any* $a \in \mathbb{R}$. *Therefore, by taking* $a_0 = \boldsymbol{Cov}(\widetilde{X}, \widetilde{Y})/\boldsymbol{var}(\widetilde{X})$, *item (1) is verified. To prove (2) and (3), note that if* $|\boldsymbol{\rho}(\widetilde{X}, \widetilde{Y})| = 1$, *then there exists a* a_0 *such that* $h(a_0) = 0$. *By Eq. (4.12), therefore:*

1. *if* $\boldsymbol{\rho}(\widetilde{X}, \widetilde{Y}) = 1$, *then* $a_0 = \sqrt{\boldsymbol{var}(\widetilde{Y})/\boldsymbol{var}(\widetilde{X})}$ *and thus* $P(d_2^2(\widetilde{Y} \oplus (a \otimes \widetilde{E}(\widetilde{X})), \widetilde{E}(\widetilde{Y}) \oplus (a \otimes \widetilde{X})) = 0) = 1$ *that is* $P(\widetilde{Y} \oplus (a_0 \otimes \widetilde{E}(\widetilde{X})) = \widetilde{E}(\widetilde{Y}) \oplus (a_0 \otimes \widetilde{X})) = 1$.

2. *If* $\boldsymbol{\rho}(\widetilde{X}, \widetilde{Y}) = -1$, *then* $a_0 = -\sqrt{\boldsymbol{var}(\widetilde{Y})/\boldsymbol{var}(\widetilde{X})}$ *and thus* $P(d_2^2(\widetilde{Y} \oplus ((-a) \otimes \widetilde{X}), \widetilde{E}(\widetilde{Y}) \oplus ((-a) \otimes \widetilde{E}(\widetilde{X}))) = 0) = 1$, *that is* $P(\widetilde{Y} \oplus (a_0 \otimes \widetilde{X}) = \widetilde{E}(\widetilde{Y}) \oplus (a_0 \otimes \widetilde{E}(\widetilde{X}))) = 1$. *Hence the necessary condition holds.*

The sufficiency could be assigned to the fact that if $P(\widetilde{Y} \oplus (a_0 \otimes \widetilde{E}(\widetilde{X})) = \widetilde{E}(\widetilde{Y}) \oplus (a_0 \otimes \widetilde{X})) = 1$ *then, we have:*

$$\begin{aligned}
\boldsymbol{Cov}(\widetilde{X}, \widetilde{Y}) &= \frac{1}{a_0} \boldsymbol{Cov}((a_0 \otimes \widetilde{X}) \oplus \widetilde{E}(\widetilde{Y}), \widetilde{Y} \oplus (a_0 \otimes \widetilde{E}(\widetilde{X}))) \\
&= \frac{1}{a_0} \boldsymbol{Cov}(\widetilde{Y} \oplus (a_0 \otimes \widetilde{E}(\widetilde{X})), \widetilde{Y} \oplus (a_0 \otimes \widetilde{E}(\widetilde{X}))) \\
&= \frac{1}{a_0} \boldsymbol{Cov}(\widetilde{Y}, \widetilde{Y}) = \sqrt{\boldsymbol{var}(\widetilde{Y}) \boldsymbol{var}(\widetilde{X})},
\end{aligned}$$

that is $\boldsymbol{\rho}(\widetilde{X}, \widetilde{Y}) = 1$. *Further, if* $P(\widetilde{Y} \oplus (a_0 \otimes \widetilde{X}) = \widetilde{E}(\widetilde{Y}) \oplus (a_0 \otimes \widetilde{E}(\widetilde{X}))) = 1$, *we have:*

$$\begin{aligned}
\boldsymbol{Cov}(\widetilde{X}, \widetilde{Y}) &= \int_0^1 E(\widetilde{X}_\alpha \widetilde{X}_\alpha) d\alpha - \int_0^1 E(\widetilde{X}_\alpha) E(\widetilde{X}_\alpha) d\alpha \\
&= \int_0^1 E((\widetilde{Y} \oplus (a_0 \otimes \widetilde{X}))_\alpha) \widetilde{X}_\alpha d\alpha - a_0 \int_0^1 E((\widetilde{X}_\alpha)^2) d\alpha \\
&- \int_0^1 E(\widetilde{X}_\alpha) E(\widetilde{X}_\alpha) d\alpha \\
&= \int_0^1 E(\widetilde{E}(\widetilde{Y}) \oplus (a_0 \otimes \widetilde{E}(\widetilde{X})))_\alpha \widetilde{X}_\alpha) d\alpha - a_0 \int_0^1 E((\widetilde{X}_\alpha)^2) d\alpha \\
&- \int_0^1 E(\widetilde{X}_\alpha) E(\widetilde{X}_\alpha) d\alpha \\
&= \int_0^1 (E(\widetilde{Y}_\alpha) + (a_0 E(\widetilde{X}_\alpha))) E(\widetilde{X}_\alpha) d\alpha - a_0 \int_0^1 E((\widetilde{X}_\alpha)^2) d\alpha \\
&- \int_0^1 E(\widetilde{X}_\alpha) E(\widetilde{X}_\alpha) d\alpha \\
&= -a_0 \boldsymbol{var}(\widetilde{X}) = -\sqrt{\boldsymbol{var}(\widetilde{Y}) \boldsymbol{var}(\widetilde{X})},
\end{aligned}$$

that is $\boldsymbol{\rho}(\widetilde{X}, \widetilde{Y}) = -1$.

Example 4.5 *Recall Example 4.4. Then:*

$$\begin{aligned}
\boldsymbol{var}(\widetilde{X}) &= \int_0^1 var(X-(1-2\alpha)U)d\alpha \\
&= \int_0^1 (var(X)+(1-2\alpha)^2 var(U) - 2(1-2\alpha)cov(X,U))d\alpha \\
&= var(X) + (\int_0^1 (1-2\alpha)^2 d\alpha) var(U) = 4 + (\frac{1}{3})(\frac{1}{3}) = \frac{37}{3}.
\end{aligned}$$

Further,

$$\begin{aligned}
\boldsymbol{var}(\widetilde{Y}) &= \int_0^1 var(Y-(1-2\alpha)U^2)d\alpha \\
&= \int_0^1 (var(Y)+(1-2\alpha)^2 var(U^2) - 2(1-2\alpha)cov(Y,U^2))d\alpha \\
&= var(Y) + (\int_0^1 (1-2\alpha)^2 d\alpha) var(U^2) = 9 + (\frac{1}{3})(\frac{64}{45}) = \frac{1279}{135}.
\end{aligned}$$

This concludes that

$$\boldsymbol{\rho}(\widetilde{X},\widetilde{Y}) = \frac{\frac{226}{45}}{\sqrt{\frac{37}{3}\times\frac{1279}{135}}} = 0.465.$$

4.1.3 Probability of an interval based on FRVs

Let X be a continuous random variable with density function $f_X(x)$. The probability that X is in the interval $[a,b]$ can be calculated by integrating the $f_X(x)$ as:

$$P(a < X < b) = \int_{\mathbb{R}} f_X(x)dx.$$

Here, a concept of $P(a < X < b)$ is developed for a **FRV** $\widetilde{X}$.

Definition 4.8 *Let $\widetilde{X}$ be a **FRV**. The probability of $\widetilde{X} \in (a,b)$ where $(a,b) \subseteq \mathbb{R}$ is defined as a **FN** of $\widetilde{P}(\widetilde{X} \in (a,b))$ with the following α-values:*

$$(\widetilde{P}(\widetilde{X} \in (a,b)))_\alpha = \begin{cases} \inf_{\beta \in I_{2\alpha}} P(a < \widetilde{X}_\beta < b), & 0 \le \alpha \le 0.5, \\ \sup_{\beta \in I_{2(1-\alpha)}} P(a < \widetilde{X}_\beta < b), & 0.5 < \alpha \le 1, \end{cases} \tag{4.13}$$

where $I_\alpha = [\alpha/2, 1-\alpha/2]$ and

$$P(a < \widetilde{X}_\beta < b) = \int_{\mathbb{R}} f_{\widetilde{X}_\beta}(x)dx. \tag{4.14}$$

Example 4.6 *A tire and rubber company is interested in the quality of a tire it has recently developed. Since, under some unexpected situations, we cannot measure the tire lifetime (KM) precisely, we can just mention the tire lifetime using approximate terms. Therefore the tire lifetime can be taken as a fuzzy quantity such as* $\widetilde{X} = (X; 3000 + \epsilon)$ *where* $X \sim N(32000, (5000)^2)$, $\epsilon \sim N(0, (200)^2)$ *in which* X *and* ϵ *are independent random variables. We wish to evaluate* $\widetilde{P}(\widetilde{X} \in (35000, \infty))$. *For this purpose, note that* $\widetilde{X}_\alpha = X - (1-2\alpha)(3000+\epsilon)$. *Therefore:*

$$P(\widetilde{X}_\beta > 35000) = 1 - F_Z\left(\frac{35000 - (32000 - 3000(1-2\beta)}{\sqrt{(5000)^2 + (200)^2(1-2\beta)^2}}\right),$$

where F_Z *is the* ***CDF*** *of standard normal distribution function. Thus, the* α*-cuts of* $\widetilde{P}(\widetilde{X} \in (35000, \infty))$ *can be evaluated as follows:*

$$(\widetilde{P}(\widetilde{X} \in (35000, \infty)))^L_\alpha = \inf_{\beta \in [\alpha/2, 1-\alpha/2]} \left(1 - F_Z\left(\frac{35000 - (32000 - 3000(1-2\beta))}{\sqrt{(5000)^2 + (200)^2(1-2\beta)^2}}\right)\right), \tag{4.15}$$

$$(\widetilde{P}(\widetilde{X} \in (35000, \infty)))^U_\alpha = \sup_{\beta \in [\alpha/2, 1-\alpha/2]} \left(1 - F_Z\left(\frac{35000 - (32000 - 3000(1-2\beta))}{\sqrt{(5000)^2 + (200)^2(1-2\beta)^2}}\right)\right). \tag{4.16}$$

To compute the α*-cut of* $\widetilde{P}(\widetilde{X} \in (35000, \infty))$, *there is a need for a optimization problem to evaluate the lower and upper limits. Some* α*-cuts of* $\widetilde{P}(\widetilde{X} \in (35000, \infty))$ *are shown in Table 4.1. By plotting the lover and upper values of* $\widetilde{P}(\widetilde{X} \in (35000, \infty))$ *for each* $\alpha = 0, 0.01, 0.02, ..., 0.99, 1$, *the membership function of* $\widetilde{P}(\widetilde{X} \in (35000, \infty))$ *is shown in Fig. 4.2. Accordingly, it can be said that the fuzzy probability of* $\widetilde{X} \in (35000, \infty)$ *is a* ***FN*** *as 'about 0.5'.*

TABLE 4.1
Some α-cuts of $\widetilde{P}(\widetilde{X} \in (35000, \infty))$ in Example 4.6.

α	$(\widetilde{P}(\widetilde{X} \in (35000, \infty)))^L_\alpha$	$(\widetilde{P}(\widetilde{X} \in (35000, \infty)))^U_\alpha$
0.07	0.1323	0.5
0.17	0.1597	0.5
0.27	0.1905	0.5
0.37	0.2248	0.5
0.47	0.2623	0.5
0.57	0.3029	0.5
0.67	0.3461	0.5
0.77	0.3913	0.5
0.87	0.4380	0.5
0.97	0.4856	0.5
1	0.5	0.5

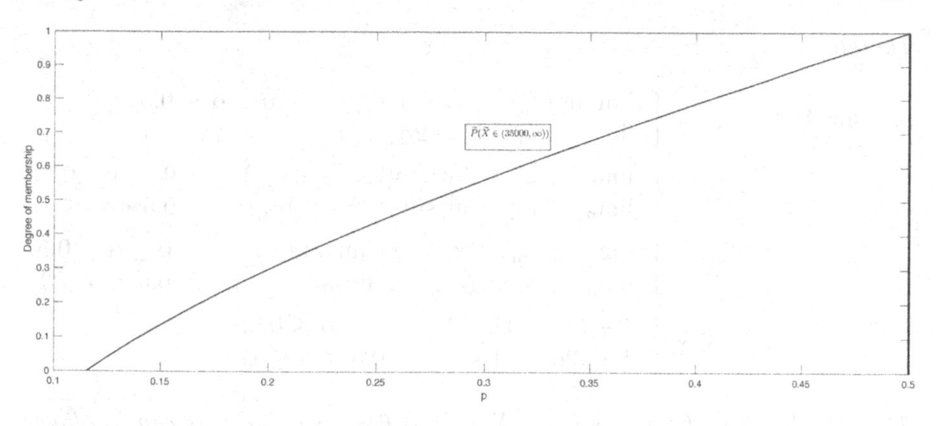

FIGURE 4.2
Plot of $\widetilde{P}(\widetilde{X} \in (35000, \infty))$ in Example 4.6.

4.1.4 Liminf and limsup for a sequence of FRVs

Here, a notion of lim inf and lim sup is introduced and discussed for **FRVs**.

Definition 4.9 *Let $\{\widetilde{X}_n\}_{n \in \mathbb{N}}$ be a sequence of **FRVs**. The* lim inf *and* lim sup *of a sequence of **FRVs** $\{\widetilde{X}_n\}_{n \geq 1}$ can be defined as **FNs** in the form of $\liminf_{n\to\infty} \widetilde{X}_n = \widetilde{\max}_{n\geq 1}\widetilde{\min}_{m\geq n}\widetilde{X}_m$ and $\limsup_{n\to\infty} \widetilde{X}_n = \widetilde{\min}_{n\geq 1}\widetilde{\max}_{m\geq n}\widetilde{X}_m$.*

Theorem 4.4 *If $\{\widetilde{X}_n\}_{n \in \mathbb{N}}$ is a sequence of **FRVs**, then:*

1) $(\liminf_{n\to\infty} \widetilde{X}_n)_\alpha = \liminf_{n\to\infty} (\widetilde{X}_n)_\alpha$.

2) $(\limsup_{n\to\infty} \widetilde{X}_n)_\alpha = \limsup_{n\to\infty} (\widetilde{X}_n)_\alpha$.

Proof *To establish (1), by definition of maximum and minimum of **FNs**, it is easy to check for every $\alpha \in [0, 1]$ that:*

$$\begin{aligned}(\liminf_{n\to\infty} \widetilde{X}_n)_\alpha &= (\widetilde{\max}_{n\geq 1}\widetilde{\min}_{m\geq n}\widetilde{X}_m)_\alpha = \max_{n\geq 1}(\widetilde{\min}_{m\geq n}\widetilde{X}_m)_\alpha \\ &= \max_{n\geq 1}\min_{m\geq n}(\widetilde{X}_m)_\alpha = \liminf_{n\to\infty}(\widetilde{X}_n)_\alpha.\end{aligned}$$

Similar reasoning holds for (2).

Example 4.7 *Consider a countable sequence of **FRVs** as $\widetilde{X}_n = X \otimes \widetilde{A}_n$ where $P(X > 0) = 1$. Moreover, let $\widetilde{A}_n = (a_n; l_{a_n}, r_{a_n})_T$ with $a_n = 2^{(1/(n+1))}$, $l_{a_n} = (n/(n+1))I(n \text{ is odd}) + (1/(n+1))I(n \text{ is even})$ and $r_{a_n} = 5 + \cos(n\pi/4)$.*

Therefore:

$$(\limsup \widetilde{X}_n)_\alpha = X \times \begin{cases} \limsup(a_n + (2\alpha - 1)l_{a_n}), & 0 \le \alpha \le 0.5, \\ \limsup(a_n + (1 - 2\alpha)r_{a_n}), & 0.5 < \alpha \le 1. \end{cases}$$
$$= X \times \begin{cases} \lim_{n\to\infty} a_n + \limsup((2\alpha - 1)l_{a_n}), & 0 \le \alpha \le 0.5, \\ \lim_{n\to\infty} a_n + \limsup((2\alpha - 1)r_{a_n}), & 0.5 < \alpha \le 1. \end{cases}$$
$$= X \times \begin{cases} \lim_{n\to\infty} a_n + (2\alpha - 1)\liminf l_{a_n}, & 0 \le \alpha \le 0.5, \\ \lim_{n\to\infty} a_n + (1 - 2\alpha)\limsup(r_{a_n}), & 0.5 < \alpha \le 1. \end{cases}$$
$$= X \times \begin{cases} 1 + (2\alpha - 1) \times 0, & 0 \le \alpha \le 0.5, \\ 1 + (2\alpha - 1) \times 6, & 0.5 < \alpha \le 1. \end{cases}$$

This concludes that $\limsup \widetilde{X}_n = X \otimes (1; 0, 6)_T$. *Similarly, it can be shown that:*

$$(\liminf \widetilde{A}_n)_\alpha = X \times \begin{cases} 1 + (2\alpha - 1) \times 1, & 0 \le \alpha \le 0.5, \\ 1 + (1 - 2\alpha) \times 4, & 0.5 < \alpha \le 1, \end{cases}$$

that is $\liminf \widetilde{X}_n = X \otimes (1; 1, 4)_T$.

Proposition 4.1 *If* $\{\widetilde{X}_n\}_{n\in\mathbb{N}}$ *and* $\{\widetilde{Y}_n\}_{n\in\mathbb{N}}$ *are two sequence of* ***FRVs*** *then:*

1) $P_d(\liminf_{n\to\infty}(\widetilde{X}_n \oplus \widetilde{Y}_n) \succeq \liminf_{n\to\infty} \widetilde{X}_n \oplus \liminf_{n\to\infty} \widetilde{Y}_n) = 1$.

2) $P_d(\limsup_{n\to\infty}(\widetilde{X}_n \oplus \widetilde{Y}_n) \preceq \limsup_{n\to\infty} \widetilde{X}_n \oplus \limsup_{n\to\infty} \widetilde{Y}_n) = 1$.

Proof *To establish (1), for every* $\alpha \in [0, 1]$, *we easily have:*

$$\begin{aligned}(\liminf_{n\to\infty}(\widetilde{X}_n \oplus \widetilde{Y}_n))_\alpha &= \liminf_{n\to\infty}((\widetilde{X}_n)_\alpha + (\widetilde{Y}_n)_\alpha) \\ &\ge \liminf_{n\to\infty}(\widetilde{X}_n)_\alpha + \liminf_{n\to\infty}(\widetilde{Y}_n)_\alpha \\ &= (\liminf_{n\to\infty} \widetilde{X}_n)_\alpha + (\liminf_{n\to\infty} \widetilde{Y}_n)_\alpha.\end{aligned}$$

This simply concludes (1). Similar reasoning holds also for (2).

4.1.5 Fuzzy cumulative distribution function

The cumulative distribution function (**CDF**) is the probability that the variable takes a value less than or equal to x. Here, a notion of the fuzzy cumulative distribution function is defined based on **FRV**s at exact points.

Definition 4.10 *[85] Let* $\widetilde{X}$ *be a* ***FRV*** *on a probability space* $(\Omega, \mathcal{A}, P)$. *The fuzzy cumulative distribution function (**FCDF**) of* $\widetilde{X}$ *at* $x \in \mathbb{R}$ *is defined as a* ***FN*** $\widetilde{F}_{\widetilde{X}}(x)$ *with the following* α*-values:*

$$(\widetilde{F}_{\widetilde{X}}(x))_\alpha = P(\widetilde{X}_{1-\alpha} \le x).$$

Example 4.8 *Consider a* ***FRV*** *as* $\widetilde{X} = (X; 0.1X)_T$ *where* X *is distributed according to an exponential distribution with a mean of* $\lambda = 80$. *Therefore:*

$$\begin{aligned}(\widetilde{F}_{\widetilde{X}}(x))_\alpha &= P(\widetilde{X}_{1-\alpha} \leq x)\\ &= P(X + 0.1(1-2\alpha)X \leq x)\\ &= F_X(\frac{x}{1.1 - 0.1(1-2\alpha)})\\ &= 1 - \exp(-\frac{1}{80} \times \frac{x}{1.1 - 0.1(1-2\alpha)}).\end{aligned}$$

Thus the α*-cut of* $\widetilde{F}_{\widetilde{X}}$ *can be evaluated as follows:*

$$\begin{aligned}(\widetilde{F}_{\widetilde{X}}(x))_\alpha^L &= 1 - \exp(-\frac{1}{80} \times \frac{x}{0.9 + 0.1\alpha}),\\ (\widetilde{F}_{\widetilde{X}}(x))_\alpha^U &= 1 - \exp(-\frac{1}{80} \times \frac{x}{1.1 - 0.1\alpha}).\end{aligned}$$

Some specified values of $(\widetilde{F}_{\widetilde{X}}(x))_\alpha^L$ *and* $(\widetilde{F}_{\widetilde{X}}(x))_\alpha^U$ *at* $x = 188, 190, 192$, *and* 194 *are listed in Table 4.2. Furthermore, the plots of* $\widetilde{F}_{\widetilde{X}}$ *at* $x = 188, 190, 192$, *and* 194 *are drown in Fig. 4.3. Accordingly, the fuzzy cumulative distribution of* $\widetilde{X}$ *at* $x = 188, 190, 192$, *and* 194 *are some* ***FNs*** *as 'about 0.90463', 'about 0.90699', 'about 0.90928' and 'about 0.91152'. In addition, the three-dimensional curve of* $\widetilde{F}_{\widetilde{X}}(x)$ *on* $[185, 200]$ *is plotted in Fig. 4.4.*

Example 4.9 *Consider the* ***FRV*** *of* $\widetilde{X} = (X; 0.1|Y|)$ *where* X *and* Y *are two independent random variables distributed according to standard normal distribution. Since* $\widetilde{X}_\alpha = X - 0.1(1-2\alpha)|Y|$, *by definition of a* ***FCDF***, *the* α*-cut of* $\widetilde{F}_{\widetilde{X}}$ *can be evaluated as follows:*

$$\begin{aligned}(\widetilde{F}_{\widetilde{X}}(x))_\alpha^L &= P(\widetilde{X}_{1-\alpha/2} \leq x)\\ &= P(X + 0.1(1-\alpha)|Y| \leq x) = \int_0^\infty F_X(x - 0.1(1-\alpha)|y|) f_Y(y)dy\\ &= \frac{1}{2\pi}\int_0^\infty \int_{-\infty}^{x-0.1(1-\alpha|y|)} \exp(-0.5(z^2+y^2))dzdy,\\ (\widetilde{F}_{\widetilde{X}}(x))_\alpha^U &= P(\widetilde{X}_{\alpha/2} \leq x)\\ &= P(X - 0.1(1-\alpha)|Y| \leq x) = \int_0^\infty F_X(x + 0.1(1-\alpha)|y|) f_Y(y)dy\\ &= \frac{1}{2\pi}\int_0^\infty \int_{-\infty}^{x+0.1(1-\alpha|y|)} \exp(-0.5(z^2+y^2))dzdy.\end{aligned}$$

Some α*-cuts of* $\widetilde{F}_{\widetilde{X}}(x)$ *at* $x = -1, 1$ *are listed in Table 4.3. The plot of* $\widetilde{F}_{\widetilde{X}}(x)$ *at* $x = -1, 1$ *are also plotted in Fig. 4.5. The results show that* $\widetilde{F}_{\widetilde{X}}(-1)$ *is 'about 0.1586' and* $\widetilde{F}_{\widetilde{X}}(+1)$ *is 'about 0.8413'.*

TABLE 4.2
Some α-cuts of $\widetilde{P}(\widetilde{X} \in (35000, \infty))$ in Example 4.8.

α	$(\widetilde{F}_{\widetilde{X}}(188))^L_\alpha$	$(\widetilde{F}_{\widetilde{X}}(188))^U_\alpha$	$(\widetilde{F}_{\widetilde{X}}(190))^L_\alpha$	$(\widetilde{F}_{\widetilde{X}}(190))^U_\alpha$	$(\widetilde{F}_{\widetilde{X}}(192))^L_\alpha$	$(\widetilde{F}_{\widetilde{X}}(192))^U_\alpha$	$(\widetilde{F}_{\widetilde{X}}(194))^L_\alpha$	$(\widetilde{F}_{\widetilde{X}}(194))^U_\alpha$
0.03	0.92591	0.8826	0.92793	0.88525	0.9299	0.88784	0.93181	0.89036
0.13	0.92376	0.8849	0.92582	0.88751	0.92783	0.89007	0.92978	0.89257
0.23	0.92161	0.88718	0.9237	0.88977	0.92574	0.8923	0.92773	0.89477
0.33	0.91944	0.88947	0.92157	0.89203	0.92364	0.89453	0.92566	0.89697
0.43	0.91726	0.89175	0.91942	0.89428	0.92153	0.89675	0.92359	0.89916
0.53	0.91507	0.89402	0.91727	0.89652	0.91941	0.89896	0.9215	0.90135
0.63	0.91286	0.89629	0.9151	0.89876	0.91727	0.90117	0.91939	0.90353
0.73	0.91065	0.89855	0.91292	0.90099	0.91513	0.90337	0.91728	0.9057
0.83	0.90843	0.90081	0.91073	0.90322	0.91297	0.90557	0.91516	0.90786
0.93	0.9062	0.90306	0.90853	0.90544	0.9108	0.90776	0.91302	0.91002
1	0.90463	0.90463	0.90699	0.90699	0.90928	0.90928	0.91152	0.91152

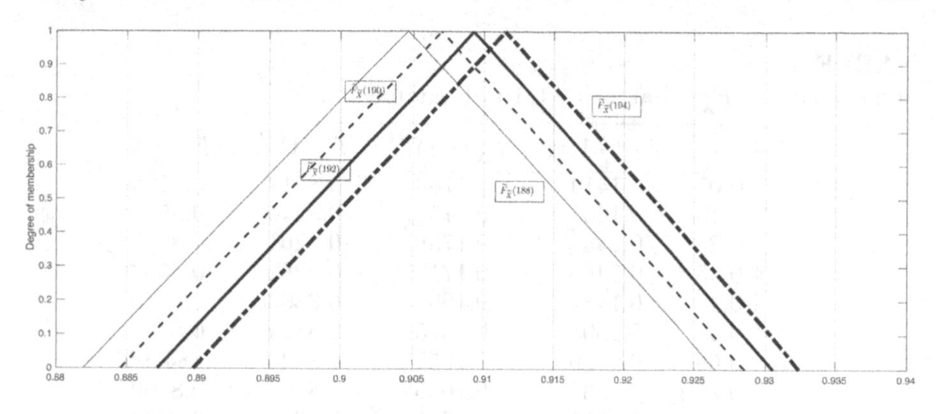

FIGURE 4.3
Plot of $\widetilde{F}_{\widetilde{X}}(x)$ at $x = 188, 190, 192, 194$ in Example 4.8.

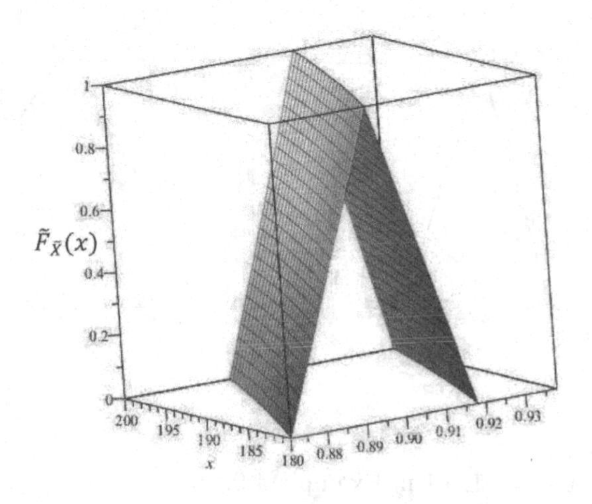

FIGURE 4.4
Plot of $\widetilde{F}_{\widetilde{X}}(x)$ on $[185, 200]$ in Example 4.8.

Theorem 4.5 *Let $\widetilde{X}$ be a **FRV** on the probability space $(\Omega, \mathcal{A}, P)$. Then $\widetilde{F}_{\widetilde{X}}$ meets the following properties:*

(i) For $t, s \in \mathbb{R}$ if $t < s$ then $P_d(\widetilde{F}_{\widetilde{X}}(t) \preceq \widetilde{F}_{\widetilde{X}}(s)) = 1$.

(ii) If $\{t_n\}_{n\in\mathbb{N}}$ is a decreasing sequence of real-valued numbers with $t_n \to t \in \mathbb{R}$ then $\lim_{n\to+\infty} d_\infty(\widetilde{F}_{\widetilde{X}}(t_n), \widetilde{F}_{\widetilde{X}}(t)) = 0$.

TABLE 4.3
Some α-cuts of $\widetilde{F}_{\widetilde{X}}(x)$ at $x = -1, 1$ in Example 4.9.

α	$(\widetilde{F}_{\widetilde{X}}(-1))_\alpha^L$	$(\widetilde{F}_{\widetilde{X}}(-1))_\alpha^U$	$(\widetilde{F}_{\widetilde{X}}(1))_\alpha^L$	$(\widetilde{F}_{\widetilde{X}}(1))_\alpha^U$
0.05	0.1414	0.1780	0.8219	0.8585
0.15	0.1431	0.1759	0.8240	0.8565
0.25	0.1448	0.1738	0.8261	0.8551
0.35	0.1466	0.1717	0.8282	0.8533
0.45	0.1484	0.1696	0.8303	0.8515
0.55	0.1502	0.1675	0.8324	0.8497
0.65	0.1520	0.1655	0.8344	0.84795
0.75	0.1539	0.1635	0.8364	0.8460
0.85	0.1557	0.1615	0.8384	0.8442
0.95	0.1576	0.1596	0.8403	0.8423
1	0.1586	0.1586	0.8413	0.8413

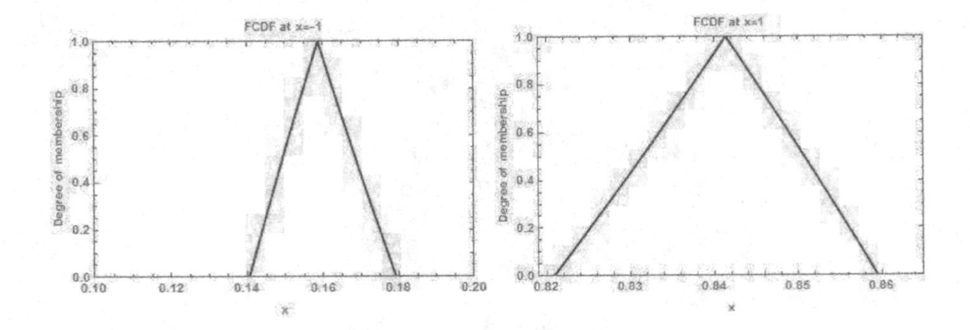

FIGURE 4.5
Plot of $\widetilde{F}_{\widetilde{X}}(x)$ at $x = -1, +1$ in Example 4.9.

(iii) If $\{t_n\}_{n\in\mathbb{N}}$ is an increasing sequence of real-valued numbers with $t_n \to +\infty$ then $\lim_{n\to+\infty} d_\infty(\widetilde{F}_{\widetilde{X}}(t_n), I\{1\})) = 0$.

(iv) If $\{t_n\}_{n\in\mathbb{N}}$ is a decreasing sequence of real-valued numbers with $t_n \to -\infty$ then $\lim_{n\to+\infty} d_\infty(\widetilde{F}_{\widetilde{X}}(t_n), I\{0\})) = 0$ where $d_\infty(\widetilde{A}, \widetilde{B}) = \sup_{\alpha\in[0,1]} |\widetilde{A}_\alpha - \widetilde{B}_\alpha|$ is the absolute error distance between two **FN**s of $\widetilde{A}$ and $\widetilde{B}$.

where I denotes the indicator function.

Proof. For two real-valued numbers of t and s ($t < s$), considering the monotonic property of a ordinary **CDF**, we have $(\widetilde{F}_{\widetilde{X}}(t))_\alpha \leq (\widetilde{F}_{\widetilde{X}}(s))_\alpha$ for any $\alpha \in [0,1]$. This provides (i). To prove (ii), let $\{t_n\}_{n\in\mathbb{N}}$ be a sequence of decreasing numbers which converges to t. Then, based on the classical properties of **CDF**, we have $\lim_{n\to+\infty} \widetilde{F}_{\widetilde{X}_\alpha}(t_n) = \widetilde{F}_{\widetilde{X}_\alpha}(t)$ for any $\alpha \in [0,1]$ and so

$\lim_{n\to+\infty} d_\infty(\widetilde{F}_{\widetilde{X}}(t_n), \widetilde{F}_{\widetilde{X}}(t)) = 0$. To establish (iii), suppose that $\{t_n\}_{n\in\mathbb{N}}$ is an increasing sequence of real-numbers with $t_n \to +\infty$. Therefore for any positive integer of M, there exists an integer N where $M < t_n$, for all $n \geq N$. Applying the classical properties of a **CDF**, for any $\alpha \in [0,1]$, it can be concluded that $\lim_{n\to+\infty} \widetilde{F}_{(\widetilde{X})_\alpha}(t_n) = 1$ and hence $\lim_{n\to+\infty} d_\infty(\widetilde{F}_{\widetilde{X}}(t_n), I\{1\})) = 0$. Part (iv) can be verified similarly.

Remark 4.2 *For $a < b$, if $(\widetilde{F}_{\widetilde{X}}(a))_1 \leq (\widetilde{F}_{\widetilde{X}}(b))_0$ then it is easy to verify that $\left(\widetilde{P}(\widetilde{X} \in (a,b))\right)_\alpha = (\widetilde{F}_{\widetilde{X}}(b))_\alpha - (\widetilde{F}_{\widetilde{X}}(a))_{1-\alpha}$. Therefore, by the fuzzy arithmetic properties, the fuzzy probability of $\widetilde{X} \in (a,b)$ can be represented by difference of two **FCDF** of $\widetilde{F}_{\widetilde{X}}(b)$ and $\widetilde{F}_{\widetilde{X}}(a)$ as:*

$$\widetilde{P}(\widetilde{X} \in (a,b)) = \widetilde{F}_{\widetilde{X}}(b) \ominus \widetilde{F}_{\widetilde{X}}(a). \tag{4.17}$$

Therefore, the α-cuts of $\widetilde{P}(\widetilde{X} \in (a,b))$ can be evaluated as follows:

$$\begin{aligned}\widetilde{P}(\widetilde{X} \in (a,b))[\alpha] \\ &= [(\widetilde{F}_{\widetilde{X}}(b))_{\alpha/2} - (\widetilde{F}_{\widetilde{X}}(a))_{1-\alpha/2}, (\widetilde{F}_{\widetilde{X}}(b))_{1-\alpha/2} - (\widetilde{F}_{\widetilde{X}}(a))_{\alpha/2}].\end{aligned}$$

To plot the membership function of $\widetilde{P}(\widetilde{X} \in (a,b))$, it is enough to evaluate $\widetilde{P}(\widetilde{X} \in (a,b))[\alpha]$ for all values of $\alpha = 0, 0.01, 0.02, ..., 1$ and then plot its lower and upper bounds points by points.

4.1.6 Location-Scale FRVs

As remarked in the introductory chapter, **TFN**s can serve as an alternative tool to easily describe the fuzziness for most experts (as well as non-experts). Another advantage of employing **TFN**s is that their simple arithmetic operations. Furthermore, in many real applications, the imprecise data are often reported by experts as non-exact observations 'about x_i', for $i = 1, 2, \ldots, n$ in form of a **TFN**. Notably, randomness is included in $\widetilde{x}_i$'s since x_1, x_2, ..., x_n can be regarded as observed random samples distributed according to a probability distribution function. On the other hand, fuzziness is relevant to the expert's point of view when he/she assigns some left and right boundaries to each x_i as the minimum and maximum imprecisions. Such values may be independently selected form X as observed values of two uniform random variables. It is worth noting that such a uniform distributions are known since they can be completely specified by expert's measurement investigation when minimum or maximum impressions were included in the experiment. Therefore, statistical inference for such cases needs to be extended when two random variables are involved in modeling a **FRV**: (1) random variable induced by the random experiment, and (2) domain of impression as uniform random variable. It should be also noted that many probability distributions can be characterized by location and scale parameters. Location scale family [125] is a family of distributions formed by translation and rescaling of a standard family member. Such location and scale parameters are frequently and specifically used in statistical inferences and reliability theory. This section proposes the notion of location and scale fuzzy random variables as the **TFN**s based on [74].

Definition 4.11 *Let X be a location-scale random variable distributed according to a family of distribution functions $\{f_X(x) = \frac{1}{\sigma}g((x-\mu)/\sigma) : x \in S_X \subseteq \mathbb{R}, \mu \in \mathbb{R}, \sigma > 0\}$. Then:*

*1) A **FRV** $\widetilde{X}$ is said to be a location-scale fuzzy random variable (**LSFRV**) if $\widetilde{X} = X \oplus (0; U_1, U_2)_T$ in which*

a) $P(0 \in S_X) = 1$,

b) U_1, U_2 are independent of X and $U_1 \sim U(a_1, b_1)$ and $U_2 \sim U(a_2, b_2)$ with $\min\{a_1, a_2\} \geq 0$.

*2) A **FRV** $\widetilde{X}$ is said to be a scale fuzzy random variable (**SFRV**) if $\widetilde{X} = X \otimes (1; U_1', U_2')_T$ in which*

a) $\mu = 0$ and $P(S_X \subseteq (0, \infty)) = 1$.

b) U_1', U_2' are independent of X and $U_1' \sim U(a_1, b_1)$ and $U_2' \sim U(a_2, b_2)$ with $0 \leq a_1 < b_1 \leq 1, 0 \leq a_2 < b_2$.

Noteworthy, **LSFRV** or **SFRV** can be defined as the fuzzy result of an imprecise mapping. Therefore, it considers both aspects of fuzziness and randomness needed for statistical inference in the fuzzy environment. Such a definition of a **FRV** seems more realistic than others since many experts often are interested to evaluate the fuzziness involved in a statistical process using uniform distribution in such situations. Moreover, if the imprecision parts of the proposed **LSFRV/SFRV** (that is $(0; U_1, U_2)_T$ and $(1; U_1', U_2')_T$) reduce to exact values of 0 and 1 then **LSFRV/SFRV** will convert into the classical location-scale random variables. In the following, some statistical features of the proposed location scale fuzzy random variables including **FCDF**, fuzzy expectation, and exact variance will be investigated.

Theorem 4.6 *Let $\widetilde{X}$ be a **SFRV**. Then, the α-values of $\widetilde{F}_{\widetilde{X}}(z)$ can be evaluated as follows:*

$$(\widetilde{F}_{\widetilde{X}}(z))_\alpha = \tag{4.18}$$

$$\begin{cases} F_X(\frac{z}{1+(1-2\alpha)b_2}) + \int_{\frac{z}{1-(1-2\alpha)b_2}}^{\frac{z}{1-(1-2\alpha)a_2}} \frac{z/x-1-(1-2\alpha)a_2}{(1-2\alpha)(b_2-a_2)} f_X(x)dx, & 0.0 \leq \alpha \leq 0.50, \\ F_X(\frac{z}{1+(1-2\alpha)a_1}) + \int_{\frac{z}{1+(1-2\alpha)a_1}}^{\frac{z}{1+(1-2\alpha)b_1}} \frac{z/x-1-(1-2\alpha)b_1}{(2\alpha-1)(b_1-a_1)} f_X(x)dx, & 0.50 < \alpha \leq 1. \end{cases}$$

Proof *First note that:*

$$\widetilde{X}_\alpha = \begin{cases} XY_1^\alpha, & 0 \leq \alpha \leq 0.5, \\ XY_2^\alpha, & 0.5 < \alpha \leq 1, \end{cases} \tag{4.19}$$

where $Y_1^\alpha \sim U(1-(1-2\alpha)b_1, 1-(1-2\alpha)a_1)$ *and* $Y_2^\alpha \sim U(1-(1-2\alpha)a_2, 1-(1-2\alpha)b_2)$. *Since* Y_1^α *and* X *for* $\alpha \in [0, 0.5]$ *and* Y_2^α *and* X *for* $\alpha \in [0.5, 1]$ *are independent random variables, by product of two non-negative, independent continuous random variables, we can get:*

$$P(\widetilde{X}_{1-\alpha} \leq z) = \begin{cases} P(XY_2^{1-\alpha} \leq z), & 0 < \alpha \leq 0.5. \\ P(XY_1^{1-\alpha} \leq z), & 0.5 \leq \alpha \leq 1. \end{cases}$$

$$= \begin{cases} \int_0^\infty F_{Y_2^{1-\alpha}}(z/x) f_X(x)dx, & 0 \leq \alpha \leq 0.5, \\ \int_0^\infty F_{Y_1^{1-\alpha}}(z/x) f_X(x)dx, & 0.5 < \alpha \leq 1, \end{cases}$$

$$= \begin{cases} F_X\left(\frac{z}{1+(1-2\alpha)b_2}\right) + \int_{\frac{z}{1-(2\alpha-1)b_2}}^{\frac{z}{1-(2\alpha-1)a_2}} \frac{z/x-1-(1-2\alpha)a_2}{(1-2\alpha)(b_2-a_2)} f_X(x)dx, & 0 \leq \alpha \leq 0.5, \\ F_X\left(\frac{z}{1+(1-2\alpha)a_1}\right) + \int_{\frac{z}{1+(1-2\alpha)a_1}}^{\frac{z}{1+(1-2\alpha)b_1}} \frac{z/x-1-(1-2\alpha)b_1}{(2\alpha-1)(b_1-a_1)} f_X(x)dx, & 0.5 < \alpha \leq 1. \end{cases}$$

Theorem 4.7 *Let* $\widetilde{X}$ *be a* ***LSFRV***. *Then:*

$$(\widetilde{F}_{\widetilde{X}}(z))_\alpha = \tag{4.20}$$

$$\begin{cases} F_X(z+(1-2\alpha)a_2) + \int_{z+(1-2\alpha)a_2}^{z+(1-2\alpha)b_2} \frac{(1-2\alpha)b_2-x+z}{(1-2\alpha)(b_2-a_2)} f_X(x)dx, & 0 \leq \alpha \leq 0.5, \\ F_X(z+(1-2\alpha)b_1) + \int_{z+(1-2\alpha)b_1}^{z+(1-2\alpha)a_1} \frac{(1-2\alpha)a_1-x+z}{(2\alpha-1)(b_1-a_1)} f_X(x)dx, & 0.5 < \alpha \leq 1. \end{cases}$$

Proof *If* $\widetilde{X}$ *is a* ***LSFRV***, *then it can be shown that:*

$$\widetilde{X}_\alpha = \begin{cases} X - Y_1^\alpha, & 0 \leq \alpha \leq 0.5, \\ X - Y_2^\alpha, & 0.5 < \alpha \leq 1, \end{cases} \tag{4.21}$$

where $Y_1^\alpha \sim U(-(1-2\alpha)b_1, -(1-2\alpha)a_1)$ *and* $Y_2^\alpha \sim U(-(1-2\alpha)a_2, -(1-2\alpha)b_2)$. *Therefore, the* α*-values of* $\widetilde{F}_{\widetilde{X}}(z)$ *can be easily evaluated using the convolution rule as follows:*

$$P(\widetilde{X}_{1-\alpha} \leq z) = \begin{cases} \int_{\mathbb{R}} P(X - Y_2^{1-\alpha} \leq z | X = x) f_X(x)dx, & 0 \leq \alpha \leq 0.5, \\ \int_{\mathbb{R}} P(X - Y_1^{1-\alpha} \leq z | X = x) f_X(x)dx, & 0.5 < \alpha \leq 1, \end{cases}$$

$$= \begin{cases} \int_{\mathbb{R}} P(Y_2^{1-\alpha} \geq x - z) f_X(x)dx, & 0 \leq \alpha \leq 0.5, \\ \int_{\mathbb{R}} P(Y_1^{1-\alpha} \geq x - z) f_X(x)dx, & 0.5 < \alpha \leq 1, \end{cases}$$

$$= \begin{cases} F_X(z+(1-2\alpha)a_2) + \int_{z+(1-2\alpha)a_2}^{z+(1-2\alpha)b_2} \frac{(1-2\alpha)b_2-x+z}{(1-2\alpha)(b_2-a_2)} f_X(x)dx, & 0 \leq \alpha \leq 0.5, \\ F_X(z+(1-2\alpha)b_1) + \int_{z+(1-2\alpha)b_1}^{z+(1-2\alpha)a_1} \frac{(1-2\alpha)a_1-x+z}{(2\alpha-1)(b_1-a_1)} f_X(x)dx, & 0.5 < \alpha \leq 1. \end{cases}$$

This completes the proof.

Proposition 4.2 *Let $\widetilde{X}$ be a* ***SFRV****. Then:*

1) $\widetilde{E}(\widetilde{X}) = E(X) \otimes (1; \frac{a_1+b_1}{2}, \frac{a_2+b_2}{2})_T$,

2) $\boldsymbol{var}(\widetilde{X}) = Var(X)(\frac{\frac{1}{3}+\frac{1}{24}(-2+a_1+b_1)^3}{a_1+b_1} + \frac{\frac{-1}{3}+\frac{1}{24}(2+a_2+b_2)^3}{a_2+b_2}) + E(X^2)(\frac{(b_1-a_1)^2}{72} + \frac{(b_2-a_2)^2}{6})$.

Proof *To prove (1), since* $Y_1^\alpha \sim U(1-(1-2\alpha)b_1, 1-(1-2\alpha)a_1)$ *and* $Y_2^\alpha \sim U(1-(1-2\alpha)a_2, 1-(1-2\alpha)b_2)$, *we have:*

$$E(\widetilde{X}_\alpha) = \begin{cases} E(XY_1^\alpha), & 0 \le \alpha \le 0.5, \\ E(XY_2^\alpha), & 0.5 < \alpha \le 1, \end{cases}$$

$$= E(X) \times \begin{cases} 1-(1-2\alpha)\frac{(a_1+b_1)}{2}, & 0 \le \alpha \le 0.5, \\ 1-(1-2\alpha)\frac{(a_2+b_2)}{2}, & 0.5 < \alpha \le 1. \end{cases}$$

Therefore, $\widetilde{E}(\widetilde{X}) = E(X) \otimes (1; \frac{a_1+b_1}{2}, \frac{a_2+b_2}{2})_T$. *To establish (2), one finds that:*

$$\begin{aligned}
\boldsymbol{var}(\widetilde{X}) &= \int_0^1 var(\widetilde{X}_\alpha) d\alpha \\
&= \int_0^1 (var(E(\widetilde{X}_\alpha|X)) + E(var(\widetilde{X}_\alpha|X)))d\alpha \\
&= \int_0^{0.5} (var(E(XY_1^\alpha|X) + E(var(XY_1^\alpha|X)))d\alpha \\
&+ \int_{0.5}^1 (var(E(XY_2^\alpha|X) + E(var(XY_2^\alpha|X)))d\alpha \\
&= \int_0^{0.5} ((E(Y_1^\alpha))^2 var(X) + E(X^2) var(Y_1^\alpha))d\alpha \\
&+ \int_{0.5}^1 ((E(Y_2^\alpha))^2 var(X) + E(X^2) var(Y_2^\alpha))d\alpha \\
&= var(X)(\int_0^{0.5} (E(Y_1^\alpha))^2 d\alpha + \int_{0.5}^1 (E(Y_2^\alpha))^2 d\alpha) \\
&+ E(X^2)(\int_0^{0.5} var(Y_1^\alpha) d\alpha + \int_{0.5}^1 var(Y_2^\alpha) d\alpha).
\end{aligned}$$

But,

$$E(Y_1^\alpha) = 1 - \frac{(1-2\alpha)(a_1+b_1)}{2},$$

$$E(Y_2^\alpha) = 1 - \frac{(1-2\alpha)(a_2+b_2)}{2},$$

and

$$var(Y_1^\alpha) = \frac{(1-2\alpha)^2(b_1-a_1)^2}{12},$$

$$var(Y_2^\alpha) = \frac{((1-2\alpha)^2(b_2-a_2)^2}{12}.$$

Therefore, it can be easily shown that

$$\int_0^{0.5}(E(Y_1^\alpha))^2 d\alpha + \int_{0.5}^1 (E(Y_2^\alpha))^2 d\alpha = \frac{\frac{1}{3}+\frac{1}{24}(-2+a_1+b_1)^3}{a_1+b_1} + \frac{\frac{-1}{3}+\frac{1}{24}(2+a_2+b_2)^3}{a_2+b_2},$$

and

$$\int_0^{0.5} var(Y_1^\alpha)d\alpha + \int_{0.5}^1 var(Y_2^\alpha)d\alpha = \frac{(b_1-a_1)^2}{72} + \frac{(b_2-a_2)^2}{6}.$$

The result is then verified.

Proposition 4.3 *Let* $\widetilde{X}$ *be a* ***LSFRV****. Then:*

1) $\widetilde{E}(\widetilde{X}) = E(X) \oplus (0; \frac{a_1+b_1}{2}, \frac{a_2+b_2}{2})_T$,

2) $\boldsymbol{var}(\widetilde{X}) = var(X) + \frac{1}{6}(\frac{(b_1-a_1)^2+(b_2-a_2)^2}{12})$.

Proof *Note that*

$$E(\widetilde{X}_\alpha) = \begin{cases} E(X - Y_1^\alpha), & 0 \le \alpha \le 0.5, \\ E(X - Y_2^\alpha), & 0.5 < \alpha \le 1, \end{cases}$$
$$= E(X) - \begin{cases} -(1-2\alpha)\frac{(a_1+b_1)}{2}, & 0 \le \alpha \le 0.5, \\ -(1-2\alpha)\frac{(a_2+b_2)}{2}, & 0.5 < \alpha \le 1. \end{cases}$$

It follows that $\widetilde{E}(\widetilde{X}) = E(X) \oplus (0; \frac{a_1+b_1}{2}, \frac{a_2+b_2}{2})_T$ *and hence (1) is verified. Moreover,*

$$\begin{aligned} \boldsymbol{var}(\widetilde{X}) &= \int_0^1 var(\widetilde{X}_\alpha)d\alpha \\ &= \int_0^{0.5} var(X - Y_1^\alpha)d\alpha + \int_{0.5}^1 var(X - Y_2^\alpha)d\alpha \\ &= var(X) + \int_0^{0.5} var(Y_1^\alpha)d\alpha + \int_{0.5}^1 var(Y_2^\alpha)d\alpha \\ &= var(X) + \int_0^{0.5}(1-2\alpha)^2\frac{(b_1-a_1)^2}{12}d\alpha + \int_{0.5}^1 (2\alpha-1)^2\frac{(b_2-a_2)^2}{12}d\alpha \\ &= var(X) + \frac{1}{6}(\frac{(b_1-a_1)^2+(b_2-a_2)^2}{12}), \end{aligned}$$

which proves (2).

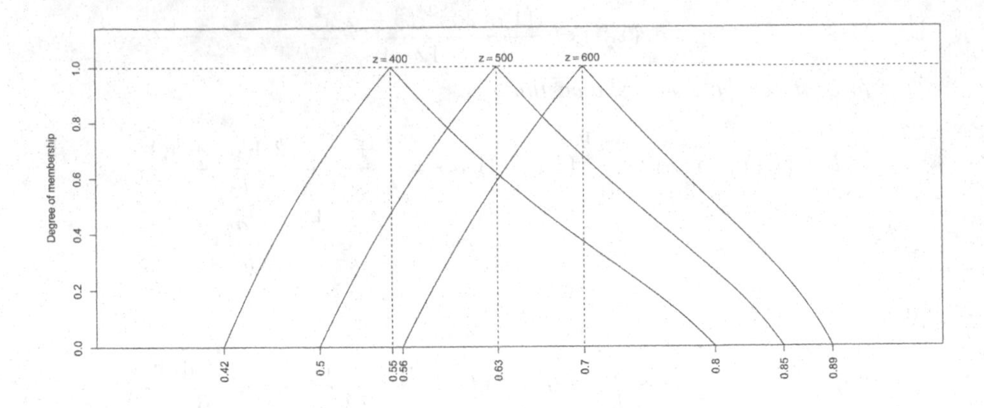

FIGURE 4.6
Plot of $\widetilde{F}_{\widetilde{X}}(z)$ at three points of 400, 500, and 600 in Example 4.10.

Example 4.10 *Let $\widetilde{X}$ be a **SFRV** where $X \sim exp(0.002)$, $U_1 \sim U(0,1)$, and $U_2 \sim U(0,1)$. According to Eq. (4.18), the α-values of $\widetilde{F}_{\widetilde{X}}(z)$ can be obtained by:*

$$(\widetilde{F}_{\widetilde{X}}(z))_\alpha = \begin{cases} F_X(z) + \int_{\frac{z}{\alpha}}^{z} \frac{z/x-1}{2\alpha-1} f_X(x)dx, & 0 \le \alpha \le 0.5, \\ F_X(\frac{z}{2(1-\alpha)}) + \int_{\frac{z}{2(1-\alpha)}}^{z} \frac{z/x-1}{2\alpha-1} f_X(x)dx, & 0.5 < \alpha \le 1, \end{cases} \tag{4.22}$$

$$= \begin{cases} (1-\exp(-0.002z)) + 0.002 \int_{\frac{z}{\alpha}}^{z} \frac{z/x-1}{2\alpha-1} \exp(-0.002x)dx, & 0 \le \alpha \le 0.5, \\ (1-\exp(\frac{-0.001z}{(1-\alpha)}) + 0.002 \int_{\frac{z}{2(1-\alpha)}}^{z} \frac{z/x-1}{2\alpha-1} \exp(-0.002x)dx, & 0.5 < \alpha \le 1. \end{cases}$$

The plots of $\widetilde{F}_{\widetilde{X}}$ at three points of 400, 500, and 600 are depicted in Fig. 4.6. Further, we get $\widetilde{E}(\widetilde{X}) = (500; 250, 250)_T$ and $\boldsymbol{var}(\widetilde{X}) = 409722.2$. Additionally, using Eq. (4.17), we have

$$\widetilde{P}(\widetilde{X} \in (200, 700)) = \widetilde{F}_{\widetilde{X}}(700) \ominus \widetilde{F}_{\widetilde{X}}(200).$$

The plot of $\widetilde{P}(\widetilde{X} \in (200, 700))$ is shown in Fig. 4.7 as 'about 0.42'.

Example 4.11 *Let $\widetilde{X}$ be a **LSFRV** where $X \sim N(6, 3.61)$, $U_1 \sim U(-1,0)$ and $U_2 \sim U(0,1)$. The α-values of $\widetilde{F}_{\widetilde{X}}(z)$ can be obtained by Theorem 4.7 as follows:*

$$(\widetilde{F}_{\widetilde{X}}(z))_\alpha = \begin{cases} F_X(z) + \int_{0}^{z+(1-2\alpha)} \frac{z-x+(1-2\alpha)}{2\alpha-1} f_X(x)dx, & 0 \le \alpha \le 0.5, \\ F_X(z) + \int_{z}^{z+(1-2\alpha)} \frac{z-x-(2\alpha-1)}{\alpha-1} f_X(x)dx, & 0.5 < \alpha \le 1. \end{cases}$$

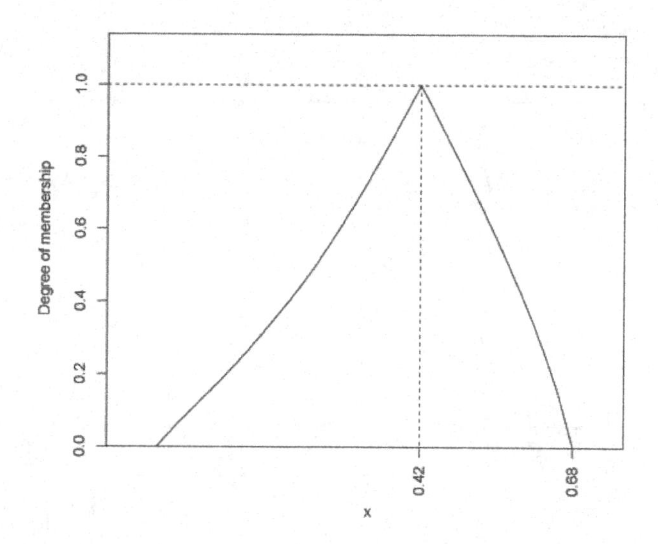

FIGURE 4.7
Plot of $\widetilde{P}(\widetilde{X} \in (200, 700))$ in Example 4.10.

*Where F_X and f_X are the **CDF** and density functions of normal distribution with mean 6 and variance 3.61. The plots of $\widetilde{F}_{\widetilde{X}}$ at three points of $-1, 0,$ and 2 are drawn as presented in Fig. 4.8. Using Lemma 4.3, the fuzzy expectation and exact variance of $\widetilde{X}$ can be calculated as $\widetilde{E}(\widetilde{X}) = (6; 0.5, 0.5)_T$ and $\boldsymbol{var}(\widetilde{X}) = 3.64$, respectively. Further, assume we wish to evaluate the fuzzy probability of $\widetilde{X} \in (7, 10)$. According to Eq. (4.17), note that $\widetilde{P}(\widetilde{X} \in (7, 10)) = \widetilde{F}_{\widetilde{X}}(10) \ominus \widetilde{F}_{\widetilde{X}}(7)$. Therefore, the α-cuts of $\widetilde{P}(\widetilde{X} \in (7, 10))$ can be evaluated as follows:*

$$\widetilde{P}(\widetilde{X} \in (7, 10))[\alpha] =$$

$$[(\widetilde{F}_{\widetilde{X}}(10))_{\alpha/2} - (\widetilde{F}_{\widetilde{X}}(7))_{1-\alpha/2}, (\widetilde{F}_{\widetilde{X}}(10))_{1-\alpha/2} - (\widetilde{F}_{\widetilde{X}}(7))_{\alpha/2}].$$

By plotting the lower and upper bounds of $\widetilde{P}(\widetilde{X} \in (7, 10))[\alpha]$ for all values of $\alpha = 0, 0.01, ..., 0.99, 1$, the membership function of $\widetilde{P}(\widetilde{X} \in (7, 10))$ can be depicted as shown in Fig. 4.9.

There are several approaches in the classical inferences to evaluate the unknown parameters based on a sample data. Assume that $\widetilde{X}$ is a **LSFRV** with unknown shape parameters. Here, a common statistical procedure is extended to estimate such parameters based on a **FRS**. For this purpose, we extend the classical moment estimator to estimate the parameters of **LSFRV**s.

Definition 4.12 *Let $\widetilde{X}$ be a **LSFRV** (**SFRV**) and $\widetilde{x}_1, \widetilde{x}_2, \ldots, \widetilde{x}_n$ be an observed **FRS**. We say $\widehat{\boldsymbol{\theta}}(\widetilde{x}_1, \widetilde{x}_2, \ldots, \widetilde{x}_n)$ is a real-valued moment estimator of*

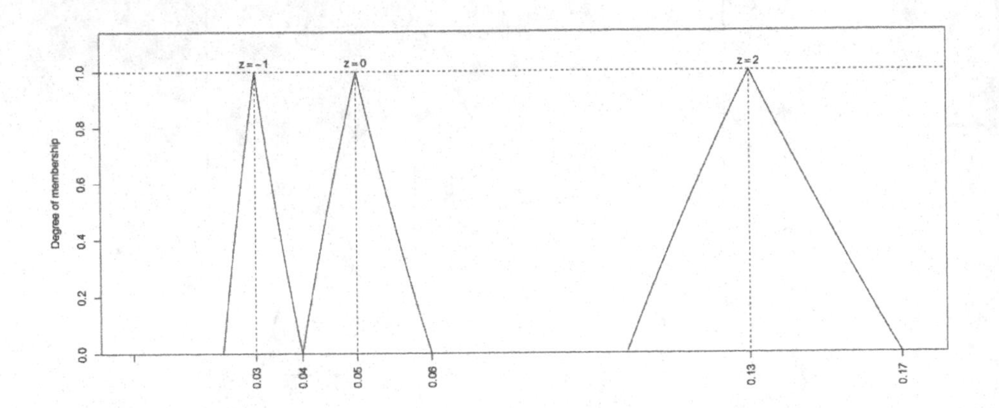

FIGURE 4.8
Plot of $\widetilde{F}_{\widetilde{X}}(z)$ at three points of $-1, 0$, and 2 in Example 4.11.

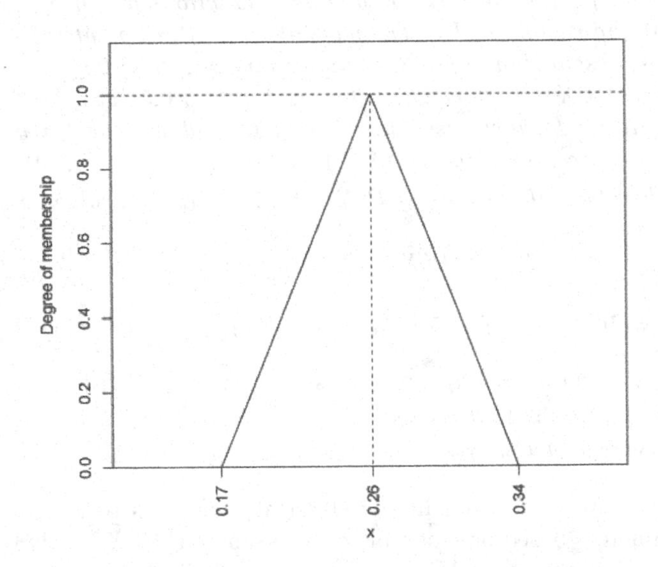

FIGURE 4.9
Plot of $\widetilde{P}(\widetilde{X} \in (7, 10))$ in Example 4.11.

an unknown vector of parameters of $\boldsymbol{\theta} = (\mu, \sigma)^\top$ if it is obtained by solving the following equations:

$$\int_0^1 E((\widetilde{X}_\alpha)^r)d\alpha = \int_0^1 (\widetilde{\overline{x^r}})_\alpha d\alpha, \tag{4.23}$$

where $(\widetilde{\overline{x^r}})_\alpha = 1/n \sum_{i=1}^n ((\widetilde{x}_i)_\alpha)^r$ *and* $r = 1, 2, \ldots$.

Example 4.12 *Let $\widetilde{X}$ be a **SFRV** with $X \sim exp(\lambda)$ and $\widetilde{x}_1, \widetilde{x}_2, \ldots, \widetilde{x}_n$ be an observed **FRS**. Since*

$$\widetilde{X}_\alpha = \begin{cases} XY_1^\alpha, & 0 \leq \alpha \leq 0.5, \\ XY_2^\alpha, & 0.5 < \alpha \leq 1, \end{cases} \tag{4.24}$$

where $Y_1^\alpha \sim U(1-(1-2\alpha)b_1, 1-(1-2\alpha)a_1)$, and $Y_2^\alpha \sim U(1-(1-2\alpha)a_2, 1-(1-2\alpha)b_2)$, we have

$$\int_0^1 E(\widetilde{X}_\alpha)d\alpha = \int_0^{0.5} E(XY_1^\alpha)d\alpha + \int_{0.5}^1 E(XY_2^\alpha)d\alpha.$$

Since Y_1^α and X for $\alpha \in [0, 0.5]$ and Y_2^α and X for $\alpha \in [0.5, 1]$ are independent random variables, therefore

$$\begin{aligned}\int_0^1 E(\widetilde{X}_\alpha)d\alpha &= E(X)(\int_0^{0.5} E(Y_1^\alpha)d\alpha + \int_{0.5}^1 E(Y_2^\alpha)d\alpha) \\ &= \lambda(\int_0^{0.5} (\frac{2-(1-2\alpha)(a_1+b_1)}{2})d\alpha \\ &+ \int_{0.5}^1 (\frac{2-(1-2\alpha)(a_2+b_2)}{2}d\alpha) \\ &= \lambda(1 + \frac{(a_2+b_2)-(a_1+b_1)}{8}).\end{aligned}$$

Therefore, the moment estimator of λ can be obtained as:

$$\widehat{\lambda} = \frac{\int_0^1 (\widetilde{\overline{x}})_\alpha d\alpha}{(1 + \frac{(a_2+b_2)-(a_1+b_1)}{8})}. \tag{4.25}$$

Example 4.13 *Let $\widetilde{X}$ be **LSFRV** where $X \sim N(\mu, \sigma^2)$. Here, the goal is to find the moment estimators of $\boldsymbol{\theta} = (\mu, \sigma^2)$ based on an **FRS** of $\widetilde{x}_1, \widetilde{x}_2, \ldots, \widetilde{x}_n$. For $r = 1$, we get: $\int_0^1 E(\widetilde{X}_\alpha)d\alpha = \int_0^{0.5} E(X - Y_1^\alpha)d\alpha + \int_{0.5}^1 E(X - Y_2^\alpha)d\alpha = \mu + (\frac{(a_1+b_1)+(a_2+b_2)}{8})$. For $r = 2$, we also have $\int_0^1 E(\widetilde{X}_\alpha)^2 d\alpha = \int_0^{0.5} E(X - Y_1^\alpha)^2 d\alpha + \int_{0.5}^1 E(X - Y_2^\alpha)^2 d\alpha$. Note that, $E(X - Y_j^\alpha)^2 = var(X - Y_j^\alpha) + (E(X - Y_j^\alpha))^2 = (\sigma^2 + var(Y_j^\alpha)) + (\mu + E(Y_j^\alpha))^2$ for $j = 1, 2$ where*

$$\begin{cases} Y_1^\alpha \sim U(-(1-2\alpha)b_1, -(1-2\alpha)a_1), & 0 \leq \alpha \leq 0.5, \\ Y_2^\alpha \sim U(-(2\alpha-1)b_2, -(2\alpha-1)a_2), & 0.5 < \alpha \leq 1. \end{cases}$$

TABLE 4.4
Data set in Example 4.14.

No.	$\widetilde{x}_i$	No.	$\widetilde{x}_i$	No.	$\widetilde{x}_i$
1	$(0.84; 0.1, 0.42)_T$	6	$(1.78; 0.1, 0.26)_T$	11	$(1.55; 0.2, 0.42)_T$
2	$(0.96; 0.05, 0.30)_T$	7	$(1.84; 0.18, 0.37)_T$	12	$(1; 0.17, 0.36)_T$
3	$(1.28; 0.22, 0.54)_T$	8	$(1.92; 0.24, 0.22)_T$	13	$(1.37; 0.15, 0.24)_T$
4	$(1.43; 0.15, 0.48)_T$	9	$(2.14; 0.27, 0.1)_T$	14	$(1.75; 0.1, 0.28)_T$
5	$(1.65; 0.25, 0.34)_T$	10	$(2.2; 0.32, 0.15)_T$	-	-

Therefore, $\int_0^1 E(\widetilde{X}_\alpha)^2 d\alpha = \mu^2 + \sigma^2 + \frac{(a_1^2+b_1^2)+(a_2^2+b_2^2)}{18} + \frac{a_1b_1+a_2b_2}{4}$. *Now, by solving the following equations:*

$$\widehat{\mu} + k_1 = \int_0^1 (\widetilde{\overline{x}})_\alpha d\alpha,$$

$$\widehat{\mu}^2 + \widehat{\sigma}^2 + k_2 = \int_0^1 (\widetilde{\overline{x^2}})_\alpha d\alpha,$$

where

$$k_1 = \frac{(a_1+b_1)+(a_2+b_2)}{8},$$

$$k_2 = \frac{(a_1^2+b_1^2)+(a_2^2+b_2^2)}{18} + \frac{a_1b_1+a_2b_2}{4},$$

the moment estimator of $\boldsymbol{\theta} = (\mu, \sigma^2)$ *can be evaluated as* $\widehat{\boldsymbol{\theta}} = (\widehat{\mu}, \widehat{\sigma}^2)$ *with:*

$$\widehat{\mu} = \int_0^1 (\widetilde{\overline{x}})_\alpha d\alpha - k_1,$$

and

$$\widehat{\sigma}^2 = \frac{1}{n}\int_0^1 \sum_{i=1}^n \left((\widetilde{x}_i)_\alpha - (\widetilde{\overline{x}})_\alpha\right)^2 d\alpha + 2k_1 \int_0^1 (\widetilde{\overline{x}})_\alpha d\alpha + k_1^2 - k_2.$$

Example 4.14 *In business and educational settings, computer-based presentations are displayed by liquid crystal display projectors. The lamp failure is the most common failure mode of these projectors. Manufacturers often include the expected lamp life in their technical specification documents. Naturally, it is assumed that the projection time of each lamp has an exponential distribution under normal operating conditions. To test this claim, a large private university placed identical lamps in three projector models. The university staff recorded the number of presentations (as measured by the projector per 1000 hours) when each lamp burned out. However, in real situations,* *the lifetime of a lamp may not be precisely measured. A lamp may work perfectly within a specific time but with irregular interruption, and finally work out. Therefore, the collected data should be reported by inexact quantities. In this regard, according to the proposed method, the fuzzy quantities were reported in the form of a* ***FRS*** *of* $\widetilde{X} = (X, U_1, U_2)_T$ *where* X *is distributed from Weibull distribution with the shape parameter* $\gamma = 2$ *and scale parameter* β ($f_X(x) = (2/\beta^2)x\exp(-(x/\beta)^2)$, $x > 0$), $U_1 \sim U(0, 0.4)$ *and* $U_2 \sim U(0, 0.6)$. *The observed fuzzy data are shown in Table 4.4. First note that*

$$E(\widetilde{X}_\alpha) = \begin{cases} E(X) + 0.4\alpha - 0.2 = \frac{\sqrt{\pi}}{2}\beta + 0.4\alpha - 0.2, & 0 \le \alpha \le 0.5, \\ E(X) + 0.6\alpha - 0.3 = \frac{\sqrt{\pi}}{2}\beta + 0.6\alpha - 0.3, & 0.5 < \alpha \le 1. \end{cases}$$

This concludes that

$$\begin{aligned} \int_0^1 E(\widetilde{X}_\alpha)d\alpha &= \int_0^{0.5} (\frac{\sqrt{\pi}}{2}\beta + 0.4\alpha - 0.2)d\alpha + \int_{0.5}^1 (\frac{\sqrt{\pi}}{2}\beta + 0.6\alpha - 0.3)d\alpha \\ &= \frac{\sqrt{\pi}}{2}\beta - 0.025. \end{aligned}$$

In addition,

$$\begin{aligned} \int_0^1 (\widetilde{\overline{x}})_\alpha d\alpha &= \overline{x} + \frac{1}{4}((\frac{1}{15})\sum_{i=1}^{15} r_{x_i} - (\frac{1}{15})\sum_{i=1}^{15} l_{x_i}) \\ &= 1.5507 + \frac{0.32 - 0.1785}{4} = 1.5860. \end{aligned}$$

According to the moment estimation procedure, the unknown parameter of β *can be evaluated from the following identity:*

$$\frac{\sqrt{\pi}}{2}\widehat{\beta} - 0.025 = 1.5860.$$

This concludes that $\widehat{\beta} = 2(1.5860 + 0.025)/\sqrt{\pi} = 1.8178$.

4.2 Probabilistic Inequalities for Fuzzy Random Variables

Probability inequalities of random variables play important roles in analytic probability theory such as limiting theorems. Some well-known probabilistic inequalities are extended for the **FRV**s in this section inspired by Hesamian and Akbari [71]. For this purpose, the main focus is on popular inequalities in probability theory namely Jensen , Hölder', and Minkowski inequalities. Fatou's lemma is also developed for a sequence of **FRV**s.

Definition 4.13 *Let $(\Omega, \mathcal{A}, P)$ be a probability space and $p > 0$. The fuzzy L^p space $(\widetilde{L}^p(P))$ is defined to be a class of all **FRVs** $\widetilde{X}$ with $\widetilde{E}(|\widetilde{X}|^p) < \infty$ in which $sup_{\alpha\in[0,1]}E((|\widetilde{X}|^p)_\alpha) < \infty$.*

Definition 4.14 *For $\widetilde{X} \in \widetilde{L}^p(P)$, the the fuzzy L^p-norm of $\widetilde{X}$ is defined by $\|\widetilde{X}\|_p = f(\widetilde{E}(\widetilde{X}^p))$ where $f(x) = x^{1/p}$.*

Theorem 4.8 *Let $\widetilde{X}$ be a **FRV** on probability space $(\Omega, \mathcal{A}, P)$. Then $P_d(|\widetilde{E}(\widetilde{X})| \preceq \widetilde{E}(|\widetilde{X}|)) = 1$.*

Proof *Since $P_d(|\widetilde{X}| \succeq \widetilde{X}) = 1$, for any $\alpha \in [0,1]$, we have $|\widetilde{X}|_\alpha \geq \widetilde{X}_\alpha$. This concludes $(\widetilde{E}(|\widetilde{X}|))_\alpha \geq (|\widetilde{E}(\widetilde{X})|)_\alpha$ for any $\alpha \in [0,1]$.*

Theorem 4.9 *For $1 < p < \infty$, assume that $\widetilde{X} \in \widetilde{L}^p(P)$ and $\widetilde{Y} \in \widetilde{L}^q(P)$ where $\widetilde{X}$ $(\widetilde{Y})$ is a positive or negative **FN** and $1 < q < \infty$ and $1/p + 1/q = 1$, then $\widetilde{X} \otimes \widetilde{Y} \in \widetilde{L}^1(P)$.*

Proof *First, assume that $\widetilde{X}$ and $\widetilde{Y}$ are positive **FRVs**. Applying the classical Hölder's inequality, we can immediately get:*

$$\begin{aligned}(\widetilde{E}|\widetilde{X} \otimes \widetilde{Y}|)_\alpha &= E(|\widetilde{X} \otimes \widetilde{Y}|_\alpha) = E(|\widetilde{X}|_\alpha \times |\widetilde{Y}|_\alpha)\\ &\leq (E(|\widetilde{X}|^p_\alpha))^{1/p} \times (E(|\widetilde{Y}|^q_\alpha))^{1/q},\end{aligned}$$

*which verifies the claim since $\widetilde{X} \in \widetilde{L}^p(P)$ and $\widetilde{Y} \in \widetilde{L}^q(P)$. Now suppose that $\widetilde{X}$ is a positive and $\widetilde{Y}$ is a negative **FRV**. It follows that:*

$$(\widetilde{E}|\widetilde{X} \otimes \widetilde{Y}|)_\alpha \leq (E((|\widetilde{X}|^\alpha)^p))^{1/p} \times (E((|\widetilde{Y}|_\alpha)^q))^{1/q},$$

*which completes the proof. A similar argument can be used for cases in which both $\widetilde{X}$ and $\widetilde{Y}$ are negative **FRVs**.*

Theorem 4.10 *Let $\widetilde{X}$ and $\widetilde{Y}$ be two positive **FRV**. If $0 < p < 1$ and $q > 0$ with $1/p + 1/q = 1$ then $P_d(|\widetilde{X} \otimes \widetilde{Y}| \succeq \|\widetilde{X}\|_p \otimes \|\widetilde{Y}\|_q) = 1$.*

Proof *For $0 < p < 1$, let $q = p/(1-p)$, $k = 1/p$ and $m = 1/(1-(1/k)) = 1/(1-p)$. Therefore, we get $1/k + 1/m = 1$ where $k > 1$ and $m > 1$. Moreover, let $\widetilde{V}_\alpha = (\widetilde{Y}_\alpha)^{-1/p}$ and $\widetilde{U}_\alpha = (\widetilde{X}_\alpha\widetilde{Y}_\alpha)^{1/p}$, $\alpha \in [0,1]$. By Holder's inequality, the following result could be obtained for every $\alpha \in [0,1]$:*

$$E(|\widetilde{U}_\alpha\widetilde{V}_\alpha|) \leq (E(\widetilde{U}^k_\alpha))^{1/k} \times (E(\widetilde{V}^m_\alpha))^{1/m}.$$

The above inequality is equivalent to:

$$E(\widetilde{X}_\alpha)^p \leq (E(\widetilde{X}_\alpha\widetilde{Y}_\alpha))^p \times (E(\widetilde{Y}_\alpha)^{-q/p})^{1-p}.$$

This simply concludes that:

$$E(|\widetilde{X}_\alpha\widetilde{Y}_\alpha|) \geq (E(\widetilde{X}_\alpha)^p)^{1/p} \times (E(\widetilde{Y}_\alpha)^q)^{1/q},$$

which leads to $P_d(|\widetilde{X} \otimes \widetilde{Y}| \succeq \|\widetilde{X}\|_p \otimes \|\widetilde{Y}\|_q) = 1$.

Theorem 4.11 *For $p > 1$, assume that $\widetilde{X}, \widetilde{Y} \in \widetilde{L}^p(P)$ where $\widetilde{X}$ and $\widetilde{Y}$ are two positive* ***FNs****. Then $P_d(\|\widetilde{X} + \widetilde{Y}\|_p \preceq \|\widetilde{X}\|_p + \|\widetilde{Y}\|_p) = 1$.*

Proof *For every $\alpha \in [0,1]$, one has the following results:*

$$
\begin{aligned}
\left(\widetilde{E}|\widetilde{X} \oplus \widetilde{Y}|^p\right)_\alpha = E\left(|\widetilde{X} \oplus \widetilde{Y}|^p\right)_\alpha = E\left(|\widetilde{X} \oplus \widetilde{Y}|_\alpha\right)^p = \\
E\left[\left(|\widetilde{X} \oplus \widetilde{Y}|_\alpha\right)^{p-1} |\widetilde{X} \oplus \widetilde{Y}|_\alpha\right] \\
\leq E\left[\left(|\widetilde{X} \oplus \widetilde{Y}|_\alpha\right)^{p-1} \left(|\widetilde{X}|_\alpha + |\widetilde{Y}|_\alpha\right)\right] \\
= E\left[\left(|\widetilde{X} \oplus \widetilde{Y}|_\alpha\right)^{p-1} |\widetilde{X}|_\alpha\right] + E\left[\left(|\widetilde{X} \oplus \widetilde{Y}|_\alpha\right)^{p-1} |\widetilde{Y}|_\alpha\right] \\
\leq \|\widetilde{X}_\alpha\|_p \|((\widetilde{X} \oplus \widetilde{Y})_\alpha)^{p-1}\|_q + \|\widetilde{Y}_\alpha\|_p \|((\widetilde{X} \oplus \widetilde{Y})_\alpha)^{p-1}\|_q \\
= (\|\widetilde{X}_\alpha\|_p + \|\widetilde{Y}_\alpha\|_p)(\|(\widetilde{X} \oplus \widetilde{Y})_\alpha\|_p)^{p/q}.
\end{aligned}
$$

The first inequality is immediately followed since $|\widetilde{A} \oplus \widetilde{B}|_\alpha \leq |\widetilde{A}|_\alpha + |\widetilde{B}|_\alpha$ for any $\alpha \in [0,1]$. The second inequality follows from the Hölder's inequality. Therefore:

$$
\left(\|\widetilde{X} \oplus \widetilde{Y}\|_p\right)_\alpha \leq \left(\|\widetilde{X}\|_p\right)_\alpha + \left(\|\widetilde{Y}\|_p\right)_\alpha,
$$

which immediately proves the claim.

Theorem 4.12 *If $0 < p < 1$ and $\widetilde{X}, \widetilde{Y} \in \widetilde{L}^p(P)$ are two positive* ***FRVs****, then $P_d(\|\widetilde{X} \oplus \widetilde{Y}\|_p \succeq \|\widetilde{X}\|_p \oplus |\widetilde{Y}\|_p) = 1$.*

Proof *Let $q = p/(1-p)$. By Theorem 4.10, we have:*

$$
\begin{aligned}
\left(\widetilde{E}(\widetilde{X} \oplus \widetilde{Y})^p\right)_\alpha &= E\left[\left((\widetilde{X} \oplus \widetilde{Y})_\alpha\right)^{p-1} \widetilde{X}_\alpha\right] + E\left[\left((\widetilde{X} \oplus \widetilde{Y})_\alpha\right)^{p-1} \widetilde{Y}_\alpha\right] \\
&\geq \left[E\left((\widetilde{X} \oplus \widetilde{Y})_\alpha\right)^{q(p-1)}\right]^{1/q} \times \left[[E(\widetilde{X}_\alpha)^p]^{1/p} + [E(\widetilde{Y}_\alpha)^p]^{1/p}\right],
\end{aligned}
$$

which implies that:

$$
\left(\|\widetilde{X} \oplus \widetilde{Y}\|_p\right)_\alpha \geq \left(\|\widetilde{X}\|_p\right)_\alpha + \left(\|\widetilde{Y}\|_p\right)_\alpha,
$$

for all $\alpha \in [0,1]$. This proves that $P_d(\|\widetilde{X} \oplus \widetilde{Y}\|_p \succeq \|\widetilde{X}\|_p \oplus |\widetilde{Y}\|_p) = 1$.

Theorem 4.13 *Let $\{\widetilde{X}_n\}_{n=1}^{\infty}$ be a sequence of* ***FRVs****. If $\widetilde{X} = \liminf \widetilde{X}_n$, then $P_d(\widetilde{E}(\widetilde{X}) \preceq \liminf(\widetilde{E}(\widetilde{X}_n))) = 1$.*

Proof *For all $\alpha \in [0,1]$, we have:*

$$(\widetilde{E}(\widetilde{X}))_\alpha = \left(\widetilde{E}(\liminf_{n\to\infty} \widetilde{X}_n)\right)_\alpha = E\left(\left(\liminf_{n\to\infty} \widetilde{X}_n\right)_\alpha\right) = E\left(\liminf_{n\to\infty} (\widetilde{X}_n)_\alpha\right)$$

$$\leq \sup_{n\geq 1} \inf_{m\geq n} E((\widetilde{X}_n)_\alpha) = \sup_{n\geq 1} \left(\inf_{m\geq n} E(\widetilde{X}_m)\right)_\alpha = \left(\sup_{n\geq 1} \inf_{m\geq n} E(\widetilde{X}_m)\right)_\alpha .$$

Note that the above inequality simply follows using Fatou's lemma [132] based on a sequence of ordinary random variables variables $(\widetilde{X}_n)_\alpha$. Therefore,

$$P_d(\widetilde{E}(\widetilde{X}) \preceq \liminf(\widetilde{E}(\widetilde{X}_n))) = 1.$$

Remark 4.3 *As discussed in this section, some well-known inequalities were investigated for **FRV**s. However, our results are mainly different from the method presented in [213] as their discussion mainly concerned the inequalities based on the chance measure. Our results however rely on a probability space.*

4.3 Limit Theorems Based on FRVs

Limit theorems in probability theory and statistics are regarded as results giving convergence of sequences of random variables or their distribution functions. Using the notion of α-values for the **FRV**s and a distance measure on the space of **FN**s, this section is aimed to discuss some limit theorems for the **FRV**s inspired by Hesamian et al. [71].

Definition 4.15 *Let $\widetilde{X}$ be a **FRV** and $\{\widetilde{X}_n\}_{n=1}^\infty$, a sequence of **FRV**s defined on the same probability space. Then, we can say that:*

1) *$\widetilde{X}_n \to \widetilde{X}$ converges completely and is denoted by $\widetilde{X}_n \to^c \widetilde{X}$, if $d_\infty(\widetilde{X}_n, \widetilde{X}) \to 0$ as $n \to \infty$.*

2) *$\widetilde{X}_n \to \widetilde{X}$ converges in probability and is denoted by $\widetilde{X}_n \to^p \widetilde{X}$, if for every $\varepsilon > 0$, $\lim_{n\to\infty} P\left(d_\infty(\widetilde{X}_n, \widetilde{X}) > \varepsilon\right) = 0$.*

3) *$\widetilde{X}_n \to \widetilde{X}$ converges almost surely and is denoted by $\widetilde{X}_n \to^{a.s} \widetilde{X}$, for every $\varepsilon > 0$, $P\left(\lim_{n\to\infty} d_\infty(\widetilde{X}_n, \widetilde{X}) > \varepsilon\right) = 0$.*

4) *$\widetilde{X}_n \to \widetilde{X}$ converges in distribution and is denoted by $\widetilde{X}_n \to^d \widetilde{X}$, if for any $x \in \mathbb{R}$, $\lim_{n\to\infty} d_\infty(\widetilde{F}_n(x), \widetilde{F}(x)) = 0$.*

5) *Assume $\widetilde{X}_n, \widetilde{X} \in \widetilde{L}^p$. Then, we say $\widetilde{X}_n \to \widetilde{X}$ and is denoted by $\widetilde{X}_n \to^{\widetilde{L}^p} \widetilde{X}$, if $\lim_{n\to\infty} E(d^p_\infty(\widetilde{X}_n, \widetilde{X})) = 0$ where $d_\infty(\widetilde{A}, \widetilde{B}) = (d^p_\infty(\widetilde{A}, \widetilde{B}))^p$.*

*Here, d_∞ is a absolute error distance between two **FNs** $\widetilde{A}$ and $\widetilde{B}$ defined as:*

$$d_\infty(\widetilde{A},\widetilde{B}) = \sup_{\alpha\in[0,1]} |\widetilde{A}_\alpha - \widetilde{B}_\alpha|.$$

Example 4.15 *Assume that $\widetilde{X}_n = (U_n; 0.01U_n)_T$ is a sequence of independent **FRVs** in which $U_n \sim U(0,1/n)$ and $\widetilde{X} \equiv \widetilde{c} = (0.5, 0.005)_T$. Then, for any $\epsilon > 0$,*

$$P(\sup_{\alpha\in[0,1]} |(\widetilde{X}_n)_\alpha - \widetilde{X}_\alpha| \geq \epsilon) \leq P(|(\widetilde{X}_n)_\alpha - \widetilde{X}_\alpha| \geq \epsilon),$$

for any $\alpha \in (0,1]$. Therefore,

$$\begin{aligned} P(|(\widetilde{X}_n)_\alpha - \widetilde{X}_\alpha| \geq \epsilon) &= P(|U_n - 0.5| \geq \frac{\epsilon}{0.99+0.02\alpha}) \\ &\leq \frac{(0.99+0.02\alpha)^2 var(U_n)}{\epsilon^2} \\ &= \frac{(0.99+0.02\alpha)^2}{12\epsilon^2 n^2} \to 0. \end{aligned}$$

This concludes that $\widetilde{X}_n \to^p \widetilde{c}$.

Example 4.16 *Assume that $\widetilde{X}_n = (1-(1/2n)) \otimes \widetilde{X}$ is a sequence of independent **FRVs** in which $\widetilde{X} = (X, 0.01X)_T$ with $P(X > 0) = 1$ and $E(X^2) < \infty$. Then, for any $\epsilon > 0$:*

$$\begin{aligned} P(\sup_{\alpha\in[0,1]} |(\widetilde{X}_n)_\alpha - \widetilde{X}_\alpha| \geq \epsilon) &\leq P(|(\widetilde{X}_n)_\alpha - \widetilde{X}_\alpha| \geq \epsilon) \\ &= P(|(1-(1/2n))\widetilde{X}_\alpha - \widetilde{X}_\alpha| \geq \epsilon), \end{aligned}$$

for any $\alpha \in [0,1]$. This simply concludes that:

$$\begin{aligned} P(|(1-(1/2n))\widetilde{X}_\alpha - \widetilde{X}_\alpha| \geq \epsilon) &= P(|\widetilde{X}_\alpha| \geq 2n\epsilon) \\ &= P(|X| \geq \frac{2n\epsilon}{0.99+0.02\alpha}) \\ &\leq \frac{(0.99+0.02\alpha)^2 E(X^2)}{4n^2\epsilon^2} \to 0 \text{ as } n \to \infty. \end{aligned}$$

This concludes that $\widetilde{X}_n \to^p \widetilde{X}$.

Lemma 4.5 *For any $\alpha \in [0,1]$, if $(\widetilde{X}_n)_\alpha \to^d \widetilde{X}_\alpha$ then $\widetilde{X}_n \to^d \widetilde{X}$.*

Proof *The proof is left for the reader.*

Proposition 4.4 *If $\widetilde{X}_n \to \widetilde{X}$ completely, then $(\widetilde{X}_n)_\alpha \to \widetilde{X}_\alpha$ for all $\alpha \in [0,1]$. The converse is also true if at least one of the following conditions is satisfied:*
i) $h(\alpha) = (\widetilde{X}_n)_\alpha$ converges uniformly on $[0,1]$.
ii) $h(\alpha) = (\widetilde{X}_n)_\alpha$ be a continuous function of α.

Proof *The proof of (i) is quite trivial. For the converse, consider the following cases:*
i) If $(\widetilde{X}_n)_\alpha \to \widetilde{X}_\alpha$ *uniformly, then* $\forall\epsilon > 0$, $\exists N > 0$ *such that for* $n > N$, $|(\widetilde{X}_n)_\alpha - \widetilde{X}_\alpha| < \epsilon$. *It concludes that* $\sup_{0\le\alpha\le1} |(\widetilde{X}_n)_\alpha - \widetilde{X}_\alpha| < \epsilon$ *which proves the claim.*
ii) If $\{(\widetilde{X}_n)_\alpha\}$ *is a sequence of continuous random variables and* $(\widetilde{X}_n)_\alpha \to \widetilde{X}_\alpha$, *then* $\widetilde{X}_\alpha$ *and* $d_\infty(\widetilde{X}_n, \widetilde{X})$ *are also continuous functions of* α. *On the other hand, if* $(\widetilde{X}_n)_\alpha \to \widetilde{X}_\alpha$, *then* $\forall\epsilon > 0$, $\exists N > 0$ *such that for* $n > N$, $|(\widetilde{X}_n)_\alpha - \widetilde{X}_\alpha| < \epsilon$. *Due to continuity of* $d_\infty(\widetilde{X}_n, \widetilde{X})$, *it follows that* $\sup_{0\le\alpha\le1} |(\widetilde{X}_n)_\alpha - \widetilde{X}_\alpha| = \max_{0\le\alpha\le1} |(\widetilde{X}_n)_\alpha - \widetilde{X}_\alpha| < \epsilon$. *This completes the proof.*

Proposition 4.5 *Complete convergence implies almost surely convergence.*

Proof *A direct proof can be done by Proposition 4.4.*

Theorem 4.14 *If* $(\widetilde{X}_n)_\alpha \to^{a.s} \widetilde{X}_\alpha$ *for all* $\alpha \in [0, 1]$ *and* $(\widetilde{X}_n)_\alpha$ *is finite-valued for all* $\alpha \in [0, 1]$, *then* $\widetilde{X}_n \to^{a.s} \widetilde{X}$.

Proof *First note that if* g *is a finite-valued monotone increasing function on* $[a, b]$. *Then* g *is a continuous except on a set of points which is at most countable [37]. Since* $(\widetilde{X}_n)_\alpha$ *is an increasing function of* α, *it is continuous function of* α *except on a set of points which are at most countable (say* F'*). Now, if* $(\widetilde{X}_n)_\alpha \to^{a.s} \widetilde{X}_\alpha$ *for all* $\alpha \in [0, 1]$, *then there exists a set of* E_α *such that* $P(E'_\alpha) = 0$. *Moreover, for every* $\epsilon_\alpha > 0$, *there exists* N_α *such that for all* $n > N_\alpha$, $|(\widetilde{X}_n)_\alpha - \widetilde{X}_\alpha| < \epsilon_\alpha$. *Now, we define the set* $E = F \cap (\cap_{\alpha\in[0,1]} E_\alpha)$. *Considering the continuity of* $d_\infty(\widetilde{X}_n, \widetilde{X})$, *for all* $n > \max N_\alpha$, *it follows that:*

$$\sup_{0\le\alpha\le1} |(\widetilde{X}_n)_\alpha - \widetilde{X}_\alpha| = \max_{0\le\alpha\le1} |(\widetilde{X}_n)_\alpha - \widetilde{X}_\alpha| < \max_\alpha \epsilon_\alpha = \epsilon'.$$

That means $\widetilde{X}_n \to \widetilde{X}$ *on* E. *To complete the proof, it is enough to show that* $P(F' \cup (\cup_{\alpha\in[0,1]} E'_\alpha)) = 0$ *or* $P(E') = P(\cup_{\alpha\in[0,1]} E'_\alpha) = 0$. *For this purpose, assume that* $P(E') > 0$. *Therefore, there exists a set of* $F \subseteq E'$ *and* $\{\alpha_1, ..., \alpha_k\}$, *such that* $F = \cup_{i=1}^k E'_{\alpha_i}$ *and* $P(F) > 0$. *However, we have:*

$$P(F) = P(\cup_{i=1}^k E'_{\alpha_i}) \le \sum_{i=1}^k P(E'_{\alpha_i}) = 0,$$

which is inconsistent with the assumption of $P(E') > 0$.

Theorem 4.15 *If* $\widetilde{X}_n \to^{a.s} \widetilde{X}$ *then* $\widetilde{X}_n \to^p \widetilde{X}$.

Proof *Let* $\Omega_0 = \Omega - N$ *where* N *is a null set* ($P(N) = 0$) *and*

$$\forall\omega \in \Omega_0 : \lim_{n\to\infty} d_\infty(\widetilde{X}_n(\omega), \widetilde{X}(\omega)) = 0.$$

For $m \geq 1$, define $A_m(\epsilon) = \cap_{n=m}^{\infty}\{d_\infty(\widetilde{X}_n, \widetilde{X}) \leq \epsilon\}$. Notice that $A_m(\epsilon)$ is an increasing function of m. For each ω_0, $\{\widetilde{X}_n(\omega_0)\} \to \widetilde{X}(\omega_0)$ implies that for any $\epsilon \geq 0$, there exists an integer $m(\omega_0, \epsilon)$ such that for $n \geq m(\omega_0, \epsilon)$, $d_\infty(\widetilde{X}_n(\omega_0), \widetilde{X}(\omega_0)) \leq \epsilon$. Hence, $\Omega_0 \subset \cup_{m=1}^{\infty} A_m(\epsilon)$ since $\omega_0 \in A_m(\epsilon)$. Therefore, according to the monotone convergence theorem [132], we get:

$$\begin{aligned}1 = P(\Omega_0) &\leq P(\cup_{m=1}^{\infty} A_m(\epsilon)) = \lim_{m\to\infty} P(A_m(\epsilon)) \\ &= \lim_{m\to\infty} P(\cap_{n=m}^{\infty}\{d_\infty(\widetilde{X}_n, \widetilde{X}) \leq \epsilon\}) \\ &= \lim_{m\to\infty} P(d_\infty(\widetilde{X}_n, \widetilde{X}) \leq \epsilon \text{ for all } n \geq m),\end{aligned}$$

or equivalently:

$$\lim_{m\to\infty} P(d_\infty(\widetilde{X}_n, \widetilde{X}) > \epsilon \text{ for some } n \geq m) = 0.$$

These conclude that $\lim_{n\to\infty} P(d_\infty(\widetilde{X}_n, \widetilde{X}) > \epsilon) = 0$ which completes the proof.

Theorem 4.16 *If $\widetilde{X}_n \to^p \widetilde{X}_1$ and $\widetilde{X}_n \to^p \widetilde{X}_2$ for any **FRVs** of $\widetilde{X}_1$ and $\widetilde{X}_2$, then $P(\widetilde{X}_1 = \widetilde{X}_2) = 1$.*

Proof *Assume that $\widetilde{X}_n \to \widetilde{X}_1$ and $\widetilde{X}_n \to \widetilde{X}_2$ in probability. Since:*

$$\sup_{0\leq\alpha\leq1} |(\widetilde{X}_1)_\alpha - (\widetilde{X}_2)_\alpha| \leq \sup_{0\leq\alpha\leq1} |(\widetilde{X}_n)_\alpha - (\widetilde{X}_1)_\alpha| + \sup_{0\leq\alpha\leq1} |(\widetilde{X}_n)_\alpha - (\widetilde{X}_2)_\alpha|,$$

for any $\varepsilon > 0$, we can easily get:

$$\begin{aligned}P(\sup_{0\leq\alpha\leq1} |(\widetilde{X}_1)_\alpha - (\widetilde{X}_2)_\alpha| > 2\varepsilon) &\leq P(\sup_{0\leq\alpha\leq1} |(\widetilde{X}_n)_\alpha - (\widetilde{X}_1)_\alpha| > \varepsilon) \\ &\quad + P(\sup_{0\leq\alpha\leq1} |(\widetilde{X}_n)_\alpha - (\widetilde{X}_2)_\alpha| > \varepsilon).\end{aligned}$$

But the right side tends to zero as n goes to infinity. Therefore, $P(d_\infty(\widetilde{X}_1, \widetilde{X}_2) = 0) = 1$ or $P(\widetilde{X}_1 = \widetilde{X}_2) = 1$.

Theorem 4.17 *If $\widetilde{X}_n \to^{\widetilde{L}^p} \widetilde{X}$, then $\widetilde{X}_n \to^p \widetilde{X}$. The converse is also true provided that there exists a **FRV** of $\widetilde{Y} \in \widetilde{L}^p$ such that $P_d(|\widetilde{X}_n| \preceq \widetilde{Y}) = 1$ for all n.*

Proof *Regarding Markov inequality [132], note that:*

$$P(\sup_{0\leq\alpha\leq1} |(\widetilde{X}_n)_\alpha - (\widetilde{X})_\alpha| \geq \varepsilon) \leq \frac{E\left(\sup_{0\leq\alpha\leq1} |(\widetilde{X}_n)_\alpha - (\widetilde{X})_\alpha|^p\right)}{\varepsilon^p}.$$

However, the right side goes to zero as $n \to \infty$, so does the left side. This proves the first assertion. Now, if $P_d(|\widetilde{X}_n| \preceq \widetilde{Y}) = 1$ for $\widetilde{Y} \in \widetilde{L}^p(P)$, then

$P_d(|\widetilde{X}_n \ominus \widetilde{X}| \preceq \widetilde{Y} \oplus |\widetilde{X}| = \widetilde{Y}') = 1$ *which is a member of* $\widetilde{L}^p$. *Moreover, it can be observed that* $Z_{n,\alpha} = \sup_{0\leq\alpha\leq 1} |(\widetilde{X}_n)_\alpha - \widetilde{X}_\alpha| \leq \sup_{0\leq\alpha\leq 1} \widetilde{Y}'_\alpha = Z'_\alpha$. *Therefore:*

$$E(\sup_{0\leq\alpha\leq 1} |(\widetilde{X}_n)_\alpha - \widetilde{X}_\alpha|^p) = \int_{\{Z_{n,\alpha}<\epsilon\}} (Z_{n,\alpha})^p dP + \int_{\{Z_{n,\alpha}\geq\epsilon\}} (Z_{n,\alpha})^p dP$$
$$\leq \epsilon^p + \int_{\{Z_{n,\alpha}\geq\epsilon\}} (Z'_\alpha)^p dP.$$

Additionally, the last integral will tend to zero since $E((Z'_\alpha)^p) < \infty$ *and* $P(Z_{n,\alpha} \geq \varepsilon)$ *tends to zero. This proves the claim.*

Theorem 4.18 *If* $\widetilde{X}_n \to^{a.s} \widetilde{X}$ *and* $\widetilde{X}_n \in \widetilde{L}^p$ *then* $\widetilde{X} \in \widetilde{L}^p$ *and* $\widetilde{X}_n \to^{\widetilde{L}^p} \widetilde{X}$.

Proof *Since* $\widetilde{X}_n \to^{a.s} \widetilde{X}$, *for sufficiently large integer of* n, *we have* $|(\widetilde{X}_n(\omega))_\alpha - (\widetilde{X}(\omega))_\alpha| < \epsilon$ *for any* $\omega \in \Omega$ *and any* $\alpha \in [0,1]$. *Therefore,* $|(\widetilde{X}(\omega))_\alpha|^p \leq 2^p(|(\widetilde{X}_n(\omega))_\alpha|^p + \epsilon^p)$ *for any* $\omega \in \Omega$ *and any* $\alpha \in [0,1]$. *This simply concludes that* $\widetilde{X} \in \widetilde{L}^p$. *In addition, for any* $\epsilon > 0$,

$$E(\sup_{\alpha\in[0,1]} |(\widetilde{X}_n)_\alpha - (\widetilde{X})_\alpha|^p) \leq$$
$$\int_{\{\sup_{\alpha\in[0,1]} |(\widetilde{X}_n)_\alpha - (\widetilde{X})_\alpha|>\epsilon\}} \sup_{\alpha\in[0,1]} |(\widetilde{X}_n)_\alpha - (\widetilde{X})_\alpha|^p dp$$
$$+\epsilon.$$

Since $\widetilde{X}_n \to^p \widetilde{X}$, *the first sentence on the right side goes to zero as* $n \to \infty$. *This completes the proof.*

Theorem 4.19 *If* $\widetilde{X}_n \to^d \widetilde{c}$, *then* $\widetilde{X}_n \to^p \widetilde{c}$.

Proof *If* $\widetilde{X}_n \to^d \widetilde{c}$ *then* $(\widetilde{X}_n)_\alpha \to^d \widetilde{c}_\alpha$ *for any* $\alpha \in [0,1]$. *This simply follows that* $(\widetilde{X}_n)_\alpha \to \widetilde{c}_\alpha$ *in probability for any* $\alpha \in [0,1]$. *This also implies that* $P(\sup_{0\leq\alpha\leq 1} |(\widetilde{X}_n)_\alpha - \widetilde{c}_\alpha| > \epsilon) \to 1$ *as* $n \to \infty$ *for any* $\epsilon > 0$ *which completes the proof.*

Theorem 4.20 $\widetilde{X}_n \to^d \widetilde{X}$ *if and only if* $\widetilde{E}(g(\widetilde{X}_n)) \to \widetilde{E}(g(\widetilde{X}))$ *for any monotone continuous and bounded function of* g.

Proof *Let* g *be a monotone continuous and bounded function. To prove the above lemma, it is enough to show that* $d_\infty(\widetilde{E}(g(\widetilde{X}_n)), \widetilde{E}(g(\widetilde{X}))) \longrightarrow 0$. *Note that* $\widetilde{X}_n \to \widetilde{X}$ *in distribution if and only if there exists an integer number of* N *such that for* $n > N$,

$$|(\widetilde{F}_n(x))_\alpha - (\widetilde{F}(x))_\alpha| < \epsilon, \quad \forall\alpha \in [0,1].$$

for all $\epsilon > 0$. *Since* g *is a continuous and bounded function , from the classical probability theory [132], we get:*

$$|E\left((g(\widetilde{X}_n))_\alpha\right) - E\left((g(\widetilde{X}))_\alpha\right)| = |E\left(g((\widetilde{X}_n)_\alpha)\right) - E\left(g(\widetilde{X})_\alpha\right)| < \epsilon.$$

This means $E\left(g((\widetilde{X}_n)_\alpha)\right) \to E(g(\widetilde{X})_\alpha)$ *uniformly on* $[0, 1]$ *which simply concludes* $d_\infty(\widetilde{E}\ (g(\widetilde{X}_n)), \widetilde{E}(g(\widetilde{X})) \longrightarrow 0$ *as* n *tends to infinity.*

Proposition 4.6 *If* $\widetilde{X}_n \to^{a.s} \widetilde{X}$, *then* $\widetilde{X}_n \to^p \widetilde{X}$.

Proof *Assume that* $\widetilde{X}_n \to \widetilde{X}$ *in probability. Therefore, for each continuous and bounded function of* g, *we get* $g(\widetilde{X}_n) \to^p g(\widetilde{X})$. *According to Theorem 4.17, note that this convergence also holds in* L^1 *space. It follows that* $d_\infty(\widetilde{E}(g(\widetilde{X}_n)), \widetilde{E}(g(\widetilde{X}))) \to 0$ *as* n *goes to infinity. This simply proves the claim from Theorem 4.20.*

Theorem 4.21 *If* $\widetilde{X}_n \to^d \widetilde{X}$ *and* $\widetilde{Y}_n \to^p I\{0\}$, *then* $\widetilde{X}_n \oplus \widetilde{Y}_n \to^d \widetilde{X}$.

Proof *Let* g *be a bounded continuous and monotone function with* $|g| \leq M$. *Therefore, for any* $\epsilon > 0$, *there exists a* $\delta > 0$ *such that* $|x - y| \leq \delta$ *implies* $|g(x) - g(y)| \leq \epsilon$. *For all* $\alpha \in [0, 1]$, *we then have:*

$$\begin{aligned}
E(|g((\widetilde{X}_n)_\alpha + (\widetilde{Y}_n)_\alpha) - g((\widetilde{X}_n)_\alpha)|) &\leq \epsilon P(|g((\widetilde{X}_n)_\alpha + (\widetilde{Y}_n)_\alpha) - g((\widetilde{X}_n)_\alpha)| \leq \epsilon) \\
&\quad + 2MP(|g((\widetilde{X}_n)_\alpha + (\widetilde{Y}_n)_\alpha) - g((\widetilde{X}_n)_\alpha)| > \epsilon) \\
&\leq \epsilon + 2MP(|(\widetilde{Y}_n)_\alpha| > \delta).
\end{aligned}$$

Since $\lim_{n\to\infty} P(|(\widetilde{Y}_n)_\alpha| > \delta) = 0$, *we also get:*

$$\lim_{n\to\infty} E(g((\widetilde{X}_n)_\alpha + (\widetilde{Y}_n)_\alpha)) = \lim_{n\to\infty} E(g((\widetilde{X}_n)_\alpha)) = E(g(\widetilde{X}_\alpha)), \quad \forall\alpha \in [0, 1].$$

By Theorem 4.20, this completes the proof.

Theorem 4.22 *Let* $\widetilde{X}$, $\widetilde{X}_n$, $\widetilde{Y}$ *and* $\widetilde{Y}_n$ *be **FRVs**. Then,*

1) if $\widetilde{X}_n \to^{a.s} \widetilde{X}$ *and* $\widetilde{Y}_n \to^{a.s} \widetilde{Y}$ *then* $\widetilde{X}_n \oplus \widetilde{Y}_n \to^{a.s} \widetilde{X} \oplus \widetilde{Y}$,

2) $\widetilde{X}_n \to^p \widetilde{X}$ *and* $\widetilde{Y}_n \to^p \widetilde{Y}$ *then* $\widetilde{X}_n \oplus \widetilde{Y}_n \to^p \widetilde{X} \oplus \widetilde{Y}$,

3) $\widetilde{X}_n \to^{L^1} \widetilde{X}$ *and* $\widetilde{Y}_n \to^{L^1} \widetilde{Y}$ *then* $\widetilde{X}_n \oplus \widetilde{Y}_n \to^{L^1} \widetilde{X} \oplus \widetilde{Y}$.

Proof *First, note that:*

$$d_\infty(\widetilde{X}_n \oplus \widetilde{Y}_n, \widetilde{X} \oplus \widetilde{Y}) \leq d_\infty(\widetilde{X}_n, \widetilde{X}) + d_\infty(\widetilde{Y}_n, \widetilde{Y}).$$

Therefore, for any $\epsilon > 0$

1) $P(d_\infty(\widetilde{X}_n \oplus \widetilde{Y}_n, \widetilde{X} \oplus \widetilde{Y}) > \epsilon) \leq P(d_\infty(\widetilde{X}_n, \widetilde{X}) > \epsilon) + P(d_\infty(\widetilde{Y}_n, \widetilde{Y}) > \epsilon)$,

2) By Minkowski inequality, we get $E(d_1^2(\widetilde{X}_n \oplus \widetilde{Y}_n, \widetilde{X} \oplus \widetilde{Y})) \leq E(d_1^2(\widetilde{X}_n, \widetilde{X}))+ E(d_1^2(\widetilde{Y}_n, \widetilde{Y}))$.

3) $E(d_\infty(\widetilde{X}_n \oplus \widetilde{Y}_n, \widetilde{X} \oplus \widetilde{Y})) \leq E(d_\infty(\widetilde{X}_n, \widetilde{X})) + E(d_\infty(\widetilde{Y}_n, \widetilde{Y}))$.

These simply verified 1, 2, and 3.

Theorem 4.23 ***(Strong Law of Large Numbers (SLLN))*** *Let* $\{\widetilde{X}_n\}$ *be a sequence of independent and identically distributed **FRV**s. For* $p \in (0, 2)$*, if* $\widetilde{X}_1 \in L^p(P)$ *then:*

$$\overline{\widetilde{X}} = (\frac{1}{n}) \otimes \bigoplus_{i=1}^{n} \widetilde{X}_i \rightarrow \widetilde{E}(\widetilde{X}_1) \quad a.s. \tag{4.26}$$

Proof *Since* $\{(\widetilde{X}_n)_\alpha\}$ *is a sequence of independent and identically distributed (non-fuzzy) random variables, according to arithmetic* α*-values based operations of **FN**s and Marcinkiewz-Zygmund strong low of large numbers [132], it concludes that:*

$$\left((\frac{1}{n}) \otimes \bigoplus_{i=1}^{n} \widetilde{X}_i \right)_\alpha = \frac{1}{n} \sum_{i=1}^{n} \widetilde{X}_{i\alpha} \rightarrow E(\widetilde{X}_{1\alpha}), \quad a.s. \quad \forall \alpha \in [0, 1].$$

On the other hand, since $\widetilde{X}_1 \in L^p(P)$*, we get* $E(\widetilde{X}_\alpha) < \infty$ *for all* $\alpha \in [0, 1]$*. Therefore,* $\widetilde{X}_\alpha$ *is almost surely finite valued which completes the proof by Theorem 4.14.*

Example 4.17 *Assume that* $\widetilde{X}_n = (U_n; 0.01U_n)_T$ *is a sequence of independent **FRV**s in which* $U_n \sim U(0, 1)$*. Let* $\widetilde{Z}_n = \bigoplus_{i=1}^{n} \widetilde{X}_n$*. For any* $\alpha \in [0, 1]$*,* $E(|(\widetilde{X}_1)_\alpha|) = E(|U_n(1 - 0.01(1 - 2\alpha))|) \leq 1.01E(U_n) < 1$*. Therefore, since* $\widetilde{X}_1 \in L^1(P)$*, according to Theorem 4.23, we get* $\overline{\widetilde{X}} \rightarrow^{a.s.} \widetilde{E}(\widetilde{X}_1)$ *where* $\widetilde{E}(\widetilde{X}_1) = (0.5, 0.005)_T$*.*

Remark 4.4 *It should be noted that many authors have studied the law of large numbers for sum of the independent **FRV**s. Some authors generalized them in Banach space for fuzzy random set [164, 137, 195, 7, 57, 190, 99]. Other studies also addressed the weak laws of large number for **FRV**s in the setting of chance theory [102, 190, 91] or possibility space [130, 213]. Therefore, our method is completely different from such methods since it is conducted based on the **FRV**s. Several authors also examined the problem of large sample properties for a sequence of **FRV**s in probability space. Such methods basically require several conditions relevant to lower and upper bounds of the of **FRV**s or probability space. For instance, a notion of the strong low of large numbers is suggested **(SLLN)** in [113] under the condition of* $E(||supp\widetilde{X}||) < \infty$*, where* $||.||$ *denotes the Euclidian norm. Under some conditions for* $P(||supp\widetilde{X}_n|| > x) < P(|\widetilde{X}| > x)$ *a version of **SLLN** is proposed* *for **FRVs** in [90]. Wu [207, 208] also replaced this condition with the uniform convergence to investigate the notion of **SLLN**. Another version of **SLLN** was introduced in [111] using continuity properties on α-cuts of **FRVs**. However, the proposed method discussed these properties for the **FRVs** using the minimum requirements.*

Definition 4.16 *Let $\{\widetilde{X}_n\}$ be a sequence of independent and identically distributed with $\widetilde{E}(\widetilde{X}_n) = \widetilde{\mu}$ and $\boldsymbol{var}(\widetilde{X}_n) = \sigma^2$. We say $\widetilde{X}_n$ is a generalized **FRV** (**GFRV**) if there exists a **FRV** $\widetilde{Z}_n$ in which $\widetilde{X}_n \ominus_G \widetilde{\mu} = \widetilde{Z}_n$ with $\widetilde{E}(\widetilde{Z}_n) = I\{0\}$ and $\boldsymbol{var}(\widetilde{Z}_n) = \sigma^2$.*

The Central Limit Theorem is an important tool in probability theory because it mathematically explains why the normal distribution is observed so commonly in nature. Here, a notion of the central limit theorem is presented for independent **FRV**s via the generalized difference operator $\ominus_G$.

Theorem 4.24 ***(Central Limit Theorem)*** *Let $\{\widetilde{X}_n\}$ be a sequence of **GFRVs**. Then:*

$$(\frac{\sqrt{n}}{\sigma}) \otimes \widetilde{\overline{Y}}_n \to^d \widetilde{Z}, \tag{4.27}$$

where $\widetilde{\overline{Y}}_n = (\frac{1}{n}) \bigoplus_{i=1}^{n} (\widetilde{X}_i \ominus_G \widetilde{\mu})$ and $\widetilde{Z}_\alpha \sim N(0,1)$.

Proof *We show that $((\frac{\sqrt{n}}{\sigma}) \otimes \widetilde{\overline{Y}}_n)_\alpha \to^d \widetilde{Z}_\alpha$ for any $\alpha \in [0,1]$. For this purpose, we have*

$$((\frac{\sqrt{n}}{\sigma}) \otimes \widetilde{\overline{Y}}_n)_\alpha = \frac{1}{n} \sum_{i=1}^{n} \frac{\sqrt{n}}{\sigma} (\widetilde{X}_i \ominus_G \widetilde{\mu})_\alpha.$$

Since $\frac{\sqrt{n}}{\sigma}(\widetilde{X}_i \ominus_G \widetilde{\mu})_\alpha =^d \widetilde{Z}_\alpha$, the result simply follows from the conventional Central Limit Theorem.

Example 4.18 *Let $\widetilde{X}_n = \widetilde{\mu} \oplus (\sigma \otimes \widetilde{Z}_n)$ where $\widetilde{E}(\widetilde{Z}_n) = I\{0\}$ and $\boldsymbol{var}(\widetilde{Z}_n) = 1$. Since, $\widetilde{X}_n \ominus_G \widetilde{\mu} = (\sigma \otimes \widetilde{Z}_n)$, it can be concluded that $\{\widetilde{X}_n\}$ is a sequence of **GFRVs**. By Theorem 4.24, we have:*

$$(\frac{\sqrt{n}}{\sigma}) \otimes \Big((\frac{1}{n}) \bigoplus_{i=1}^{n} (\widetilde{X}_i \ominus_G \widetilde{\mu}) \Big) \to^d \widetilde{Z}.$$

Remark 4.5 *It should be noted that Wu [207] also proposed a notion of central limit theorems of **FRVs** based on Hukuhara difference [165] called H-**FRVs**. He provided some additional conditions that an H-**FRV** can be well defined. Compared to his method, our method needs no additional conditions since $\ominus_G$ always exists.*

4.4 Exercise

Exercise 4.1 *Let $\widetilde{X} = (X; U)$ be a* ***LSFRV*** *where X is distributed according to double exponential distribution with density function of $f_X(x) = (1/2\theta)\exp(-|\mu - x|/\theta)$ and $U \sim U(1,3)$. Based on a given* ***FRS*** *of $\widetilde{x}_1, \widetilde{x}_2, \ldots, \widetilde{x}_n$:*

(1) *Evaluate the α-cuts of $\widetilde{F}_{\widetilde{X}}(x)$.*

(2) *Find the moment estimations of μ and θ.*

(3) *Assume that $\mu = 0$ and $\theta = 1$. Find the α-cuts of $\widetilde{P}(\widetilde{X} \in (-1, 1)$.*

Exercise 4.2 *Let $\widetilde{X} = (X; UX)_T$ be a* ***SFRV*** *where $X \sim \chi^2_r$ and $U \sim U(0.25, 0.75)$. According to a* ***FRS*** *of $\widetilde{x}_1, \widetilde{x}_2, \ldots, \widetilde{x}_n$:*

(1) *Evaluate the α-cuts of $\widetilde{F}_{\widetilde{X}}(x)$.*

(2) *Find the moment estimations of r.*

(3) *Assume that $r = 2$. Find the α-cuts of $\widetilde{P}(\widetilde{X} \in (1, \infty)$ and plot its membership function.*

Exercise 4.3 *Consider a* ***FRV*** *of $\widetilde{X} = (X; Z)$ where $X \sim N(\mu, \sigma^2)$, $Z \sim N_{0.5}(0, \tau^2)$ in which X and Z are two independent random variables and $N_{0.5}$ shows the half-normal distribution. Find $\widetilde{E}(\widetilde{X})$ and* ***var***$(\widetilde{X})$.

Exercise 4.4 *Prove Lemma 4.1.*

Exercise 4.5 *Let $\widetilde{X} = (X_1; U_1)_T$ and $\widetilde{Y} = (X_2; U_2)_T$ be two* ***FRVs****. Compute* ***corr***$(\widetilde{X}, \widetilde{Y})$ *for the cases where*

1) $\mathbf{X} = (X_1, X_2)$ *is a bivariate normal random variable with $Var(X_1) = 2$, $Var(X_2) = 3$, and $Corr(X_1, X_2) = -0.75$.*

2) $\boldsymbol{U} = (U_1, U_2)$ *is a bivariate random variable with the following joint density function:*

$$f(u_1, u_2) = \begin{cases} 2, & u_1, u_2 > 0, u_1 + u_2 < 1 \\ 0, & o.w, \end{cases} \tag{4.28}$$

3) $\mathbf{X}$ *and* $\boldsymbol{U}$ *are independent.*

Exercise 4.6 *Let $\widetilde{X}$ be a* ***FRV*** *with fuzzy mean of $\widetilde{E}(\widetilde{X})$. For any $\epsilon > 0$, prove that $P(d_2^2(\widetilde{X}, \widetilde{E}(\widetilde{X})) > \epsilon) \leq$* ***var***$(\widetilde{X})/\epsilon^2$.

Exercise 4.7 *Let $\{\widetilde{X}_n\}_{n\in\mathbb{N}}$ and $\{\widetilde{Y}_n\}_{n\in\mathbb{N}}$ be two sequences of **FRVs** where $\widetilde{Y}_n \to^c \widetilde{Y}$. Prove that*

$$\limsup_{n\to\infty}(\widetilde{X}_n \oplus \widetilde{Y}_n) = \limsup_{n\to\infty} \widetilde{X}_n \oplus \widetilde{Y}.$$

Exercise 4.8 *For two **FRVs** of $\widetilde{X}$ and $\widetilde{Y}$, define*

$$d^*(\widetilde{X},\widetilde{Y}) = E(\frac{d_\infty(\widetilde{X},\widetilde{Y})}{1+d_\infty(\widetilde{X},\widetilde{Y})}).$$

(1) *Prove that $d^* : \mathcal{F}(\mathbb{R}) \times \mathcal{F}(\mathbb{R}) \to [0,1]$ is a distance measure.*

(2) *Prove $\widetilde{X}_n \to^p \widetilde{X}$ if and only if $P(\lim_{n\to\infty} d^*(\widetilde{X_n},\widetilde{X}) = 0) = 1$.*

Exercise 4.9 *Let $\{\widetilde{X}_n\}_{n\in\mathbb{N}}$ be a sequence of **FRVs**. For any $\epsilon > 0$, prove that if $\sum_{n=1}^{\infty} P(d_\infty(\widetilde{X_n},\widetilde{X}) \geq \epsilon) < \infty$, then $\widetilde{X}_n \to^{a.s.} \widetilde{X}$.*

Exercise 4.10 *Construct a sequence of **FRVs** $\{\widetilde{X}_n\}_{n=1}^{\infty}$ and a **FRV** of $\widetilde{X}$ in which $\widetilde{X}_n \to^p \widetilde{X}$ but $\widetilde{X}_n \nrightarrow^{L^1} \widetilde{X}$.*

Exercise 4.11 *Construct a sequence of **FRVs** $\{\widetilde{X}_n\}_{n=1}^{\infty}$ and a **FRV** $\widetilde{X}$ in which $\widetilde{X}_n \to^p \widetilde{X}$ but $\widetilde{X}_n \nrightarrow^{a.s.} \widetilde{X}$.*

Exercise 4.12 *For a sequence of **FRVs** $\{\widetilde{X}_n\}_{n=1}^{\infty}$ and a **FRV** $\widetilde{X}$, assume that $\widetilde{X}_n \to^p \widetilde{X}$. Prove that there exists a subsequence of **FRVs** $\{\widetilde{X}_{n_k}\}$ such that $\widetilde{X}_{n_k} \to^{a.s.} \widetilde{X}$.*

Exercise 4.13 *Assume that $\widetilde{X}_n = (U_n; 0.01U_n)_T$ is a sequence of independent **FRVs** in which $U_n \sim U(0,1)$. Let $\widetilde{Z}_n = \bigotimes_{i=1}^{n} \sqrt[n]{\widetilde{X}_n}$. Show that there exists a **FRV** of $\widetilde{Z}$ such that $\widetilde{Z}_n \to^{a.s.} \widetilde{Z}$.*

Exercise 4.14 *For a sequence of **FRVs** $\{\widetilde{X}_n\}$ and a **FRV** of $\widetilde{X}$, prove that $\widetilde{X}_n \to^{a.s.} \widetilde{X}$ if and only if $\lim_{n\to\infty} P(\cup_{k=n}^{\infty}\{d_\infty(\widetilde{X}_k,\widetilde{X}) > \epsilon\}) = 0$ for any $\epsilon > 0$.*

Exercise 4.15 *Assume that $\{\widetilde{X}_n\}$, $\{\widetilde{Y}_n\}$ (as well as $\widetilde{X}$ and $\widetilde{Y}$) are independent **FRVs**. If $\widetilde{X}_n \to^d \widetilde{X}$ and $\widetilde{Y}_n \to^d \widetilde{Y}$ then $\widetilde{X}_n \oplus \widetilde{Y}_n \to^d \widetilde{X} \oplus \widetilde{Y}$.*

Exercise 4.16 *Prove Lemma 1.9.*

Exercise 4.17 *Let $\{\widetilde{X}_n\}$ be a sequence of **HFRVs**. Prove that*

$$(\frac{\sqrt{n}}{S_n}) \otimes \overline{\widetilde{Y}}_n \to^d \widetilde{Z},$$

where $S_n^2 = \frac{1}{n-1}\sum_{i=1}^{n} d_2^2(\widetilde{X}_i,\overline{\widetilde{X}})$, $\overline{\widetilde{Y}}_n = (\frac{1}{n}\bigoplus(\widetilde{X}_i \ominus_H \widetilde{\mu}))$ and $\widetilde{Z}_\alpha \sim N(0,1)$.

Exercise 4.18 *Prove Lemma 4.5.*

4.5 Glossary

Convergence of a sequence of a fuzzy random variable: An extension of limit theorems based on fuzzy random variables.

Fuzzy cumulative distribution function: A generalization of chance that the fuzzy random variable is less than or equal to x.

Fuzzy correlation: A measure of linear association between two fuzzy random variable.

Fuzzy expectation: The fuzzy mean-valued of a fuzzy random variable.

Fuzzy empirical cumulative distribution function: An estimated cumulative distribution function directly from a fuzzy sample data without assuming an underlying algebraic form.

Fuzzy moment estimator: An extended value of the average value of the elements raised to the rth power based on α-values of fuzzy data.

Fuzzy Random variable: An assignment of fuzzy numbers to possible outcomes of a random experiment.

Variance of a fuzzy random variable: An extension of the second moment about the fuzzy mean of a fuzzy random variable.

5

Fuzzy Hypothesis Testing for ***FRV****s*

Hypothesis testing, also known as significance testing, is the most frequent statistical method to guard against making claims unjustified by data [147, 181, 188]. This method has been used to evaluate the strength of evidence from the sample and provides a framework for population-related decision-making, i.e, it provides a method for understanding how reliably one can extrapolate observed findings in a sample under study to the larger population from which the sample was drawn. A statistical hypothesis (H) is a statement about the population (populations) from which one or more samples are drawn. The hypothesis H_0 under test is called the null hypothesis. A statistical procedure which enables one to make a decision whether or not H_0 should be rejected is called a test. The classical approaches usually depend on certain basic assumptions about the underlying population such as crisp observations or crisp hypotheses. In practical studies, as discussed in Chapter 4, it is frequently difficult to assume that the outcomes of a random variable are recorded as precise values. For such cases, there are two approaches to model the parameters, for which the distribution of a **FRV** is determined; exact parameters or fuzzy-valued parameters. For the second case, it is assumed that the hypotheses of interest are presented as exact relations. This chapter assumes that imprecision, brought about by the imprecise meaning of the data and hypotheses can be modeled by **FN**s.

Therefore, if fuzziness involves to record observations or hypotheses we face quite challengeable problem to model type I error and type II error, p-values and method of making decision at a specified significance level. To achieve suitable testing statistical methods dealing with fuzzy information, we need to model such imprecision and extend the usual approaches to fuzzy domain. For this purpose, the fuzzy set theory has been developed and applied in some statistical contexts to deal with fuzzy conditions. Especially, statistical tests based on fuzzy information have been extensively studied. The proposed approaches mainly rely on fuzzy data, fuzzy hypotheses, and both fuzzy data and fuzzy hypotheses for parametric [6, 18, 65, 84, 95, 141, 152, 169, 197, 209] and non-parametric cases [29, 39, 56, 60, 61, 62, 86, 89, 94, 105, 128]. For a comprehensive review on past studies on fuzzy hypothesis tests, see Chukhrova and Johannssen [30] until 2020.

This chapter also extends the one-way analysis of variance (ANOVA) to fuzzy domain. As discussed in Chapter 4, there are several real-life populations in which imprecise values can be assigned to their experimental outcomes. For

DOI: 10.1201/9781003248644-5

instance, a food company wished to test four different package designs for a new product. The sales volumes for Some stores are recorded as the experimental units. In this way, 'sales volumes' is a fuzzy quantity and **TFN**s may be suitable models to formalize and handle a sales volume of a store. For such a case, a combination of fuzzy set theory and the classical statistical inference can help to produce a fuzzy ANOVA for such imprecise cases. In this regard, it should be noted that ANOVA with fuzzy information has been studied by many authors. Degaribay [38] developed the one-way ANOVA by comparing the behavior of the one-way ANOVA and linear regression based on fuzzy data. Montenegro et al. [138] considered the one-way ANOVA based on measuring differences between fuzzy numbers. Gil et al. [56] presented asymptotic multi-sample testing of means for simple fuzzy random variables based on a bootstrap approach. Konishi et al. [116] proposed the method of ANOVA for the fuzzy interval data by using the concept of fuzzy sets. Wu [210] proposed a method for the traditional ANOVA model based on a definition of the normal distribution with fuzzy parameters. Lubiano and Trutschnig [134] provided an *R* package for testing the equality of means of two or more fuzzy random variables with exact parameters. Rodríguez et al. [168] proposed an one-way ANOVA by using a random mechanism producing fuzzy data and relevant associated parameters by using concepts for Hilbert space-valued random elements. Kalpanapriy proposed a one-way ANOVA approach using a random mechanism producing fuzzy data and relevant associated parameters based on the concepts of Hilbert space-valued random elements. Kalpanapriya and Pandian [106] examined the one-way ANOVA model with triangular fuzzy data and fuzzy hypothesis. Andrés-Sánchez [36] proposed the fuzzy least-squares regression that combines fuzzy regression with the classical statistical scheme based on two-way of ANOVA. Nourbakhsh et al. [145] extended the one-way ANOVA using Zadeh's extension principle based on **TFN**s. Jiryaei et al. [101], used the least-square method to extend the one-way analysis of variance based on **FRV**s where all random errors and parameters are considered as **FN**s. Pandian and Kalpanapriya [151] proposed a quality statistical hypothesis tests for fuzzy data with respect to membership grade of fuzzy set(s) using analysis of variance technique for analyzing the effect of factors of classification of a population.

5.1 Testing Fuzzy Hypotheses Based on Normal FRVs

In this section, some common parametric hypothesis tests are extended based on **FRV**s and fuzzy hypothesis. For this purpose, the focus is on triangular fuzzy random variables defined as follows.

Definition 5.1 *Let* $(\Omega, \mathbb{F}(\mathbb{R}), P)$ *be a probability space. A fuzzy-valued mapping of* $\widetilde{X} : \Omega \to \mathcal{F}(\mathbb{R})$ *are said to be a triangular fuzzy random variable*

*(**TFRV**) if $\widetilde{X} = (X^L, X, X^U)_T$ where $P(X^L < X < X^U) = 1$. Moreover, two **TFRVs** of $\widetilde{X}$ and $\widetilde{Y}$ is said to be independent and identically distributed if X^L, Y^L, X, Y and X^U, Y^U are independent and identically distributed. Such an idea can be extended for a set of n **TFRVs** $\widetilde{X}_1, \ldots, \widetilde{X}_n$. In this regard, we say that $\widetilde{\mathbf{X}} = (\widetilde{X}_1, \ldots, \widetilde{X}_n)$ is a **FRS** and its observed value can be denoted by $\widetilde{\boldsymbol{x}} = (\widetilde{x}_1, \ldots, \widetilde{x}_n)$.*

Let $\widetilde{X}$ be a **TFRV**. In this regard, the fuzzy expectation of $\widetilde{X} = (X^L, X, X^U)_T$ is considered as $\widetilde{E}(\widetilde{X}) = (E(X^L), E(X), E(X^U))_T$ with the variance of $\mathbf{var}(\widetilde{X}) = E(d_2^2(\widetilde{X}, \widetilde{E}(\widetilde{X})))$ where $d_2^2(\widetilde{A}, \widetilde{B}) = ((a^L - b^L)^2 + (a - b)^2 + (a^U - b^U)^2)/3$ is the square error distance measure between two **TFNs** of $\widetilde{A}$ and $\widetilde{B}$. This simply concludes that $\mathbf{var}(\widetilde{X}) = (var(X^L) + var(X) + var(X^U))/3$. Furthermore, according to a **FRS** of $\widetilde{\mathbf{X}} = (\widetilde{X}_1, \ldots, \widetilde{X}_n)$, the sample variance can be defined as $S_n^2 = (1/(n-1)) \sum_{i=1}^n d_2^2(\widetilde{X}_i, \overline{\widetilde{X}})^2$. Therefore, the sample variance can be shown by $S_n^2 = (S_L^2 + S^2 + S_U^2)/3$ where

1) $S_L^2 = (1/(n-1)) \sum_{i=1}^n (X_i^L - \overline{X}^L)^2$,
2) $S^2 = (1/(n-1)) \sum_{i=1}^n (X_i - \overline{X})^2$,
3) $S_U^2 = (1/(n-1)) \sum_{i=1}^n (X_i^U - \overline{X}^U)^2$.

To make **TFRVs** applicable to statistical hypothesis tests, a notion of normal **TFRV** is given.

Definition 5.2 *A **TFRV** $\widetilde{X} = (X^L, X, X^U)_T$ is said to be a triangular normal fuzzy random variable (**TNFRV**) denoted by $\widetilde{X} \sim N(\widetilde{\mu}, \sigma^2)$ if*

1) $\widetilde{\mu} = (\mu^L, \mu, \mu^U)_T$,

2) $\mu^L = \mu - l_\mu$ and $\mu^U = \mu + r_\mu$ where $l_\mu, r_\mu > 0$,

3) $X^L = Y - l_\mu + \epsilon$, $X = Y + \epsilon$ and $X^U = Y + r_\mu + \epsilon$,

4) $Y \sim N(\mu, \sigma^2/2)$ and $\epsilon \sim N(0, \sigma^2/2)$ are independent random variables.

Lemma 5.1 *If $\widetilde{X}$ is a **TNFRV** then $\widetilde{E}(\widetilde{X}) = \widetilde{\mu}$ and $\mathbf{var}(\widetilde{X}) = \sigma^2$.*

Proof *The proof is left as an exercise.*

Lemma 5.2 *Let $\widetilde{X}$ be a **TNFRV** and $\widetilde{\mathbf{X}} = (\widetilde{X}_1, \ldots, \widetilde{X}_n)$ be **FRS**. Then,*

1) $\frac{M_{\overline{\widetilde{X}}} - M_{\widetilde{\mu}}}{\sigma/\sqrt{3n}} \sim N(0,1)$.

2) $\frac{3(n-1)S_n^2}{\sigma^2} \sim \chi^2_{3(n-1)}$.

3) $\frac{M_{\overline{\widetilde{X}}} - M_{\widetilde{\mu}}}{S_n/\sqrt{n}} \sim t_{3(n-1)}$.

Proof *To prove, it is enough to note that if* $\widetilde{\mathbf{X}} = (\widetilde{X}_1, \ldots, \widetilde{X}_n)$ *is a normal* ***FRS*** *then* $M_{\widetilde{X}_1}, M_{\widetilde{X}_2}, \ldots, M_{\widetilde{X}_n}$ *is also a random sample of* $N(M_{\widetilde{\mu}}, \sigma^2)$.

Here, a notion of fuzzy hypotheses of the fuzzy mean of a **TNFRV** is introduced.

Definition 5.3 *Let* $\widetilde{X}$ *be a* ***TNFRV*** *with fuzzy mean* $\widetilde{\mu} = (\mu^L, \mu, \mu^U)_T$. *Then,*

1. *Any hypothesis of the form*

$$\widetilde{H} : \widetilde{\mu} \simeq_M \widetilde{\mu}_0 = (\mu_0^L, \mu_0, \mu_0^U)_T \qquad \equiv \qquad \widetilde{H} : M_{\widetilde{\mu}} = M_{\widetilde{\mu}_0},$$

   is called a null fuzzy hypothesis.

2. *Any hypothesis of the form*

$$\widetilde{H} : \widetilde{\mu} \succeq_M \widetilde{\mu}_0 \qquad \equiv \qquad \widetilde{H} : M_{\widetilde{\mu}} \geq M_{\widetilde{\mu}_0},$$

   is called a right one-sided fuzzy null hypothesis.

3. *Any hypothesis of the form*

$$\widetilde{H} : \widetilde{\mu} \preceq_M \widetilde{\mu}_0 \qquad \equiv \qquad \widetilde{H} : M_{\widetilde{\mu}} \leq M_{\widetilde{\mu}_0},$$

   is called a left one-sided fuzzy null hypothesis.

4. *Any hypothesis of the form*

$$\widetilde{H} : \widetilde{\mu} \succ_M \widetilde{\mu}_0 \qquad \equiv \qquad \widetilde{H} : M_{\widetilde{\mu}} > M_{\widetilde{\mu}_0},$$

   is called a right one-sided fuzzy alternative hypothesis.

5. *Any hypothesis of the form*

$$\widetilde{H} : \widetilde{\mu} \prec_M \widetilde{\mu}_0 \qquad \equiv \qquad \widetilde{H} : M_{\widetilde{\mu}} < M_{\widetilde{\mu}_0},$$

   is called a left one-sided fuzzy alternative hypothesis.

6. *Any hypothesis of the form*

$$\widetilde{H} : \widetilde{\mu} \neq_M \widetilde{\mu}_0 \qquad \equiv \qquad \widetilde{H} : M_{\widetilde{\mu}} \neq_M M_{\widetilde{\mu}_0},$$

   is called a two-sided fuzzy alternative hypothesis,

5.1.1 One-sample testing fuzzy hypotheses based on normal FRVs

Here, based on **FRV** $\widetilde{\mathbf{X}} = (\widetilde{X}_1, \ldots, \widetilde{X}_n)$ of a **TNFRV**, a procedure is established for one-sided hypothesis test of fuzzy mean of **TNFRV**s.

Definition 5.4 *Let $\widetilde{\boldsymbol{X}} = (\widetilde{X}_1, \ldots, \widetilde{X}_n)$ be a* ***FRS****. For testing $\widetilde{H}$ defined in Definition 5.3, the test statistic $T : (\mathcal{F}(\mathbb{R}))^n \to \mathbb{R}$ can be defined as $T(\widetilde{\boldsymbol{X}}) = M_{\widetilde{\overline{X}}}$.*

Definition 5.5 *Let $\widetilde{\boldsymbol{X}} = (\widetilde{X}_1, \ldots, \widetilde{X}_n)$ be a* ***FRS****. For testing $\widetilde{H}_0$ versus $\widetilde{H}_1$ defined in Definition 5.3, a test $\varphi : (\mathcal{F}(\mathbb{R}))^n \to \{0, 1\}$ is defines as:*

$$\varphi(\widetilde{\boldsymbol{X}}) = \begin{cases} 1, & M_{\widetilde{\overline{X}}} \in \mathrm{R}, \\ 0, & M_{\widetilde{\overline{X}}} \notin \mathrm{R}, \end{cases} \tag{5.1}$$

where R *denotes a specified rejection region.*

Definition 5.6 *Let $\widetilde{\boldsymbol{X}} = (\widetilde{X}_1, \ldots, \widetilde{X}_n)$ be an observed* ***FRS*** *from* ***TNFRV****. Then, For testing $\widetilde{H}_0$ versus $\widetilde{H}_1$ defined in Definition 5.3, the p-value is defined by $p - value = \inf\{\delta \in (0, 1) : \varphi(\widetilde{\boldsymbol{X}}) = 1 | M_{\widetilde{\mu}} = M_{\widetilde{\mu}_0}\}$.*

Definition 5.7 *Let $\widetilde{\boldsymbol{X}} = (\widetilde{X}_1, \ldots, \widetilde{X}_n)$ be a* ***FRS****. For testing $\widetilde{H}_0$ versus $\widetilde{H}_1$ defined in Definition 5.3,*

1) The type I error is defined by $\alpha = E_{\widetilde{H}_0 : \widetilde{\mu} \simeq_M \widetilde{\mu}_0}(\varphi(\widetilde{\boldsymbol{X}}))$,

2) The type II error is defined by $\beta = E_{\widetilde{H}_1 : \widetilde{\mu} \simeq_M \widetilde{\mu}_1}(1 - \varphi(\widetilde{\boldsymbol{X}}))$,

3) Power of test at $\widetilde{\mu}$ is defined as $\pi = E_{\widetilde{H}_1 : \widetilde{\mu} \simeq_M \widetilde{\mu}^*}(\varphi(\widetilde{\boldsymbol{X}}))$ where $\widetilde{\mu}^* \neq \widetilde{\mu}_0$.

According to Lemma 5.2, we can construct some (exact) tests for testing one-sided and two-sided fuzzy hypotheses of fuzzy mean via the conventional likelihood-ratio-test for known or unknown σ^2. Thus, the conventional tests can be used to accept or reject the null hypothesis $\widetilde{H}_0$ as follows.

Lemma 5.3 *Let $\widetilde{X}$ be a* ***TNFRV*** *with fuzzy mean of $\widetilde{\mu}$ and known variance of σ^2 and $\widetilde{\boldsymbol{X}} = (\widetilde{X}_1, \ldots, \widetilde{X}_n)$ be a* ***FRS****. Then, at the significance level of δ:*

1)

$$\varphi(\widetilde{\boldsymbol{X}}) = \begin{cases} 1, & M_{\widetilde{\overline{X}}} \geq M_{\widetilde{\mu}_0} + Z_\delta \sigma / \sqrt{3n}, \\ 0, & M_{\widetilde{\overline{X}}} < M_{\widetilde{\mu}_0} + Z_\delta \sigma / \sqrt{3n}, \end{cases} \tag{5.2}$$

is a likelihood ratio test for testing $\widetilde{H}_0 : \widetilde{\mu} \preceq_M \widetilde{\mu}_0$ versus $\widetilde{H}_1 : \widetilde{\mu}_M \succ \widetilde{\mu}_0$.

2)

$$\varphi(\widetilde{\boldsymbol{X}}) = \begin{cases} 1, & M_{\widetilde{\overline{X}}} \leq M_{\widetilde{\mu}_0} - Z_{1-\delta} \sigma / \sqrt{3n}, \\ 0, & M_{\widetilde{\overline{X}}} > M_{\widetilde{\mu}_0} - Z_{1-\delta} \sigma / \sqrt{3n}, \end{cases} \tag{5.3}$$

is a likelihood ratio test for testing $\widetilde{H}_0 : \widetilde{\mu} \succeq_M \widetilde{\mu}_0$ versus $\widetilde{H}_1 : \widetilde{\mu} \prec_M \widetilde{\mu}_0$.

3)

$$\varphi(\widetilde{\boldsymbol{X}}) = \tag{5.4}$$

$$\begin{cases} 1, & M_{\widetilde{\overline{X}}} \geq M_{\widetilde{\mu}_0} + Z_{\delta/2}\sigma/\sqrt{3n} \text{ or } M_{\widetilde{\overline{X}}} \leq M_{\widetilde{\mu}_0} - Z_{\delta/2}\sigma/\sqrt{3n}, \\ 0, & M_{\widetilde{\overline{X}}} \in (M_{\widetilde{\mu}_0} - Z_{\delta/2}\sigma/\sqrt{3n}, M_{\widetilde{\mu}_0} + Z_{\delta/2}\sigma/\sqrt{3n}), \end{cases}$$

is a likelihood ratio test for testing $\widetilde{H}_0 : \widetilde{\mu} \simeq_M \widetilde{\mu}_0$ *versus* $\widetilde{H}_1 : \widetilde{\mu} \neq_M \widetilde{\mu}_0$,

where $P(Z > Z_\delta) = \delta$ *and* Z *follows the standard normal distribution.*

Proof *The proof is similar to the conventional methods used for one-sided hypothesis test based on normal distribution function.*

The procedure to accept or reject the fuzzy null hypothesis can be carried out via the p-value concept.

Lemma 5.4 *Let* $\widetilde{\boldsymbol{x}} = (\widetilde{x}_1, \ldots, \widetilde{x}_n)$ *be an observed **FRS** from **TNFRV** with fuzzy mean of* $\widetilde{\mu}$ *and known variance of* σ^2*. Then:*

1) For testing $\widetilde{H}_0 : \widetilde{\mu} \preceq_M \widetilde{\mu}_0$ *versus* $\widetilde{H}_1 : \widetilde{\mu} \succ_M \widetilde{\mu}_0$,

$$p_1 - value = 1 - F_Z\left(\frac{M_{\widetilde{\overline{x}}} - M_{\widetilde{\mu}_0}}{\sigma/\sqrt{3n}}\right). \tag{5.5}$$

2) For testing $\widetilde{H}_0 : \widetilde{\mu} \succeq_M \widetilde{\mu}_0$ *versus* $\widetilde{H}_1 : \widetilde{\mu} \prec_M \widetilde{\mu}_0$,

$$p_2 - value = F_Z\left(\frac{M_{\widetilde{\overline{x}}} - M_{\widetilde{\mu}_0}}{\sigma/\sqrt{3n}}\right). \tag{5.6}$$

3) For testing $\widetilde{H}_0 : \widetilde{\mu} \simeq_M \widetilde{\mu}_0$ *versus* $\widetilde{H}_1 : \widetilde{\mu} \neq_M \widetilde{\mu}_0$,

$$p - value = 2\min\{p_1 - value, p_2 - value\}. \tag{5.7}$$

where $\widetilde{\overline{x}} = (1/n) \oplus_{i=1}^{n} \widetilde{x}_i$ *and* F_Z *stands for the **CDF** of the standard normal distribution function.*

Proof *The proof is left as an exercise.*

We also have the following results.

Lemma 5.5 *Let* $\widetilde{X}$ *be a **TNFRV** and* $\widetilde{\boldsymbol{X}} = (\widetilde{X}_1, \ldots, \widetilde{X}_n)$ *be a **FRS** from* $N(\widetilde{\mu}, \sigma^2)$ *with unknown* σ^2*. Then, at the significance level of* δ*:*

1)

$$\varphi(\widetilde{\boldsymbol{X}}) = \begin{cases} 1, & M_{\widetilde{\overline{X}}} \geq M_{\widetilde{\mu}_0} + t_{3n-3,\delta} S_n/\sqrt{3n}, \\ 0, & M_{\widetilde{\overline{X}}} < M_{\widetilde{\mu}_0} + t_{3n-3,\delta} S_n/\sqrt{3n}, \end{cases} \tag{5.8}$$

is a likelihood ratio test for testing $\widetilde{H}_0 : \widetilde{\mu} \preceq_M \widetilde{\mu}_0$ *versus* $\widetilde{H}_1 : \widetilde{\mu} \succ_M \widetilde{\mu}_0$.

2)

$$\varphi(\widetilde{\boldsymbol{X}}) = \begin{cases} 1, & M_{\widetilde{\overline{X}}} \leq M_{\widetilde{\mu}_0} - t_{3n-3,1-\delta} S_n/\sqrt{3n}, \\ 0, & M_{\widetilde{\overline{X}}} > M_{\widetilde{\mu}_0} - t_{3n-3,1-\delta} S_n/\sqrt{3n}, \end{cases} \tag{5.9}$$

is a likelihood ratio test for testing $\widetilde{H}_0 : \widetilde{\mu} \succeq_M \widetilde{\mu}_0$ versus $\widetilde{H}_1 : \widetilde{\mu} \prec_M \widetilde{\mu}_0$.

3)

$$\varphi(\widetilde{\boldsymbol{X}}) = \tag{5.10}$$

$$\begin{cases} 1, & M_{\widetilde{\overline{X}}} \geq M_{\widetilde{\mu}_0} + t_{3n-3,\delta/2} S_n/\sqrt{3n} \text{ or } M_{\widetilde{\overline{X}}} \leq M_{\widetilde{\mu}_0} - t_{n-1,\delta/2} S_n/\sqrt{3n}, \\ 0, & M_{\widetilde{\overline{X}}} \in (M_{\widetilde{\mu}_0} - t_{3n-3,\delta/2} S_n/\sqrt{n}, M_{\widetilde{\mu}_0} + t_{n-1,\delta/2} S_n/\sqrt{3n}), \end{cases}$$

is a likelihood-ratio test for testing $\widetilde{H}_0 : \widetilde{\mu} \simeq_M \widetilde{\mu}_0$ versus $\widetilde{H}_1 : \widetilde{\mu} \neq_M \widetilde{\mu}_0$, where $P(t_\nu > t_{\nu,\delta/2}) = \delta$ and t_ν denotes the t-student distribution with ν degree of freedom.

Proof *The proof is left as an exercise.*

Lemma 5.6 *Let $\widetilde{x} = (\widetilde{x}_1, \ldots, \widetilde{x}_n)$ be an observed **FRS** from **TNFRV** with fuzzy mean $\widetilde{\mu}$ and unknown variance of σ^2. Then:*

1) For testing $\widetilde{H}_0 : \widetilde{\mu} \preceq_M \widetilde{\mu}_0$ versus $\widetilde{H}_1 : \widetilde{\mu} \succ_M \widetilde{\mu}_0$,

$$p_1 - value = 1 - F_{t_{3n-3}}\left(\frac{M_{\widetilde{\overline{x}}} - M_{\widetilde{\mu}_0}}{\sigma/\sqrt{3n}}\right). \tag{5.11}$$

2) For testing $\widetilde{H}_0 : \widetilde{\mu} \succeq_M \widetilde{\mu}_0$ versus $\widetilde{H}_1 : \widetilde{\mu} \prec_M \widetilde{\mu}_0$,

$$p_2 - value = F_{t_{3n-3}}\left(\frac{M_{\widetilde{\overline{x}}} - M_{\widetilde{\mu}_0}}{\sigma/\sqrt{3n}}\right). \tag{5.12}$$

3) For testing $\widetilde{H}_0 : \widetilde{\mu} \simeq_M \widetilde{\mu}_0$ versus $\widetilde{H}_1 : \widetilde{\mu} \neq_M \widetilde{\mu}_0$,

$$p - value = 2\min\{p_1 - value, p_2 - value\}, \tag{5.13}$$

*where $F_{t_{3n-3}}$ denotes the **CDF** of t-student distribution with $3n - 3$ degrees of freedom.*

Proof *The proof is left as an exercise.*

Therefore, let $\widetilde{\mathbf{x}} = (\widetilde{x}_1, \ldots, \widetilde{x}_n)$ be an observed **FRS** from **TNFRV** and $\delta \in (0, 1)$ be a specified significance level. For testing $\widetilde{H}_0$ versus $\widetilde{H}_1$ in Definition 5.3, we can reject the null hypothesis of $\widetilde{H}_0$ if $p - value < \delta$ otherwise; we accept it.

TABLE 5.1
Data set in Example 5.1.

$\tilde{x}_1 = (3, 3.5, 3.9)_T$	$\tilde{x}_2 = (1.6, 2.3, 2.6)_T$
$\tilde{x}_3 = (2.6, 3.7, 3.8)_T$	$\tilde{x}_4 = (2.6, 3.4, 3.7)_T$
$\tilde{x}_5 = (2.4, 3.8, 4.6)_T$	$\tilde{x}_6 = (2.1, 3.2, 3.8)_T$
$\tilde{x}_7 = (1.1, 2.5, 3.7)_T$	$\tilde{x}_8 = (2.5, 3.3, 3.8)_T$
$\tilde{x}_9 = (1.3, 2.6, 4)_T$	$\tilde{x}_{10} = (1.8, 2.8, 3.6)_T$

TABLE 5.2
The fuzzy failure time observations (in projection hours) for the 31 lamps in Example 5.2.

No.	$\tilde{x}_i$	No.	$\tilde{x}_i$	No.	$\tilde{x}_i$	No.	$\tilde{x}_i$
1	$(505, 520, 530)_T$	9	$(585, 500, 515)_T$	17	$(518, 550, 582)_T$	25	$(565, 582, 599)_T$
2	$(473, 482, 491)_T$	10	$(666, 698, 737)_T$	18	$(524, 534, 546)_T$	26	$(450, 473, 488)_T$
3	$(432, 444, 452)_T$	11	$(555, 584, 613)_T$	19	$(461, 485, 509)_T$	27	$(571, 581, 596)_T$
4	$(520, 540, 555)_T$	12	$(530, 560, 593)_T$	20	$(548, 558, 565)_T$	28	$(539, 554, 569)_T$
5	$(506, 527, 548)_T$	13	$(555, 570, 583)_T$	21	$(456, 474, 492)_T$	29	$(491, 507, 523)_T$
6	$(516, 532, 548)_T$	14	$(527, 540, 553)_T$	22	$(428, 445, 462)_T$	30	$(519, 530, 544)_T$
7	$(505, 518, 538)_T$	15	$(462, 474, 482)_T$	23	$(538, 555, 572)_T$	31	$(471, 486, 501)_T$
8	$(561, 580, 599)_T$	16	$(465, 471, 477)_T$	24	$(495, 505, 518)_T$	32	$(431, 445, 459)_T$

Example 5.1 *A survey showed the of time per week (in hour) spent by adults to exercise is less than* $(2.5, 3.5, 4.5)_T$. *The data set are shown in Table 5.1. To check it up, we selected 10 adult. Here, it was assumed that* $\sigma^2 = 1.5$. *At the significance level of* $\delta = 0.05$, *we wish to test:*

$$\begin{cases} \widetilde{H}_0: & \tilde{\mu} \succeq_M \tilde{\mu}_0 = (2, 3, 4)_T, \\ \widetilde{H}_1: & \tilde{\mu} \prec_M \tilde{\mu}_0. \end{cases}$$

For this purpose, one can verify that $\widetilde{\overline{x}} = (2.10, 3.11, 3.84)_T$ *and thus* $M_{\widetilde{\overline{x}}} = 3.0167$. *Since:*

$$\begin{aligned} 3.0167 = M_{\widetilde{\overline{x}}} &> M_{\tilde{\mu}_0} - Z_{1-\delta}\sigma/\sqrt{3n} \\ &= 3 - 1.645\sqrt{1.5/30} = 2.2632, \end{aligned}$$

we accept the null hypothesis at the significance level of $\delta = 0.05$.

Example 5.2 *In business and educational settings, computer presentations use liquid crystal display projectors. The most common failure mode of these projectors is their lamp failure. Manufacturers often include the expected lamp lifetime in their technical specification documents. Naturally, it was assumed that the projection time from each lamp used under normal operating conditions has a normal distribution. To test this claim, a large private university* *placed identical lamps in three projector models. The university staff recorded the number of projection hours (as measured by the projector) when each lamp burned out. The data set in Table 5.2 presents the fuzzy failure times (in projection hours) for the lamps. The collected data are reported by inexact quantities. In this regard, according to the proposed method, the observations are reported as **STFNs**. At the significance level of $\delta = 0.05$, we wish to test the following fuzzy hypotheses:*

$$\begin{cases} \widetilde{H}_0: & \tilde{\mu} \preceq_M \tilde{\mu}_0 = (490, 500, 510)_T, \\ \widetilde{H}_1: & \tilde{\mu} \succ_M \tilde{\mu}_0. \end{cases}$$

For this purpose, we have $\widetilde{\overline{x}} = (508.16, 525.13, 541.85)_T$ *and thus* $M_{\widetilde{\overline{x}}} = 525.047$. *According to Table 5.3, we get:*

$$\begin{aligned} s_n^2 &= \frac{1}{31}\sum_{i=1}^{32}\Big(\frac{(x_i^L - 508.16)^2 + (x_i - 525.13)^2 + (x_i^U - 541.85)^2}{3}\Big) \\ &= \frac{1}{31 \times 3}(2457.62 + 2797.08 + 3313.8) = 92.1344. \end{aligned}$$

This concludes that $M_{\widetilde{\mu}_0} - t_{93,0.05}S_n/\sqrt{90} = 501.67 + 1.6614(9.59867/\sqrt{96}) = 503.298$. *since*

$$M_{\widetilde{\overline{x}}} = 525.047 > 503.298,$$

we reject the null hypothesis at the significance level of $\delta = 0.05$.

Lemma 5.7 *Let* $\widetilde{X}$ *be a **TNFRV** and* $\widetilde{\boldsymbol{X}} = (\widetilde{X}_1, \ldots, \widetilde{X}_n)$ *be a **FRS** from* $N(\widetilde{\mu}, \sigma^2)$. *Then, at the significance level of* δ:

1)

$$\varphi(\widetilde{\boldsymbol{X}}) = \begin{cases} 1, & \frac{3(n-1)S_n^2}{\sigma_0^2} \geq \chi^2_{3(n-1),\delta}, \\ 0, & \frac{3(n-1)S_n^2}{\sigma_0^2} < \chi^2_{3(n-1),\delta}, \end{cases} \tag{5.14}$$

is a likelihood ratio test for testing $H_0: \sigma^2 = \sigma_0^2$ *versus* $H_1: \sigma^2 > \sigma_0^2$.

2)

$$\varphi(\widetilde{\boldsymbol{X}}) = \begin{cases} 1, & \frac{3(n-1)S_n^2}{\sigma_0^2} \leq \chi^2_{3(n-1),1-\delta}, \\ 0, & \frac{3(n-1)S_n^2}{\sigma_0^2} > \chi^2_{3(n-1),1-\delta}, \end{cases} \tag{5.15}$$

is a likelihood ratio test for testing $H_0: \sigma_0^2 = \sigma_0^2$ *versus* $H_1: \sigma_0^2 < \sigma_0^2$.

3)

$$\varphi(\widetilde{\boldsymbol{X}}) = \tag{5.16}$$

$$\begin{cases} 1, & \frac{3(n-1)S_n^2}{\sigma_0^2} \geq \chi^2_{3(n-1),\delta/2} \text{or} \frac{3(n-1)S_n^2}{\sigma_0^2} \leq \chi^2_{3(n-1),1-\delta/2}, \\ 0, & \frac{3(n-1)S_n^2}{\sigma_0^2} \in (\chi^2_{3(n-1),1-\delta/2}, \chi^2_{3(n-1),\delta/2}), \end{cases}$$

TABLE 5.3
Computing the elements of s_n^2 in Example 5.2.

i	$(\overline{x}_i^L - 508.16)$	$(\overline{x}_i - 525.13)$	$(\overline{x}_i^U - 541.85)$
1	3.156	5.125	11.844
2	35.156	43.125	50.844
3	76.156	81.125	89.844
4	11.844	14.875	13.156
5	2.156	1.875	6.156
6	7.8447	6.875	6.156
7	3.156	7.125	3.844
8	52.844	54.875	57.156
9	23.156	25.125	26.844
10	150.844	172.875	195.156
11	46.844	58.875	71.156
12	21.844	34.875	51.156
13	46.844	44.875	41.156
14	18.844	14.875	11.156
15	46.156	51.125	59.844
16	43.156	54.125	64.844
17	9.844	24.875	40.156
18	15.844	8.875	4.156
19	47.156	40.125	32.844
20	39.844	32.875	23.156
21	52.156	51.125	49.844
22	80.156	80.125	79.844
23	29.844	29.875	30.156
24	13.156	20.125	23.844
25	56.844	56.875	57.156
26	58.156	52.125	53.844
27	62.844	55.875	54.156
28	30.844	28.875	27.156
29	17.156	18.125	18.844
30	10.844	4.875	2.156
31	37.156	39.125	40.844
32	77.156	80.125	82.844

is a likelihood-ratio test for testing $H_0 : \sigma_0^2 = \sigma_0^2$ *versus* $H_1 : \sigma_0^2 \neq \sigma_0^2$, *where* $P(\chi^2_{3(n-1)} > \chi^2_{3(n-1),\delta}) = \delta$ *and* $\chi^2_{3(n-1)}$ *denotes the Chi-square distribution with* $3(n-1)$ *degree of freedom.*

Proof *The proof is a simple consequence of the classical cases.*

Lemma 5.8 *Let* $\widetilde{\boldsymbol{x}} = (\widetilde{x}_1, \ldots, \widetilde{x}_n)$ *be an observed **FRS** from **TNFRV** with fuzzy mean of* $\widetilde{\mu}$ *and known variance of* σ^2*. Then:*

1) For testing $H_0 : \sigma^2 = \sigma_0^2$ *versus* $H_1 : \sigma^2 > \sigma_0^2$,

$$p_1 - value = F_{\chi^2_{3(n-1)}}(\frac{3(n-1)s^2}{\sigma_0^2}). \quad (5.17)$$

TABLE 5.4
Data set in Example 5.3.

No.	$\widetilde{ph}$	No.	$\widetilde{ph}$
1	$(8.19, 9.45, 10.40)_T$	9	$(9.49, 11.08, 12.30)_T$
2	$(9.18, 9.9910.99)_T$	10	$(8.73, 10.38, 12.48)_T$
3	$(6.73, 9.29, 11.85)_T$	11	$(9.37, 11.62, 13.02)_T$
4	$(9.65, 11.66, 13.33)_T$	12	$(8.02, 11.31, 14.16)_T$
5	$(10.48, 12.16, 12.9)_T$	13	$(8.23, 10.52, 13.20)_T$
6	$(8.31, 10.18, 12.18)_T$	14	$(7.66, 9.69, 11.9)_T$
7	$(7.98, 9.84, 11.84)_T$	15	$(7.32, 10.23, 13.23)_T$
8	$(10.71, 11.46, 13.87)_T$	16	$(9.37, 10.47, 11.4)_T$

2) For testing $H_0 : \sigma^2 = \sigma_0^2$ versus $H_1 : \sigma^2 < \sigma_0^2$,

$$p_2 - value = 1 - F_{\chi^2_{3(n-1)}}(\frac{3(n-1)s^2}{\sigma_0^2}). \tag{5.18}$$

3) For testing $H_0 : \sigma^2 = \sigma_0^2$ versus $H_1 : \sigma^2 \neq \sigma_0^2$,

$$p - value = 2\min\{p_1 - value, p_2 - value\}. \tag{5.19}$$

*where $F_{\chi^2_{3(n-1)}}$ stands for the **CDF** of the Chi-square distribution function with $3(n-1)$ degrees of freedom.*

Proof *The proof is left for the reader.*

Example 5.3 *Consider a study on the presence of ph content in the basic solution of industrial dye production in a fuzzy environment. A random of size 16 ph content were selected randomly as shown in Table 5.4. At the significance level of $\delta = 0.05$, we want to examine the following fuzzy hypothesis:*

$$\begin{cases} \widetilde{H}_0 : & \tilde{\mu} \simeq_M \tilde{\mu}_0 = (8, 9.5, 11)_T, \\ \widetilde{H}_1 : & \tilde{\mu} \neq_M \tilde{\mu}_0. \end{cases}$$

For this purpose, one can finds that $\widetilde{\overline{x}} = (8.736, 10.583, 12.412)_T$ and $s_n^2 = 2.152$. Since:

$$\begin{aligned} 10.568 &= M_{\widetilde{\overline{x}}} \\ &> M_{\tilde{\mu}_0} + t_{0.025,42} S_n/\sqrt{45} \\ &= 9.67 + 2.0181\sqrt{2.152/45} = 10.1113, \end{aligned}$$

we reject the null hypothesis at the significance level of $\delta = 0.05$. Furthermore, consider the following fuzzy hypothesis about the population's variance:

$$\begin{cases} H_0 : & \sigma^2 = \sigma_0^2 = 2.5, \\ H_1 : & \sigma^2 \neq \sigma_0^2 = 2.5. \end{cases}$$

The test statistics is $(45 \times 2.152)/2.5 = 38.736$. *Since* $\chi^2_{42,0.975} = 25.9986$ *and* $\chi^2_{42,0.025} = 61.7767$, *it is seen that the test statistics belongs to* $(25.9986, 61.7767)$. *Therefore, we can accept the null hypothesis at the significance level of* $\delta = 0.05$.

5.1.2 Two-samples testing fuzzy hypotheses based on normal FRVs

Let $\widetilde{\boldsymbol{X}} = (\widetilde{X}_1, \widetilde{X}_2, \ldots, \widetilde{X}_{n_1})$ and $\widetilde{\boldsymbol{Y}} = (\widetilde{Y}_1, \widetilde{Y}_2, \ldots, \widetilde{Y}_{n_2})$ be two independent **FRSs** from $N(\widetilde{\mu}_1, \sigma_1^2)$ and $N(\widetilde{\mu}_2, \sigma_2^2)$. In this section, the one-sample fuzzy hypothesis tests will be developed to compare two populations.

Definition 5.8 *Let* $\widetilde{X}$ *and* $\widetilde{Y}$ *be two independent **FRVs** of* $N(\widetilde{\mu}_1, \sigma_1^2)$ *and* $N(\widetilde{\mu}_2, \sigma_2^2)$. *Then:*

1. *Any hypothesis of the form*
$$\widetilde{H} : \widetilde{\mu}_1 \simeq_M \widetilde{\mu}_2,$$
is called a simple fuzzy null hypothesis.

2. *Any hypothesis of the form*
$$\widetilde{H} : \widetilde{\mu}_1 \succeq_M \widetilde{\mu}_2,$$
is called a right one-sided fuzzy null hypothesis.

3. *Any hypothesis of the form*
$$\widetilde{H} : \widetilde{\mu}_1 \prec_M \widetilde{\mu}_2,$$
is called a left one-sided fuzzy hypothesis.

4. *Any hypothesis of the form*
$$\widetilde{H} : \widetilde{\mu}_1 \neq_M \widetilde{\mu}_2,$$
is called to be a two-sided fuzzy alternative hypothesis.

Definition 5.9 *Let* $\widetilde{\boldsymbol{X}} = (\widetilde{X}_1, \widetilde{X}_2, \ldots, \widetilde{X}_{n_1})$ *and* $\widetilde{\boldsymbol{Y}} = (\widetilde{Y}_1, \widetilde{Y}_2, \ldots, \widetilde{Y}_{n_2})$ *be two independent **FRSs** from* $N(\widetilde{\mu}_1, \sigma_1^2)$ *and* $N(\widetilde{\mu}_2, \sigma_2^2)$. *For testing* $\widetilde{H}_0$ *defined in Definition 5.8, the test statistic is defined as* $T(\widetilde{\boldsymbol{X}}, \widetilde{\boldsymbol{Y}}) = M_{\widetilde{\overline{X}}} - M_{\widetilde{\overline{Y}}}$.

Definition 5.10 *Let* $\widetilde{\boldsymbol{X}} = (\widetilde{X}_1, \widetilde{X}_2, \ldots, \widetilde{X}_{n_1})$ *and* $\widetilde{\boldsymbol{Y}} = (\widetilde{Y}_1, \widetilde{Y}_2, \ldots, \widetilde{Y}_{n_2})$ *be two independent **FRSs** from* $N(\widetilde{\mu}_1, \sigma_1^2)$ *and* $N(\widetilde{\mu}_2, \sigma_2^2)$. *For testing* $\widetilde{H}_0$ *versus* $\widetilde{H}_1$ *defined in Definition 5.9, the test* $\varphi : \mathcal{F}(\mathbb{R})\times \to \{0, 1\}$ *is defined as:*

$$\varphi(\widetilde{\boldsymbol{X}}, \widetilde{\boldsymbol{Y}}) = \begin{cases} 1, & M_{\widetilde{\overline{X}}} - M_{\widetilde{\overline{Y}}} \in \mathrm{R}, \\ 0, & M_{\widetilde{\overline{X}}} - M_{\widetilde{\overline{Y}}} \notin \mathrm{R}, \end{cases} \tag{5.20}$$

where R *denotes a specified rejection region.*

Therefore, applying the conventional likelihood-ratio tests for two independent random samples $M_{\widetilde{X}_1}, M_{\widetilde{X}_2}, ..., M_{\widetilde{X}_{n_1}}$ and $M_{\widetilde{Y}_1}, M_{\widetilde{Y}_2}, ..., M_{\widetilde{Y}_{n_2}}$, the following tests can be reached to compare the fuzzy means of two **TNFRVs**.

Lemma 5.9 *Let* $\widetilde{\boldsymbol{X}} = (\widetilde{X}_1, \widetilde{X}_2, \ldots, \widetilde{X}_{n_1})$ *and* $\widetilde{\boldsymbol{Y}} = (\widetilde{Y}_1, \widetilde{Y}_2, \ldots, \widetilde{Y}_{n_2})$ *be two independent* ***FRSs*** *from* $N(\widetilde{\mu}_1, \sigma_1^2)$ *and* $N(\widetilde{\mu}_2, \sigma_2^2)$ *where* σ_1^2 *and* σ_2^2 *are known quantities. Then:*

1)

$$\varphi(\widetilde{\boldsymbol{X}}, \widetilde{\boldsymbol{Y}}) = \begin{cases} 1, & M_{\widetilde{\overline{X}}} - M_{\widetilde{\overline{Y}}} \geq Z_\delta / \sqrt{\frac{\sigma_1^2}{3n_1} + \frac{\sigma_2^2}{3n_2}}, \\ 0, & M_{\widetilde{\overline{X}}} - M_{\widetilde{\overline{Y}}} < Z_\delta / \sqrt{\frac{\sigma_1^2}{3n_1} + \frac{\sigma_2^2}{3n_2}}, \end{cases} \tag{5.21}$$

is a likelihood-ratio test for testing $\widetilde{H}_0 : \widetilde{\mu}_1 \preceq_M \widetilde{\mu}_2$ *versus* $\widetilde{H}_1 : \widetilde{\mu}_1 \succ_M \widetilde{\mu}_2$, *at the significance level of* δ.

2)

$$\varphi(\widetilde{\boldsymbol{X}}, \widetilde{\boldsymbol{Y}}) = \begin{cases} 1, & M_{\widetilde{\overline{X}}} - M_{\widetilde{\overline{Y}}} \leq -Z_{1-\delta} / \sqrt{\frac{\sigma_1^2}{3n_1} + \frac{\sigma_2^2}{3n_2}}, \\ 0, & M_{\widetilde{\overline{X}}} - M_{\widetilde{\overline{Y}}} > -Z_{1-\delta} / \sqrt{\frac{\sigma_1^2}{3n_1} + \frac{\sigma_2^2}{3n_2}}, \end{cases} \tag{5.22}$$

is a likelihood-ratio test for testing $\widetilde{H}_0 : \widetilde{\mu}_1 \succeq_M \widetilde{\mu}_2$ *versus* $\widetilde{H}_1 : \widetilde{\mu}_1 \prec_M \widetilde{\mu}_2$, *at the significance level of* δ.

3)

$$\varphi(\widetilde{\boldsymbol{X}}, \widetilde{\boldsymbol{Y}}) = \begin{cases} 1, & |M_{\widetilde{\overline{X}}} - M_{\widetilde{\overline{Y}}}| \geq Z_{\delta/2} / \sqrt{\frac{\sigma_1^2}{3n_1} + \frac{\sigma_2^2}{3n_2}}, \\ 0, & |M_{\widetilde{\overline{X}}} - M_{\widetilde{\overline{Y}}}| < Z_{\delta/2} / \sqrt{\frac{\sigma_1^2}{3n_1} + \frac{\sigma_2^2}{3n_2}}, \end{cases} \tag{5.23}$$

is a likelihood-ratio test for testing $\widetilde{H}_0 : \widetilde{\mu} \simeq_M \widetilde{\mu}_0$ *versus* $\widetilde{H}_1 : \widetilde{\mu} \neq_M \widetilde{\mu}_0$, *at the significance level of* δ.

Proof *The proof is left for the reader.*

Lemma 5.10 *Let* $\widetilde{\boldsymbol{x}} = (\widetilde{x}_1, \widetilde{x}_2, \ldots, \widetilde{x}_{n_1})$ *and* $\widetilde{\boldsymbol{y}} = (\widetilde{y}_1, \widetilde{y}_2, \ldots, \widetilde{y}_{n_2})$ *be observed values of two independent* ***FRSs*** *from* $N(\widetilde{\mu}_1, \sigma_1^2)$ *and* $N(\widetilde{\mu}_2, \sigma_2^2)$ *where* σ_1^2 *and* σ_2^2 *are known. Then:*

1) For testing $\widetilde{H}_0 : \widetilde{\mu}_1 \preceq_M \widetilde{\mu}_2$ *versus* $\widetilde{H}_1 : \widetilde{\mu}_1 \succ_M \widetilde{\mu}_2$,

$$p_1 - value = 1 - F_Z\left(\frac{M_{\widetilde{\overline{x}}} - M_{\widetilde{\overline{y}}}}{\sqrt{\frac{\sigma_1^2}{3n_1} + \frac{\sigma_2^2}{3n_2}}}\right). \tag{5.24}$$

2) For testing $\widetilde{H}_0 : \widetilde{\mu}_1 \succeq_M \widetilde{\mu}_2$ *versus* $\widetilde{H}_1 : \widetilde{\mu}_1 \prec_M \widetilde{\mu}_2$,

$$p_2 - value = F_Z\left(\frac{M_{\widetilde{\overline{x}}} - M_{\widetilde{\overline{y}}}}{\sqrt{\frac{\sigma_1^2}{3n_1} + \frac{\sigma_2^2}{3n_2}}}\right). \tag{5.25}$$

3) For testing $\widetilde{H}_0 : \widetilde{\mu} \simeq_M \widetilde{\mu}_0$ *versus* $\widetilde{H}_1 : \widetilde{\mu} \neq_M \widetilde{\mu}_0$,

$$p - value = 2\min\{p_1 - value, p_2 - value\}, \tag{5.26}$$

where F_Z *is the* ***CDF*** *of standard normal distribution.*

Proof *The proof is left as an exercise.*

Lemma 5.11 *Let* $\widetilde{\boldsymbol{X}} = (\widetilde{X}_1, \widetilde{X}_2, \ldots, \widetilde{X}_{n_1})$ *and* $\widetilde{\boldsymbol{Y}} = (\widetilde{Y}_1, \widetilde{Y}_2, \ldots, \widetilde{Y}_{n_2})$ *be two independent* ***FRSs*** *from* $N(\widetilde{\mu}_1, \sigma^2)$ *and* $N(\widetilde{\mu}_2, \sigma^2)$ *with unknown variance* σ^2. *Then:*

1)

$$S_p^2 = \frac{(n_1 - 1)S_1^2 + (n_2 - 1)S_2^2}{n_1 + n_2 - 2} \rightarrow^{a.s.} \sigma^2 \text{ as } n_1, n_2 \rightarrow \infty.$$

2)

$$T = \frac{M_{\widetilde{\overline{X}}} - M_{\widetilde{\overline{Y}}} - (M_{\mu_1} - M_{\mu_2})}{S_P\sqrt{\frac{1}{3n_1} + \frac{1}{3n_2}}} \sim t_{3(n_1+n_2-2)}.$$

Proof *The proof can be easily verified since* $M_{\widetilde{\overline{X}}} - M_{\widetilde{\overline{Y}}} - (M_{\mu_1} - M_{\mu_2}) \sim N(0, \sigma^2(\frac{1}{3n_1} + \frac{1}{3n_2}))$ *and* $3(n_1+n_2-2)S_p^2/\sigma^2 \sim \chi^2_{3(n_1+n_2-2)}$ *where* χ^2_ν *denotes the chi-square distribution with* ν *degrees of freedom.*

Lemma 5.12 *Let* $\widetilde{\boldsymbol{X}} = (\widetilde{X}_1, \widetilde{X}_2, \ldots, \widetilde{X}_{n_1})$ *and* $\widetilde{\boldsymbol{Y}} = (\widetilde{Y}_1, \widetilde{Y}_2, \ldots, \widetilde{Y}_{n_2})$ *be two independent* ***FRSs*** *from* $N(\widetilde{\mu}_1, \sigma^2)$ *and* $N(\widetilde{\mu}_2, \sigma^2)$ *with unknown variance* σ^2. *Then:*

1)

$$\varphi(\widetilde{\boldsymbol{X}}, \widetilde{\boldsymbol{Y}}) = \begin{cases} 1, & M_{\widetilde{\overline{X}}} - M_{\widetilde{\overline{Y}}} \geq t_{\delta,3(n_1+n_2-2)} S_p / \sqrt{\frac{1}{3n_1} + \frac{1}{3n_2}}, \\ 0, & M_{\widetilde{\overline{X}}} - M_{\widetilde{\overline{Y}}} < t_{\delta,3(n_1+n_2-2)} S_p / \sqrt{\frac{1}{3n_1} + \frac{1}{3n_2}}, \end{cases} \tag{5.27}$$

is a likelihood-ratio test for testing $\widetilde{H}_0 : \widetilde{\mu}_1 \preceq_M \widetilde{\mu}_2$ *versus* $\widetilde{H}_1 : \widetilde{\mu}_1 \succ_M \widetilde{\mu}_2$, *powerful test at the significance level of* δ.

2)

$$\varphi(\widetilde{\boldsymbol{X}}, \widetilde{\boldsymbol{Y}}) = \begin{cases} 1, & M_{\widetilde{\overline{X}}} - M_{\widetilde{\overline{Y}}} \leq -t_{1-\delta,3(n_1+n_2-2)} S_p / \sqrt{\frac{1}{3n_1} + \frac{1}{3n_2}}, \\ 0, & M_{\widetilde{\overline{X}}} - M_{\widetilde{\overline{Y}}} > -t_{1-\delta,3(n_1+n_2-2)} S_p / \sqrt{\frac{1}{3n_1} + \frac{1}{3n_2}}, \end{cases} \tag{5.28}$$

is a likelihood-ratio test for testing $\widetilde{H}_0 : \widetilde{\mu}_1 \succeq_M \widetilde{\mu}_2$ *versus* $\widetilde{H}_1 : \widetilde{\mu}_1 \prec_M \widetilde{\mu}_2$, *at the significance level of* δ.

3)

$$\varphi(\widetilde{\mathbf{X}}, \widetilde{\mathbf{Y}}) = \begin{cases} 1, & |M_{\widetilde{\overline{X}}} - M_{\widetilde{\overline{Y}}}| \geq t_{\delta/2,3(n_1+n_2-2)} S_p/\sqrt{\frac{1}{n_1} + \frac{1}{n_2}}, \\ 0, & |M_{\widetilde{\overline{X}}} - M_{\widetilde{\overline{Y}}}| < t_{\delta/2,3(n_1+n_2-2)} S_p/\sqrt{\frac{1}{3n_1} + \frac{1}{3n_2}}, \end{cases} \tag{5.29}$$

is a likelihood-ratio test for testing $\widetilde{H}_0 : \widetilde{\mu} \simeq_M \widetilde{\mu}_0$ versus $\widetilde{H}_1 : \widetilde{\mu} \neq_M \widetilde{\mu}_0$, at the significance level of δ.

Proof *The proof is left for the reader.*

Remark 5.1 *In cases where the populations are not normal and σ_1^2 and σ_2^2 are unknown if $n_1 + n_2 \geq 30$ then we could replace σ_1^2 and σ_2^2 by the sample variances of S_1^2 and S_2^2.*

Lemma 5.13 *Let $\widetilde{\boldsymbol{x}} = (\widetilde{x}_1, \widetilde{x}_2, \ldots, \widetilde{x}_{n_1})$ and $\widetilde{\boldsymbol{y}} = (\widetilde{y}_1, \widetilde{y}_2, \ldots, \widetilde{y}_{n_2})$ be observed values of two independent **FRSs** from $N(\widetilde{\mu}_1, \sigma^2)$ and $N(\widetilde{\mu}_2, \sigma^2)$ where σ^2 is unknown. Then:*

1) For testing $\widetilde{H}_0 : \widetilde{\mu}_1 \preceq_M \widetilde{\mu}_2$ versus $\widetilde{H}_1 : \widetilde{\mu}_1 \succ_M \widetilde{\mu}_2$,

$$p_1 - value = 1 - F_{t_{3(n_1+n_2-2)}}\left(\frac{M_{\widetilde{\overline{x}}} - M_{\widetilde{\overline{y}}}}{S_P\sqrt{\frac{1}{3n_1} + \frac{1}{3n_2}}}\right). \tag{5.30}$$

2) For testing $\widetilde{H}_0 : \widetilde{\mu}_1 \succeq_M \widetilde{\mu}_2$ versus $\widetilde{H}_1 : \widetilde{\mu}_1 \prec_M \widetilde{\mu}_2$,

$$p_2 - value = F_{t_{3(n_1+n_2-2)}}\left(\frac{M_{\widetilde{\overline{x}}} - M_{\widetilde{\overline{y}}}}{S_P\sqrt{\frac{1}{3n_1} + \frac{1}{3n_2}}}\right). \tag{5.31}$$

3) For testing $\widetilde{H}_0 : \widetilde{\mu} \simeq_M \widetilde{\mu}_0$ versus $\widetilde{H}_1 : \widetilde{\mu} \neq_M \widetilde{\mu}_0$,

$$p - value = 2\min\{p_1 - value, p_2 - value\}. \tag{5.32}$$

Proof *The proof is left for the reader.*

We also have the following results.

Lemma 5.14 *Let $\widetilde{\mathbf{X}} = (\widetilde{X}_1, \widetilde{X}_2, \ldots, \widetilde{X}_{n_1})$ and $\widetilde{\mathbf{Y}} = (\widetilde{Y}_1, \widetilde{Y}_2, \ldots, \widetilde{Y}_{n_2})$ be two independent **FRSs** from $N(\widetilde{\mu}_1, \sigma_1^2)$ and $N(\widetilde{\mu}_2, \sigma_2^2)$. Then:*

1)

$$\varphi(\widetilde{\mathbf{X}}, \widetilde{\mathbf{Y}}) = \begin{cases} 1, & \frac{S_1^2}{S_2^2} > F_{\delta,3(n_1-1),3(n_2-1)}, \\ 0, & \frac{S_1^2}{S_2^2} \leq F_{\delta,3(n_1-1),3(n_2-1)}, \end{cases} \tag{5.33}$$

is a likelihood-ratio test for testing $H_0 : \sigma_1^2 \leq \sigma_2^2$ versus $H_1 : \sigma_1^2 > \sigma_2^2$, at the significance level of δ.

2)

$$\varphi(\widetilde{\boldsymbol{X}}, \widetilde{\boldsymbol{Y}}) = \begin{cases} 1, & \frac{S_1^2}{S_2^2} < F_{1-\delta,3(n_1-1),3(n_2-1)}, \\ 0, & \frac{S_1^2}{S_2^2} \geq F_{1-\delta,3(n_1-1),3(n_2-1)}, \end{cases} \tag{5.34}$$

is likelihood-ratio test for testing $H_0 : \sigma_1^2 \geq \sigma_2^2$ versus $H_1 : \sigma_1^2 < \sigma_2^2$, at the significance level of δ.

3)

$$\varphi(\widetilde{\boldsymbol{X}}, \widetilde{\boldsymbol{Y}}) = \begin{cases} 1, & \frac{S_1^2}{S_2^2} > F_{\delta/2,3(n_1-1),3(n_2-1)} \text{ or } \frac{S_1^2}{S_2^2} < F_{1-\delta/2,3(n_1-1),3(n_2-1)}, \\ 0, & \frac{S_1^2}{S_2^2} \in [F_{1-\delta/2,3(n_1-1),3(n_2-1)}, F_{\delta/2,3(n_1-1),3(n_2-1)}], \end{cases} \tag{5.35}$$

is a likelihood-ratio test for testing $H_0 : \sigma_1^2 = \sigma_2^2$ versus $H_1 : \sigma_1^2 \neq \sigma_2^2$, at the significance level of δ,

where F_δ, ν_1, ν_2 represents the quantiles of F-fisher distribution function with ν_1 and ν_1 degrees of freedoms.

Lemma 5.15 *Let $\widetilde{\boldsymbol{X}} = (\widetilde{X}_1, \widetilde{X}_2, \ldots, \widetilde{X}_{n_1})$ and $\widetilde{\boldsymbol{Y}} = (\widetilde{Y}_1, \widetilde{Y}_2, \ldots, \widetilde{Y}_{n_2})$ be two independent **FRSs** from $N(\widetilde{\mu}_1, \sigma_1^2)$ and $N(\widetilde{\mu}_2, \sigma_2^2)$. Then:*

1) For testing $H_0 : \sigma_1^2 \leq \sigma_2^2$ versus $H_1 : \sigma_1^2 > \sigma_2^2$,

$$p_1 - value = 1 - F_{3(n_1-1),3(n_1-1)}(\frac{s_1^2}{s_2^2}). \tag{5.36}$$

2) For testing $H_0 : \sigma_1^2 \geq \sigma_2^2$ versus $H_1 : \sigma_1^2 < \sigma_2^2$,

$$p_2 - value = F_{3(n_1-1),3(n_1-1)}(\frac{s_1^2}{s_2^2}). \tag{5.37}$$

3) For testing $H_0 : \sigma_1^2 = \sigma_2^2$ versus $H_1 : \sigma_1^2 \neq \sigma_2^2$,

$$p - value = 2\min\{p_1 - value, p_2 - value\}, \tag{5.38}$$

where s_1^2 and s_2^2 represent the observed values of S_1^2 and S_2^2, respectively.

Example 5.4 *A psychologist studied the degree of happiness in people at various stages in life. His measure of general happiness varies from 0 to 60. In a certain study, he compared the happiness of married and single men aged 25. Each response was reported in **TFNs** as shown in Table 5.5. We wish to test if singles are happier than marries i.e.,*

$$\begin{cases} \widetilde{H}_0 : & \widetilde{\mu}_1 \simeq_M \widetilde{\mu}_2, \\ \widetilde{H}_1 : & \widetilde{\mu}_1 \neq_M \widetilde{\mu}_2. \end{cases}$$

Here, it is assumed that the variances of two populations are known values as $\sigma_1^2 = 3$ and $\sigma_2^2 = 2$. From Table 5.6, one can find $M_{\widetilde{\overline{x}}} = (\sum_{i=1}^{7}(x_i^L + x_i +$

TABLE 5.5
Data set in Example 5.4.

Married ($\widetilde{x}$)	Single ($\widetilde{y}$)
$(53, 55, 59)_T$	$(52, 54, 56)_T$
$(55, 59, 63)_T$	$(47, 49, 52)_T$
$(54, 58, 61)_T$	$(55, 56, 57)_T$
$(45, 47, 52)_T$	$(49, 53, 55)_T$
$(48, 50, 55)_T$	$(48, 52, 56)_T$
$(54, 57, 58)_T$	$(45, 48, 52)_T$
$(43, 46, 50)_T$	$(44, 45, 47)_T$
–	$(48, 51, 52)_T$

TABLE 5.6
Data set in Example 5.5

No.	x_i	y_i	No.	x_i	y_i	No.	x_i	y_i
1	65	72	11	83	79	21	68	72
2	82	80	12	68	71	22	73	67
3	75	70	13	75	72	23	70	65
4	74	73	14	83	81	24	78	73
5	73	67	15	71	68	25	71	68
6	67	70	16	86	80	26	66	69
7	80	72	17	75	81	27	82	79
8	79	74	18	83	78	28	83	76
9	77	70	19	82	81	29	84	83
10	69	82	20	74	78	30	64	68

$x_i^U))/21 = 19.905$ *and* $M_{\widetilde{y}} = (\sum_{i=1}^{8} y_i^L + y_i + y_i^U))/24 = 18.667$. *Therefore, the observed test statistics is* $|M_{\widetilde{x}} - M_{\widetilde{y}}| = 1.238$ *while the critical value is* $1.96\sqrt{\frac{3}{21} + \frac{2}{24}} = 0.932$. *We must therefore conclude that the null hypothesis* $\widetilde{H}_0$ *should be rejected at the significance level of* $\delta = 0.05$.

Example 5.5 *A random sample of twelve identical twins underwent psychological tests to measure the amount of aggressiveness. We are interested in comparing the twins to see if the firstborn twin tends to be more aggressive than the other. Assume that, because of some limitations in psychological measurements, the results of evaluations are reported as* ***TFNs*** *for the first as* $(0.99x, x, 1.01x)_T$ *and second-born as* $(0.98y, y, 1.01y)_T$ *where* x_i *and* y_i *are shown in Table 5.6. We want to test is there a significant difference between the aggressiveness of the twins at* $\delta = 0.05$, *i.e.,*

$$\begin{cases} \widetilde{H}_0: & \widetilde{\mu}_1 \simeq_M \widetilde{\mu}_2, \\ \widetilde{H}_1: & \widetilde{\mu}_1 \neq_M \widetilde{\mu}_2. \end{cases}$$

TABLE 5.7
Data set in Example 5.6.

No.	$\widetilde{x}_i$	$\widetilde{y}_i$	No.	$\widetilde{x}_i$	$\widetilde{y}_i$
1	$(68, 73, 75)_T$	$(71, 75, 77)_T$	13	$(78, 79, 82)_T$	$(71, 75, 77)_T$
2	$(67, 69, 72)_T$	$(68, 71, 74)_T$	14	$(79, 81, 84)_T$	$(74, 78, 81)_T$
3	$(79, 84, 86)_T$	$(81, 83, 87)_T$	15	$(77, 80, 85)_T$	$(74, 77, 78)_T$
4	$(75, 78, 79)_T$	$(81, 84, 85)_T$	16	$(63, 67, 69)_T$	$(68, 70, 72)_T$
5	$(72, 75, 79)_T$	$(71, 73, 74)_T$	17	$(68, 70, 71)_T$	$(71, 73, 75)_T$
6	$(66, 71, 75)_T$	$(72, 75, 76)_T$	18	$(68, 73, 76)_T$	$(65, 68, 72)_T$
7	$(75, 77, 79)_T$	$(76, 79, 81)_T$	19	$(64, 67, 70)_T$	$(64, 66, 71)_T$
8	$(71, 74, 78)_T$	$(77, 79, 82)_T$	20	$(65, 69, 73)_T$	$(71, 74, 79)_T$
9	$(65, 66, 70)_T$	$(65, 68, 71)_T$	21	$(81, 82, 87)_T$	$(74, 78, 82)_T$
10	$(79, 82, 87)_T$	$(76, 80, 82)_T$	22	$(69, 73, 77)_T$	$(72, 75, 77)_T$
11	$(76, 78, 81)_T$	$(73, 78, 81)_T$	23	$(72, 75, 79)_T$	$(76, 78, 80)_T$
12	$(74, 77, 80)_T$	$(74, 76, 77)_T$	24	$(74, 76, 80)_T$	$(77, 78, 81)_T$
	-	-	25	$(77, 82, 86)_T$	$(79, 80, 84)_T$

The direct calculations revealed that $\widetilde{\overline{x}} = (74.58, 75.333, 79.1)_T$, $\widetilde{\overline{y}} = (70.268, 73.966, 74.706)_T$, *and* $s_1^2 = (41.3219 + 42.16091 + 46.4824)/3 = 43.3217$ *and* $s_2^2 = (26.0469 + 28.8609 + 29.4410)/3 = 28.1163$. *Therefore, the observed value of the test statistics can be evaluated as* $M_{\widetilde{\overline{x}}} - M_{\widetilde{\overline{y}}} = 76.3377 - 72.98 = 3.3577$. *Moreover, for* $\delta = 0.05$, *the critical value is* $1.96\sqrt{\frac{43.3217}{90} + \frac{28.1163}{90}} = 1.74622$. *Because the observed test statistic falls in the rejection region we reject the null hypothesis for* $\delta = 0.05$.

Example 5.6 *A marketing department for a tire and rubber company laboratory tested two new formulations of A and B. Six cars were fitted with formulation A tires on the left side and formulation B tires on the right side. Tires in each side were front-rear switched every month so that there was no preferential roadside treatment. The tread lifetime (1000 miles) of each 24 tires was measured and recorded as shown in Table 5.7. First, we wish to test that two new formulations of A and B have the same variances:*

$$\begin{cases} \widetilde{H}_0: & \sigma_1^2 = \sigma_2^2, \\ \widetilde{H}_1: & \sigma_1^2 \neq \sigma_1^2. \end{cases}$$

According to Table 5.8, we get:

$$\begin{aligned} s_1^2 &= \frac{1}{3}\sum_{i=1}^{25}\left(\frac{(\overline{x}_i^L - \overline{x}^L)^2 + (\overline{x}_i - \overline{x})^2 + (\overline{x}_i^U - \overline{x}^U)^2}{24}\right) \\ &= \frac{29.91 + 27.78 + 31.25}{3} = 29.65, \end{aligned}$$

and

$$s_2^2 = \frac{1}{3}\sum_{i=1}^{25}\left(\frac{(\overline{y}_i^L - \overline{y}^L)^2 + (\overline{y}_i - \overline{y})^2 + (\overline{y}_i^U - \overline{y}^U)^2}{24}\right)$$
$$= \frac{1}{24}\left(\frac{22.107 + 20.657 + 19.69}{3}\right) = 20.818.$$

Thus $s_1^2/s_2^2 = 1.424$. *For* $\delta = 0.05$, *the critical values are* $F_{0.025,72,72} = 1.59$ *and* $F_{0.975,72,72} = 0.63$. *Since* $s_1^2/s_2^2 \in [0.63, 1.59]$ *we accept the null hypothesis* $\widetilde{H}_0 : \sigma_1^2 = \sigma_2^2$. *Thus, we can assume that two formulations of* A *and* B *have the same variances for* $\delta = 0.05$. *At the significance level of* $\delta = 0.05$, *now we want to test whether there is a significant difference between two fuzzy means of formulations:*

$$\begin{cases} \widetilde{H}_0 : & \widetilde{\mu}_1 \simeq_M \widetilde{\mu}_2, \\ \widetilde{H}_1 : & \widetilde{\mu}_1 \neq_M \widetilde{\mu}_2. \end{cases}$$

Further $\widetilde{\overline{x}} = (72.08, 75.12, 75.36)_T$, $\widetilde{\overline{y}} = (72.76, 75.64, 84.27)_T$, *and and* $s_p^2 = 5.212$ *from Table 5.8. Therefore,* $M_{\widetilde{\overline{x}}} = 75.5$ *and* $M_{\widetilde{\overline{y}}} = 75.547$. *It can be seen that:*

$$0.347 = |M_{\widetilde{\overline{x}}} - M_{\widetilde{\overline{y}}}| < 1.9766 \times \sqrt{5.212} \times \sqrt{\frac{1}{75} + \frac{1}{75}} = 0.7368.$$

At the significance level of $\delta = 0.05$, *therefore we can accept that there is no significant difference between the fuzzy mean of two populations.*

5.2 Statistical Non-parametric Tests Based on FRVs

Several conditions of validity should be met so that the result of a parametric test is reliable. In contrast, non-parametric tests are methods of statistical analysis whose distribution does not have to meet the required assumptions [115, 53]. In this section, some common non-parametric tests will be developed based on **FRVs**.

5.2.1 Sign test

Suppose that random sample of $\mathbf{X} = (X_1, X_2, \ldots, X_n)$ is drawn from a continuous **CDF** F_X with an unknown unique median of M. The hypotheses to be tested concern statements about the population median as $H_0 : M = M_0$ against $H_1 : M > M_0$ (or against $H_1 : M < M_0$, or $H_1 : M \neq M_0$), where M_0 is a specified value. The total number of observations greater than M_0 are used as a test statistic to test the validity of the hypothesis H_0 against H_1.

TABLE 5.8
Computing the elements of s^2 in Example 5.6.

i	$(\overline{x}_i^L - \overline{x}^L)$	$(\overline{x}_i - \overline{x})$	$(\overline{x}_i^U - \overline{x}^U)$	$(\overline{y}_i^L - \overline{y}^L)$	$(\overline{y}_i - \overline{y})$	$(\overline{y}_i^U - \overline{y}^U)$
1	4.08	2.12	3.4	1.76	0.64	1.24
2	5.08	6.12	6.4	4.76	4.64	4.24
3	6.92	8.88	7.6	8.24	7.36	8.76
4	2.92	2.88	0.6	8.24	8.36	6.76
5	0.08	0.12	0.6	1.76	2.64	4.24
6	6.08	4.12	3.4	0.76	0.64	2.24
7	2.92	1.88	0.6	3.24	3.36	2.76
8	1.08	1.12	0.4	4.24	3.36	3.76
9	7.08	9.12	8.4	7.76	7.64	7.24
10	6.92	6.88	8.6	3.24	4.36	3.76
11	3.92	2.88	2.6	0.24	2.36	2.76
12	1.92	1.88	1.6	1.24	0.36	1.24
13	5.92	3.88	3.6	1.76	0.64	1.24
14	6.92	5.88	5.6	1.24	2.36	2.76
15	4.92	4.88	6.6	1.24	1.36	0.24
16	9.08	8.12	9.4	6.76	5.64	6.24
17	4.08	5.12	7.4	1.76	2.64	3.24
18	4.08	2.12	2.4	7.76	7.64	6.24
19	8.08	8.12	8.4	8.76	9.64	7.24
20	7.08	6.12	5.4	1.76	1.64	0.76
21	8.92	6.88	8.6	1.24	2.36	3.76
22	3.08	2.12	1.4	0.76	0.64	1.24
23	0.08	0.12	0.6	3.24	2.36	1.76
24	1.92	0.88	1.6	4.24	2.36	2.76
25	4.92	6.88	7.6	6.24	4.36	5.76
s_i^2	29.91	27.78	31.25	22.107	20.657	19.69

The test statistic is:

$$T(\mathbf{X}) = \sum_{i=1}^{N} I(X_i > M_0).$$

Note that $T(\mathbf{X})$ counts the number of observations from the real-valued random sample $X_1, \ldots, X_n$, which exceed M_0 and reject H_0 in favor of H_1 if $T(\mathbf{X})$ is large enough. At a nominal significance level of δ, the null hypothesis of H_0 against H_1 is rejected if $t \in R_\delta$, otherwise, it is accepted (i.e. $t(\mathbf{X}) \notin R_\delta$) where t is the observed value of $T(\mathbf{X})$. The corresponding critical values for testing the null hypothesis of $H_0 : M = M_0$ against the alternative hypotheses are summarized in Table 5.9 [54]. Now, a procedure is suggested for the Sign test based on **FRV**s.

Definition 5.11 *Let $\widetilde{X}$ be a **FRV** with **FCDF** of $\widetilde{F}_{\widetilde{X}}$. The fuzzy median of the population is defined to be a **FN** with the following α-values:*

$$\widetilde{M}_\alpha = \inf\{x : (\widetilde{F}_{\widetilde{X}}(x))_{1-\alpha} \geq 0.5\}.$$

TABLE 5.9
The critical values for testing H_0 versus H_1 for the Sign test.

H_0 :	H_1 :	Critical region
$M = M_0$	$M > M_0$	$R^1_\delta = \{c^1_\delta, c^1_\delta + 1, ..., n\}$ where $c^1_\delta = \min\{k : \delta \geq \sum_{i=k}^{n} \binom{n}{i}(0.5)^n\}$
$M = M_0$	$M < M_0$	$R^2_\delta = \{0, 1, 2, ..., c^1_\delta\}$ where $c^2_\delta = \max\{k : \delta \leq \sum_{i=k}^{n} \binom{n}{i}(0.5)^n\}$
$M = M_0$	$M \neq M_0$	$R^1_{\delta/2} \cup R^2_{\delta/2}$

Example 5.7 *Let $\widetilde{X} = ((1-a)X, X, (1+a)X)_T$, where $P(X > 0) = 1$ and $a \in (0,1)$. Therefore,*

$$(\widetilde{F}_{\widetilde{X}}(x))_{1-\alpha} = F_X(x + a(1-2\alpha)).$$

This concludes that:

$$\begin{aligned}\widetilde{M}_\alpha &= \inf\{x : F_X(x + a(1-2\alpha)) \geq 0.5\}\\ &= \inf\{x : x \geq F_X^{-1}(0.5) - a(1-2\alpha)\} = M_X - a(1-2\alpha),\end{aligned}$$

where M_X is the median of X. Therefore, $\widetilde{M} = (M_X - a, M_X, M_X + a)_T$. For instance, assume that X is distributed according to an exponential distribution with the mean of λ. Therefore, the fuzzy median of $\widetilde{X}$ is a ***STFN*** *as $\widetilde{M} = (\lambda \ln 2 - 0.1, \lambda \ln 2, \lambda \ln 2 + 0.1)_T$.*

Definition 5.12 *Let $\widetilde{M}$ denotes the fuzzy median of a* ***FRV****. The fuzzy hypothesis relevant to $\widetilde{M}$ are defined as follows:*

$$(1)\begin{cases}\widetilde{H}_0 : \widetilde{M} \preceq_{P_d} \widetilde{M}_0 & \equiv \quad P_d(\widetilde{M} \preceq_M \widetilde{M}_0) \geq 0.5,\\ \widetilde{H}_1 : \widetilde{M} \succ_{P_d} \widetilde{M}_0 & \equiv \quad P_d(\widetilde{M} \preceq \widetilde{M}_0) < 0.5 \equiv P_d(\widetilde{M} \succ \widetilde{M}_0) > 0.5,\end{cases} \tag{5.39}$$

$$(2)\begin{cases}\widetilde{H}_0 : \widetilde{M} \succeq_{P_d} \widetilde{M}_0 & \equiv \quad P_d(\widetilde{M} \succeq \widetilde{M}_0) \geq 0.5,\\ \widetilde{H}_1 : \widetilde{M} \prec_{P_d} \widetilde{M}_0 & \equiv \quad P_d(\widetilde{M} \succeq \widetilde{M}_0) < 0.5 \equiv P_d(\widetilde{M} \prec \widetilde{M}_0) > 0.5,\end{cases} \tag{5.40}$$

$$(3)\begin{cases}\widetilde{H}_0 : \widetilde{M} =_{P_d} \widetilde{M}_0 & \equiv \quad P_d(\widetilde{M} \succeq \widetilde{M}_0) = P_d(\widetilde{M} \preceq \widetilde{M}_0) = 0.5,\\ \widetilde{H}_1 : \widetilde{M} \neq_{P_d} \widetilde{M}_0 & \equiv \quad P_d(\widetilde{M} \succ \widetilde{M}_0) > 0.5 \text{ or } P_d(\widetilde{M} \prec \widetilde{M}_0) > 0.5.\end{cases} \tag{5.41}$$

For extending the classical test statistic of $T(\mathbf{X})$ to fuzzy data and fuzzy hypotheses, the test statistic can be defined as the total number of the imprecise observations of $\widetilde{x}_1, \ldots, \widetilde{x}_N$ that exceed $\widetilde{M}_0$. The expression 'larger than' is not univocal in a fuzzy environment, so we apply the proposed ranking method for determining the values of $\widetilde{x}_1, \ldots, \widetilde{x}_n$, for which the expression '$\widetilde{x}_i$ is larger than $\widetilde{M}_0$' are valid.

Definition 5.13 *Let $\widetilde{x}_1, \ldots, \widetilde{x}_n$ be an observed **FRS**. The fuzzy test statistics is defined as a fuzzy set with the following membership function:*

$$\widetilde{t}(t) = \sup\{\alpha \in [0,1] : t \in \widetilde{t}[\alpha]\},$$

$$\begin{aligned}\widetilde{t}[\alpha] = \Big[& \inf_{\beta \in [\alpha/2, 1-\alpha/2]} \sum_{i=1}^{n} I(P_d(\widetilde{x}_i \succ \widetilde{M}_0) \geq 1 - \beta/2), \\ & \sup_{\beta \in [\alpha/2, 1-\alpha/2]} \sum_{i=1}^{n} I(P_d(\widetilde{x}_i \succ \widetilde{M}_0) \geq 1 - \beta/2)\Big] \\ = & \Big[\sum_{i=1}^{n} I(P_d(\widetilde{x}_i \succ \widetilde{M}_0) \geq 1 - \alpha/2), \sum_{i=1}^{n} I(P_d(\widetilde{x}_i \succ \widetilde{M}_0) \geq \alpha/2)\Big].\end{aligned} \tag{5.42}$$

It can be shown that $\widetilde{t}$ is a normal discrete fuzzy set.

Since the observed test statistics is a fuzzy quantity, at a given significance level, it is natural to get a fuzzy set (fuzzy test) to accept or reject $\widetilde{H}_0$ as

$$\widetilde{\varphi}_\delta = \{\frac{Accept\ \widetilde{H}_0}{D_A}, \frac{Reject\ \widetilde{H}_0}{D_R}\},$$

where $D_A, D_R \in [0,1]$. In such case, we are not convinced completely neither to reject nor to accept $\widetilde{H}_0$ but we could only indicate a degree of conviction that one should reject or accept the null hypothesis $\widetilde{H}_0$. For this purpose, the measure of belonging introduced in Definition 1.7 can be used.

Definition 5.14 *Let $\widetilde{T}$ be a fuzzy statistics based on an observed **FRS** $(\widetilde{x}_1, \ldots, \widetilde{x}_n)$. At the significance level of δ, we reject $\widetilde{H}_0$ if $D_R \geq D_A$ and we accept it if $D_A > D_R$ in which*

1) For testing $\widetilde{H}_0 : \widetilde{M} \preceq_{P_d} \widetilde{M}_0$ versus $\widetilde{H}_1 : \widetilde{M} \succ_{P_d} \widetilde{M}_0$,

$$D_R = d(\widetilde{t} \in R_\delta^1),\ D_A = 1 - D_R,$$

2) For testing $\widetilde{H}_0 : \widetilde{M} \succeq_{P_d} \widetilde{M}_0$ versus $\widetilde{H}_1 : \widetilde{M} \prec_{P_d} \widetilde{M}_0$,

$$D_R = d(\widetilde{t} \in R_\delta^2),\ D_A = 1 - D_R,$$

3) For testing $\widetilde{H}_0 : \widetilde{M} =_{P_d} \widetilde{M}_0$ versus $\widetilde{H}_1 : \widetilde{M} \neq_{P_d} \widetilde{M}_0$,

$$D_R = d(\widetilde{t} \in R_{\delta/2}^1 \cup R_{\delta/2}^2),\ D_A = 1 - D_R,$$

where

$$d(\widetilde{A} \in I) = \frac{\sum_{x \in I} \widetilde{A}(x)}{\sum_x \widetilde{A}(x)}.$$

denotes the the degree that $\widetilde{A}$ belongs to I.

TABLE 5.10
Preference degrees of observed **FRS** over $\widetilde{M}_0$ in Example 5.8.

j	1	2	3	4	5	6	7	8	9	10
$P_d(\widetilde{x}_i \succ \widetilde{M}_0)$	1	1	1	0	1	0	0.906	0.885	0.924	0

At the significance level of δ, according to the above definition, $\widetilde{H}_0$ is rejected if and only if $D_R > 0.5$; otherwise it is accepted. Contrary to the classical crisp test, therefore the proposed method does not lead to a binary decision to accept or reject the null hypothesis but to a fuzzy decision.

Example 5.8 *In Example 5.1, instead of a fuzzy mean, assume we wish to test the non-parametric test about the population median as:*

$$\begin{cases} \widetilde{H}_0: & \widetilde{M} \succeq_{P_d} \widetilde{M}_0 = (2,3,4)_T, \\ \widetilde{H}_1: & \widetilde{M} \prec_{P_d} \widetilde{M}_0. \end{cases}$$

The preference degree to which each observation is greater than $\widetilde{M}_0$ is evaluated in Table 5.10. Therefore, the α-cuts of the fuzzy test statistic can be evaluated by Eq. (5.42) as follows:

$$\widetilde{t}[\alpha] = \begin{cases} \{2,3,4,5,6,7\}, & 0 \leq \alpha \leq 0.149, \\ \{3,4,5,6,7\}, & 0.149 < \alpha \leq 0.188, \\ \{4,5,6,7\}, & 0.188 < \alpha \leq 0.230, \\ \{5,6,7\}, & 0.230 < \alpha \leq 0.677, \\ \{5,6\}, & 0.677 < \alpha \leq 0.682, \\ \{5\}, & 0.682 < \alpha \leq 1. \end{cases} \quad (5.43)$$

Therefore, the membership function of the fuzzy test statistic is:

$$\widetilde{t} = \{\frac{0.149}{2}, \frac{0.188}{3}, \frac{0.230}{4}, \frac{1}{5}, \frac{0.682}{6}, \frac{0.667}{7}\}.$$

Since $R^1_{0.05} = \{7,8,9,10\}$, we have $D_R = d(\widetilde{t} \in \{7,8,9,10\}) = \frac{0.677}{2.926} = 0.2314$. Since $D_R < 0.5$, we then reject $\widetilde{H}_0 : \widetilde{M} \succeq \widetilde{M}_0 = (2,3,4)_T$ at the significance level of $\delta = 0.05$.

5.2.2 Wilcoxon test

Suppose that two independent random samples of $X_1, \ldots, X_n$ and $Y_1, \ldots, Y_m$ are drawn from populations with the continuous cumulative distribution functions of F_X and F_Y, respectively. Many statistical tests applicable to the two-sample problem are based on the rank-sum statistics for the combined samples of size $N = n + m$ as $W_1 = X_1, W_2 = X_2, \ldots, W_n = X_n, W_{n+1} =$

$Y_1, \ldots, W_N = Y_m$. The alternative hypothesis is that the populations are of the same form but with a different measure of central tendency. This can be expressed symbolically by:

$$\begin{cases} H_0: \ F_Y(x) = F_X(x) \\ H_1: F_Y(x) = F_X(x-\theta), \quad \theta \neq 0. \end{cases}$$

Note that, the random variable of Y is stochastically larger than X when $\theta > 0$, and Y is stochastically smaller than X when $\theta < 0$. Thus, for example, when $\theta < 0$, the median of X (M_X) is larger than the median of Y (M_Y). The test statistic is given by:

$$W = \sum_{i=1}^{m}\sum_{j=1}^{N} I(X_i \geq W_j), \tag{5.44}$$

with probability mass function of

$$P(W = k) = \frac{w(k)}{\binom{N}{m}},$$

where $w(k)$ is the number of arrangements of m X and n Y random variables such that $W = k$. If $m \leq n$ then W has a minimum value of $m(m+1)/2$ and a maximum value of $m(2N - m + 1)/2$.

Now, let $\widetilde{x}_1, \widetilde{x}_2, \ldots, \widetilde{x}_m$ and $\widetilde{y}_1, \widetilde{y}_2, \ldots, \widetilde{y}_n$ be two independent **FRS**s with **FCDF**s $\widetilde{F}_{\widetilde{X}}$ and $\widetilde{F}_{\widetilde{Y}}$ ($n_1 \leq n_2$). Assume that $\widetilde{\boldsymbol{w}} = (\widetilde{w}_1 = \widetilde{x}_1, \widetilde{w}_2 = \widetilde{x}_2, \ldots, \widetilde{w}_m = \widetilde{x}_m, \widetilde{w}_{m+1} = \widetilde{y}_1, \ldots, \widetilde{w}_N = \widetilde{y}_n)$ is the combined **FRS** of $\widetilde{x}_1, \widetilde{x}_2, \ldots, \widetilde{x}_m$ and $\widetilde{y}_1, \widetilde{y}_2, \ldots, \widetilde{y}_n$. The interesting hypothesis is that the populations are of the same form but with a different fuzzy median. This can be expressed symbolically by:

$$\begin{cases} H_0: \ \widetilde{F}_{\widetilde{Y}}(x) = \widetilde{F}_{\widetilde{X}}(x) \\ H_1: \widetilde{F}_{\widetilde{Y}}(x) = F_{\widetilde{X}}(x-\theta), \quad \theta \neq 0. \end{cases}$$

The fuzzy test statistic is defined by a fuzzy set as:

$$\widetilde{w}(w) = \sup\{\alpha \in [0,1] : w \in \widetilde{w}[\alpha]\},$$

where

$$\begin{aligned} \widetilde{w}[\alpha] &= \Big[\inf_{\beta\in[\alpha/2,1-\alpha/2]} \sum_{i=1}^{m}\sum_{j=1}^{N} I(P_d(\widetilde{x}_i \geq \widetilde{w}_j) \geq \beta), \qquad (5.45) \\ &\quad \sup_{\beta\in[\alpha/2,1-\alpha/2]} \sum_{i=1}^{m}\sum_{j=1}^{N} I(P_d(\widetilde{x}_i \geq \widetilde{w}_j) \geq \beta)\Big], \\ &= [\sum_{i=1}^{m}\sum_{j=1}^{N} I(P_d(\widetilde{X}_i \geq \widetilde{W}_j) \geq 1-\alpha/2), \sum_{i=1}^{m}\sum_{j=1}^{N} I(P_d(\widetilde{x}_i \geq \widetilde{w}_j) \geq \alpha/2)]. \end{aligned}$$

It is easy to verify that $\widetilde{w}$ is a normal discrete fuzzy set.

TABLE 5.11
Preference degrees in Example 5.9.

j	1	2	3	4	5	6	7	8	9	10	11	12	13	14	15
$P_d(\widetilde{x}_1 \succ \widetilde{w_j})$	0.5	0	0	1	1	0.07	1	1	1	0.28	1	1	1	1	1
$P_d(\widetilde{x}_2 \succ \widetilde{w_j})$	1	0.5	1	1	1	1	1	1	1	1	1	1	1	1	1
$P_d(\widetilde{x}_3 \succ \widetilde{w_j})$	1	0	0.5	1	1	1	1	1	1	0.95	1	1	1	1	1
$P_d(\widetilde{x}_4 \succ \widetilde{w_j})$	0	0	0	0.5	0	0	1	0	0	0	0	0	0	1	0
$P_d(\widetilde{x}_5 \succ \widetilde{w_j})$	0	0	0	1	0.5	0	1	0	1	0	0	0	1	1	0.64
$P_d(\widetilde{x}_6 \succ \widetilde{w_j})$	0.93	0	0	1	1	0.5	1	1	1	0.83	1	1	1	1	1
$P_d(\widetilde{x}_7 \succ \widetilde{w_j})$	0	0	0	0	0	0	0.5	0	0	0	0	0	0	0.9	0

Definition 5.15 *Let $\widetilde{W}$ be the (Wilcoxon) fuzzy statistics based on the observed **FRSs** $\widetilde{x}_1, \ldots, \widetilde{x}_m$ and $\widetilde{y}_1, \ldots, \widetilde{y}_m$. At the significance level of δ, we reject $\widetilde{H}_0$ if $D_R \geq D_A$ and we accept it if $D_A > D_R$ in which*

1) For testing $H_0 : \theta = 0$ versus $H_1 : \theta < 0$,

$$D_R = d(\widetilde{w} \in R_\delta^1 = \{c_\delta^1, c_\delta^1 + 1, ..., m(2N - m + 1)/2\}),\ D_A = 1 - D_R,$$

2) For testing $H_0 : \theta = 0$ versus $H_1 : \theta > 0$,

$$D_R = d(\widetilde{w} \in R_\delta^2 = \{m(m + 1)/2, m(m + 1)/2 + 1, ..., c_\delta^2\}),\ D_A = 1 - D_R,$$

3) For testing $H_0 : \theta = 0$ versus $H_1 : \theta \neq 0$,

$$D_R = d(\widetilde{w} \in R_{\delta/2}^1 \cup R_{\delta/2}^2),\ D_A = 1 - D_R.$$

where

$$d(\widetilde{A} \in I) = \frac{\sum_{x \in I} \widetilde{A}(x)}{\sum_x \widetilde{A}(x)},$$

denotes the the degree that $\widetilde{A}$ belongs to I. The decision-making procedure to accept or reject the null hypothesis can be made similar to that of sign test.

Example 5.9 *Consider Example 5.4. At the significance level of $\delta = 0.05$, we wish to test $H_0 : \theta = 0$ versus $H_1 : \theta \neq 0$. According to Table 5.11, the α-cuts of fuzzy statistics $\widetilde{w}$ can be evaluated (by Eq. 5.45) as follows:*

$$\widetilde{w}[\alpha] = \begin{cases} \{53, 54, ..., 67\}, & 0 < \alpha \leq 0.0.8, \\ \{54, ..., 67\}, & 0.08 < \alpha \leq 0.14, \\ \{55, ..., 66\}, & 0.14 < \alpha \leq 0.20, \\ \{56, ..., 66\}, & 0.20 < \alpha \leq 0.33, \\ \{57, ..., 66\}, & 0.33 < \alpha \leq 0.57, \\ \{57, ..., 65\}, & 0.57 < \alpha \leq 0.71, \\ \{58, ..., 65\}, & 0.71 < \alpha \leq 0.99, \\ \{65\}, & 0.99 < \alpha \leq 1. \end{cases} \tag{5.46}$$

Therefore, the membership function of $\widetilde{W}$ *is:*

$$\widetilde{w} = \{\frac{0.08}{53}, \frac{0.14}{54}, \frac{0.2}{55}, \frac{0.33}{56}, \frac{0.71}{57}, \frac{0.99}{\{58, 59, ..., 64\}}, \frac{1}{65}, \frac{0.57}{66}, \frac{0.4}{67}\}.$$

At the significance level of $\delta = 0.05$, *the upper and lower critical regions are* $R^1_{0.025} = \{28, 29, , ..., 38\}$ *and* $R^2_{0.025} = \{74, 75, , ..., 84\}$. *Therefore, it is readily seen that* $D_R = d(\widetilde{w} \in R^1_{0.025} \cup R^2_{0.025}) = 0$. *This concludes that the null hypothesis should be accepted.*

5.2.3 Kolmogorov-Smirnov test based on FRVs

The present section is aimed to develop the one-sample and two-sample Kolmogorov-Smirnov test for **STFRVs**. For this purpose, a unified approach is suggested to simplify the method of Hesamian and Chachi [86] and Hesamian and Taheri [89] for two-sided hypotheses.

5.2.3.1 One-sample Kolmogorov-Smirnov test

Let $X_1, X_2, \ldots, X_n$ be a random sample from a population with a continuous, but unknown **CDF** of $F_X(x)$. The Kolmogorov-Smirnov one-sample test statistic is based on the differences between the **CDF** $F_X(x)$ and the empirical cumulative distribution function (**ECDF**) of $\widehat{F}_n(x)$. Due to the strong law of large numbers, $\widehat{F}_n(x) \to F_X(x)$ with has a probability of one for each x. Concerning the large sample behavior of the $\widehat{F}_n(x)$, we have also the following property (see, e.g. [54]):

$$\mathbf{P}(\lim_{n\to\infty} \sup_{x\in\mathbb{R}} |\widehat{F}_n(x) - F_X(x)| = 0) = 1.$$

In other words, with the probability of one, $\widehat{F}_n(x) \to F_X(x)$ uniformly in x. So, as n increases, the step function of $\widehat{F}_n(x)$ approaches the true distribution $F_X(x)$ for all x, so that for large n, $| \widehat{F}_n(x) - F_X(x) |$ should be small for all values of x. If n is large enough, the so-called Kolomogorov-Smirnov one-sample statistic can be defined as:

$$\sqrt{n}D_n = \sqrt{n} \sup_{x\in\mathbb{R}} | \widehat{F}_n(x) - F_X(x) | . \tag{5.47}$$

The sampling distribution of $\sqrt{n}D_n$ is given by:

$$P(\sqrt{n}D_n < \frac{1}{2\sqrt{n}} + c) = \begin{cases} 0, & v \le 0, \\ p(n), & 0 < c < \frac{2n-1}{2\sqrt{n}}, \\ 1, & c \ge \frac{2n-1}{2\sqrt{n}}, \end{cases} \tag{5.48}$$

where

$$p(n) = \int_{1/2n-\frac{c}{\sqrt{n}}}^{1/2n+\frac{c}{\sqrt{n}}} \cdots \int_{(2n-1)/2n-\frac{c}{\sqrt{n}}}^{(2n-1)/2n+\frac{c}{\sqrt{n}}} f(u_1, \ldots, u_n) du_1 \ldots du_n,$$

in which, $f(u_1, u_2, \ldots, u_n) = n!\ I(u_1 < u_2 < \ldots u_n)$, and

$$P(\sqrt{n}D_n \le c) = 1 - 2\sum_{i=1}^{\infty}(-1)^{i-1}e^{-2i^2c^2},\ c > 0, \tag{5.49}$$

for large sample n (see [54], p. 117). In the Kolmogorov-Smirnov one-sample test, we wish to test a two-sided hypothesis of $H_0 : F_X(x) = F_X^0(x)$ versus $H_1 : F_X(x) \neq F_X^0(x)$ for all x, where $F_X^0(x)$ is a specified continuous **CDF**. If $\sqrt{n}d_n$ denotes the observed value of $\sqrt{n}D_n$, then for the alternative $p-value = P_{H_0}(\sqrt{n}D_n^- > (\sqrt{n}d_n^-)) < \delta$, and for the alternative of

$$H_1 : F_X(x) \neq F_X^0(x)\ \ for\ some\ x,$$

we reject the null hypothesis when $p - value = P_{H_0}(\sqrt{n}D_n > \sqrt{n}d_n) < \delta$ ([54]).

Now, let $\widetilde{X} = (X^L, X, X^U)_T$ be a **TFRV**. Here, the fuzzy **CDF** of $\widetilde{X}$ is also considered as a **TFN**:

$$\widetilde{F}_{\widetilde{X}}(x) = (F^L(x), F(x), F^U(x))_T,$$

where $F^L(x) = P(X^U \le x)$, $F(x) = P(X \le x)$ and $F^U(x) = P(X^L \le x)$. Assume that $\widetilde{x}_1, \widetilde{x}_2, \ldots, \widetilde{x}_n$ with $\widetilde{x}_i = (x_i^L, x_i, x_i^U)_T$ is an observed **FRS**. Therefore, the fuzzy empirical cumulative distribution function (**FECDF**) of $\widetilde{X}_1, \widetilde{X}_2, \ldots, \widetilde{X}_n$ can be evaluated as a **TFN**:

$$\widetilde{\widehat{F}}_n(x) = (\widehat{F}_n^L(x), \widehat{F}_n(x), \widehat{F}_n^U(x))_T,$$

where $\widehat{F}_n^L(x) = (1/n)\sum_{i=1}^n I(x_i^U \le x)$, $F_n(x) = (1/n)\sum_{i=1}^n I(x_i \le x)$ and $F^U(x) = (1/n)\sum_{i=1}^n I(x_i^L \le x)$. The test statistics based on $\widetilde{X}_1, \widetilde{X}_2, \ldots, \widetilde{X}_n$ is defined as follows:

$$\sqrt{n}\widetilde{D}_n = \sqrt{n}\sup_{x\in\mathbb{R}} d(\widetilde{\widehat{F}}_n(x), \widetilde{F}_{\widetilde{X}}(x)),$$

where for two **TFNs** $\widetilde{A} = (a^L, a, a^U)_T$ and $\widetilde{B} = (b^L, b, b^U)_T$, the absolute distance is given by:

$$d(\widetilde{A}, \widetilde{B}) = \max\{|a^L - b^L|, |a - b|, |a^U - b^U|\}.$$

Therefore, the test statistics can be evaluated as follows:

$$\sqrt{n}\widetilde{D}_n = \max\{\sqrt{n}\widetilde{D}_n^L, \sqrt{n}\widetilde{D}_n, \sqrt{n}\widetilde{D}_n^U\},$$

where

$$\sqrt{n}\widetilde{D}_n^L = \sqrt{n}\sup_{x\in\mathbb{R}}|\widehat{F}_n^L(x) - F^L(x)|,$$

$$\sqrt{n}\widetilde{D}_n = \sqrt{n}\sup_{x\in\mathbb{R}}|\widehat{F}_n(x) - F(x)|,$$

$$\sqrt{n}\widetilde{D}_n^U = \sqrt{n}\sup_{x\in\mathbb{R}}|\widehat{F}_n^U(x) - F^U(x)|.$$

The following lemma provides a method of evaluating $\widetilde{F}_{\widetilde{X}}$ using the $\widetilde{\widehat{F}}_n$ based on observed **FRSs**.

Lemma 5.16 *Let $\widetilde{X}$ be a **NFRV** with **FCDF** of $\widetilde{F}_{\widetilde{X}}$ and **FECDF** of $\widetilde{F}_{\widetilde{X}}$. Then,*

$$P(\lim_{n\to\infty}\sup_{x\in\mathbb{R}} d_3(\widetilde{\widehat{F}}_n(x), \widetilde{F}_{\widetilde{X}}(x)) = 0) = 1.$$

Proof *First, note that:*

$$\lim_{n\to\infty}\sup_{x\in\mathbb{R}} d_3(\widetilde{\widehat{F}}_n(x), \widetilde{F}_{\widetilde{X}}(x)) = \max\{\lim_{n\to\infty}\sup_{x\in\mathbb{R}}|\widehat{F}_n^L(x) - F^L(x)|,$$

$$\lim_{n\to\infty}\sup_{x\in\mathbb{R}}|\widehat{F}_n^U(x) - F^L(x)|, \lim_{n\to\infty}\sup_{x\in\mathbb{R}}|\widehat{F}_n^U(x) - F^L(x)|\}.$$

Therefore, by the classical Kolmogorov-Smirnov theorem, we get:

$$P(\lim_{n\to\infty}\sup_{x\in\mathbb{R}}|\widehat{F}_n^L(x) - F^L(x)| = 0) = 1,$$
$$P(\lim_{n\to\infty}\sup_{x\in\mathbb{R}}|\widehat{F}_n(x) - F(x)| = 0) = 1,$$
$$P(\lim_{n\to\infty}\sup_{x\in\mathbb{R}}|\widehat{F}_n^U(x) - F(x)^U| = 0) = 1.$$

This completes the proof.

Here, we wish to test $H_0 : \widetilde{F}_{\widetilde{X}}(x) = \widetilde{F}_{\widetilde{X}}^0(x)$ for all x versus $H_1 : \widetilde{F}_{\widetilde{X}}(x) \neq \widetilde{F}_{\widetilde{X}}^0(x)$ for some x where $\widetilde{F}_{\widetilde{X}}^0$ is a specified continuous **FCDF**. If $\sqrt{n}\widetilde{d}_n$ denotes the observed value of $\sqrt{n}\widetilde{D}_n$, we reject the null hypothesis if $p - value = P_{H_0}(\sqrt{n}\widetilde{D}_n > \sqrt{n}\widetilde{d}_n) < \delta$. For this purpose, note that:

$$\begin{aligned} P_{H_0}(\sqrt{n}\widetilde{D}_n > \sqrt{n}\widetilde{d}_n) &= P_{H_0}(\max\{\sqrt{n}\widetilde{D}_n^L, \sqrt{n}\widetilde{D}_n, \sqrt{n}\widetilde{D}_n^U\} > \sqrt{n}\widetilde{d}_n) \\ &= 1 - \left(1 - P(\sqrt{n}D_n > \sqrt{n}\widetilde{d}_n)\right)^3. \end{aligned}$$

Example 5.10 *Recall the data set in Example 5.3. Consider the Kolmogorov-Smirnov two-sample test $H_0 : \widetilde{F}_{\widetilde{X}}(x) = \widetilde{F}_{\widetilde{X}_0}(x)$ for all x versus $H_1 : \widetilde{F}_{\widetilde{X}}(x) \neq \widetilde{F}_{\widetilde{X}_0}(x)$ for some x, where $\widetilde{X}_0$ is a **TNFRV** with mean $\widetilde{\mu} = (8.5, 10.5, 12.5)_T$ and $\sigma^2 = 1$. First, the **FECDF** can be evaluated as $\widetilde{\widehat{F}}_n(x) = (\widehat{F}_n^L(x), \widehat{F}_n(x),$*

$\widehat{F}_n^U(x))_T$ *with:*

$$\widehat{F}_n^L(x) = \begin{cases} 0, & x < 6.73, \\ 0.0625, & 6.73 \leq x < 7.32, \\ 0.125, & 7.32 \leq x < 7.66, \\ 0.1875, & 7.66 \leq x < 7.90, \\ 0.25, & 7.90 \leq x < 7.93, \\ 0.3125, & 7.93 \leq x < 8.19, \\ 0.3750, & 8.19 \leq x < 8.23, \\ 0.4375, & 8.23 \leq x < 8.31, \\ 0.500, & 8.31 \leq x < 8.73, \\ 0.5625, & 8.73 \leq x < 9.18, \\ 0.625, & 9.18 \leq x < 9.37, \\ 0.75, & 9.37 \leq x \leq 9.50, \\ 0.8125, & 9.50 \leq x \leq 9.65, \\ 0.8750, & 9.65 \leq x \leq 10.48, \\ 0.9375, & 10.48 \leq x \leq 10.71. \\ 1, & x \geq 10.71, \end{cases} \tag{5.50}$$

$$\widehat{F}_n(x) = \begin{cases} 0, & x < 9.25, \\ 0.0625, & 9.29 \leq x < 9.45, \\ 0.1250, & 9.45 \leq x < 9.69, \\ 0.1875, & 9.69 \leq x < 9.84, \\ 0.2500, & 9.84 \leq x < 9.99, \\ 0.3125, & 9.99 \leq x < 10.18, \\ 0.3750, & 10.18 \leq x < 10.23, \\ 0.4375, & 10.23 \leq x < 10.38, \\ 0.500, & 10.38 \leq x < 10.47, \\ 0.5625, & 10.47 \leq x < 10.52, \\ 0.6250, & 10.52 \leq x < 11.08, \\ 0.6875, & 11.08 \leq x < 11.31, \\ 0.7500, & 11.31 \leq x < 11.46, \\ 0.8125, & 11.46 \leq x < 11.62, \\ 0.8750, & 11.62 \leq x < 11.66, \\ 0.9375, & 11.66 \leq x < 12.16, \\ 1, & x \geq 12.6, \end{cases}$$

$$\widehat{F}_n^U(x) = \begin{cases} 0, & x < 10.40, \\ 0.0625, & 10.40 \leq x < 10.99, \\ 0.1250, & 10.99 \leq x < 11.40, \\ 0.1875, & 11.40 \leq x < 11.84, \\ 0.2500, & 11.84 \leq x < 11.85, \\ 0.3125, & 11.85 \leq x < 11.90, \\ 0.3750, & 11.90 \leq x < 12.03, \\ 0.4375, & 12.03 \leq x < 12.18, \\ 0.500, & 12.18 \leq x < 12.30, \\ 0.5625, & 12.30 \leq x < 12.90, \\ 0.6250, & 12.90 \leq x < 13.02, \\ 0.6875, & 13.02 \leq x < 13.20, \\ 0.7500, & 13.20 \leq x < 13.23, \\ 0.8125, & 13.23 \leq x < 13.33, \\ 0.8750, & 13.33 \leq x < 13.87, \\ 0.9375, & 13.87 \leq x < 14.16, \\ 1. & x \geq 14.16. \end{cases}$$

According to Table 5.12, the test statistics is:

$$\begin{aligned} \sqrt{n}\widetilde{d}_n &= 4 \sup_{x \in \mathbb{R}} d(\widetilde{\widehat{F}}_n(x) - \widetilde{F}_{\widetilde{X}}(x)) \\ &= 4 \sup_{x \in \mathbb{R}} \max\{|\widehat{F}_n^L(x) - F^L(x)|, |\widehat{F}_n(x) - F(x)|, |\widehat{F}_n^U(x) - F^U(x)|\} \\ &= 4 \max\{0.1418, 0.11702, 0.12675\} = 0.507. \end{aligned}$$

Therefore, the observed p-value is $P_{H_0}(\sqrt{n}\widetilde{D}_n > 0.507) = 1 - \left(2\sum_{i=1}^{\infty} (-1)^{i-1} e^{-2i^2(0.514)^2}\right)^3 = 0.959$ (> 0.05). *This concludes that the null hypothesis* $\widetilde{H}_0$ *can be accepted.*

5.2.3.2 Two-sample Kolmogorov-Smirnov test for TFRVs

The Kolmogorov-Smirnov two-sample test is a goodness-of-fit test used to determine whether two underlying one-dimensional probability distributions are the same or not. In this regard, let $X_1, X_2, \ldots, X_m$, and $Y_1, Y_2, \ldots, Y_n$, be two independent random samples from populations with unknown continuous **CDF**s of F_X and G_Y, respectively, and $\widehat{F}_m$ and $\widehat{G}_n$ be their corresponding **ECDF**s, respectively. Consider a Kolmogorov-Smirnov two-sample test to evaluate the following two-sided hypotheses:

$$\begin{cases} H_0: & F_X(x) = G_Y(x), \quad \text{for all } x \in \mathbb{R}, \\ H_1: & F_X(x) \neq G_Y(x), \quad \text{for some } x \in \mathbb{R}. \end{cases}$$

The two-sided Kolmogorov-Smirnov test statistic, denoted by $\sqrt{\frac{mn}{m+n}} D_{mn}$, is based on the maximum absolute difference between two empirical cumulative

TABLE 5.12
Calculation the elements of **FECDF** in Example 5.10.

sample	$\lvert\widehat{F}_n^L(x) - F^L(x)\rvert$	$\lvert\widehat{F}_n(x) - F(x)\rvert$	$\lvert\widehat{F}_n^U(x) - F^U(x)\rvert$
1	0.044636	0.050639	0.024136
2	0.059478	0.021859	0.006000
3	0.051834	0.021470	0.012954
4	0.004627	0.004627	0.020931
5	0.054654	0.007474	0.003114
6	0.100747	0.000516	0.003280
7	0.118322	0.043920	0.043920
8	0.125516	0.047758	0.075345
9	0.141760	0.074466	0.028454
10	0.030422	0.117022	0.126748
11	0.010968	0.031543	0.120350
12	0.008036	0.041030	0.057850
13	0.045195	0.018972	0.028845
14	0.078269	0.006357	0.000072
15	0.022843	0.060524	0.038648
16	0.048457	0.048457	0.013553

distribution functions of $\widehat{F}_m$ and $\widehat{G}_n$, i.e.,

$$\begin{aligned}\sqrt{\frac{mn}{m+n}} D_{mn} &= \sqrt{\frac{mn}{m+n}} \sup_{x\in\mathbb{R}} \left|\widehat{F}_m(x) - \widehat{G}_n(x)\right| \\ &= \sqrt{\frac{mn}{m+n}} \max_i \left|\widehat{F}_m(Z_i) - \widehat{G}_n(Z_i)\right|,\end{aligned}$$

where $Z_1, Z_2, \ldots, Z_{m+n}$ is the combined sample of $X_1, X_2, \ldots, X_m$, and $Y_1, Y_2, \ldots, Y_n$ [54]. Under the continuity assumption of F_X and G_Y, the distribution of $\sqrt{\frac{mn}{m+n}} D_{mn}$ does not depend on F_X or G_Y. Therefore, at a significance level of $\delta \in (0,1]$, the null hypothesis of H_0 is rejected if $p -$ value $< \delta$ where

$$p - value = P_{H_0}\left(\sqrt{\frac{mn}{m+n}} D_{mn} \geq \sqrt{\frac{mn}{m+n}} d_{mn}^0\right),$$

in which $\sqrt{\frac{mn}{m+n}} d_{mn}^0$ denotes the observed two-sample Kolmogorov-Smirnov test statistic. The method of calculating the exact distribution of $\sqrt{\frac{mn}{m+n}} D_{mn}$ under H_0 is presented in [54]. Especially if $m = n$, the exact distribution of $\sqrt{\frac{mn}{m+n}} D_{mn}$ under H_0 simplifies to:

$$P_{H_0}\left(\sqrt{\frac{n}{2}} D_{nn} \geq d\right) = \frac{\binom{2n}{\lfloor \frac{d}{\sqrt{n/2}} + 1 \rfloor}}{\binom{2n}{n}}, \qquad d \in [-\sqrt{\frac{n}{2}}, +\sqrt{\frac{n}{2}}],$$

where, $\lfloor a \rfloor$ denotes the greatest integer of k that $k \leq a$. If m and n are large enough and $mn/(m+n) > 4$, then the following approximation is valid:

$$P_{H_0}\left(\sqrt{\frac{mn}{m+n}}D_{mn} \geq d\right) \simeq 2\sum_{k=1}^{\infty}(-1)^{k-1}e^{-2k^2d^2}.$$

Now, let $\widetilde{X} = (X^L, X, X^U)_T$ and $\widetilde{Y} = (Y^L, Y, Y^U)_T$ be two **FRV**s with **FCDF** of $\widetilde{F}(x) = (F^L(x), F(x), F^U(x))_T$, and $\widetilde{G}(x) = (G^L(x), G(x), G^U(x))_T$, respectively. Based on the two independent **FRS**s of $\widetilde{X}_1, \widetilde{X}_2, \ldots, \widetilde{X}_m$ and $\widetilde{Y}_1, \widetilde{Y}_2, \ldots, \widetilde{Y}_n$, the two-sample Kolmogrov-Smirnov is defined as follows:

$$\sqrt{\frac{mn}{m+n}}D_{mn} = \sqrt{\frac{mn}{m+n}}\sup_{x\in\mathbb{R}} d_3(\widehat{\widetilde{F}}_m(x), \widehat{\widetilde{G}}_n(x)),$$

where

1. $\widehat{\widetilde{F}}_m(x) = (\widehat{F}^L_m(x), \widehat{F}_m(x), \widehat{F}^U_m(x))_T$ and $\widehat{\widetilde{G}}_n(x) = (\widehat{G}^L_n(x), \widehat{G}_n(x), \widehat{G}^U_n(x))_T$ are **FECDF**s,

2. d_3 is a distance measure between two **TFN**s of $\widetilde{A} = (a^L, a, a^U)_T$ and $\widetilde{B} = (b^L, b, b^U)_T$ defined by

$$d_3(\widetilde{A}, \widetilde{B}) = \max\{|a^L - b^L|, |a - b|, |a^U - b^U|\}.$$

Therefore, the test statistics can be rewritten as:

$$\sqrt{\frac{mn}{m+n}}D_{mn} = \max\{\sqrt{\frac{mn}{m+n}}D^L_{mn}, \sqrt{\frac{mn}{m+n}}D_{mn}, \sqrt{\frac{mn}{m+n}}D^U_{mn}\},$$

where

$$\sqrt{\frac{mn}{m+n}}D^L_{mn} = \sqrt{\frac{mn}{m+n}}\max_i\left|\widehat{F}^U_m(Z^U_i) - \widehat{G}^U_n(Z^U_i)\right|,$$

$$\sqrt{\frac{mn}{m+n}}D_{mn} = \sqrt{\frac{mn}{m+n}}\max_i\left|\widehat{F}_m(Z_i) - \widehat{G}_n(Z_i)\right|,$$

$$\sqrt{\frac{mn}{m+n}}D^U_{mn} = \sqrt{\frac{mn}{m+n}}\max_i\left|\widehat{F}^L_m(Z^L_i) - \widehat{G}^L_n(Z^L_i)\right|,$$

in which $Z^L_1, Z^L_2, \ldots, Z^L_{m+n}$, $Z_1, Z_2, \ldots, Z_{m+n}$ and $Z^U_1, Z^U_2, \ldots, Z^U_{m+n}$ denote the combined sample of $(X^L_1, X^L_2, \ldots, X^L_m, Y^L_1, Y^L_2, \ldots, Y^L_n)$, $(X_1, X_2, \ldots, X_m, Y_1, Y_2, \ldots, Y_n)$, and $(X^U_1, X^U_2, \ldots, X^U_m, Y^U_1, Y^U_2, \ldots, Y^U_n)$, respectively.
At a significance level of δ, we wish to test the following hypothesis test:

$$\begin{cases} H_0: & \widetilde{F}_m(x) = \widetilde{G}_n(x), & \text{for all } x \in \mathbb{R}, \\ H_1: & \widetilde{F}_m(x) \neq \widetilde{G}_n(x), & \text{for some } x \in \mathbb{R}. \end{cases}$$

TABLE 5.13
Data set in Example 5.11.

No.	Diet A	Diet B
1	$(18, 19, 20)_T$	$(22, 24, 26)_T$
2	$(16, 17, 18)_T$	$(27, 28, 29)_T$
3	$(30, 31, 32)_T$	$(28, 26, 27)_T$
4	$(28, 29, 30)_T$	$(15, 16, 17)_T$
5	$(14, 15, 16)_T$	$(17, 18, 19)_T$
6	$(20, 21, 22)_T$	$(24, 25, 26)_T$
7	$(28, 30, 32)_T$	$(19, 21, 23)_T$
8	$(25, 26, 27)_T$	$(24, 25, 26)_T$
9	$(15, 16, 17)_T$	$(20, 21, 22)_T$
10	$(26, 27, 28)_T$	$(29, 30, 31)_T$

We reject the null hypothesis when $p - value = P_{H_0}(\sqrt{\frac{mn}{m+n}}D_{mn} > \sqrt{\frac{mn}{m+n}}d^0_{mn}) < \delta$ where $\sqrt{\frac{mn}{m+n}}d^0_{mn}$ denotes the observed value of $\sqrt{\frac{mn}{m+n}}D_{mn}$. Note that:

$$
\begin{aligned}
&P_{H_0}(\sqrt{\frac{mn}{m+n}}D_{mn} > \sqrt{\frac{mn}{m+n}}d^0_{mn}) \\
&= P_{H_0}(\max\{\sqrt{\frac{mn}{m+n}}D^L_{mn}, \sqrt{\frac{mn}{m+n}}D_{mn}, \sqrt{\frac{mn}{m+n}}D^U_{mn}\} > \sqrt{\frac{mn}{m+n}}d^0_{mn}) \\
&= 1 - \left(1 - P(\sqrt{\frac{mn}{m+n}}D_{mn} > \sqrt{\frac{mn}{m+n}}d^0_{mn})\right)^3.
\end{aligned}
$$

Example 5.11 *The* ***STFNs*** *data in Table 5.13 show the weight gain (in lbs) of pet dogs fed on two types of diets (A and B). At a significance level of* 0.05*, we wish to test the following hypotheses:*

$$
\begin{cases} H_0: & \widetilde{F}_m(x) = \widetilde{G}_n(x), & \text{for all } x \in \mathbb{R}, \\ H_1: & \widetilde{F}_m(x) \neq \widetilde{G}_n(x), & \text{for some } x \in \mathbb{R}. \end{cases}
$$

First note that:

$$
\widehat{F}^U_m(x) = \begin{cases} 0, & x < 14, \\ 0.1, & 14 \leq x < 15, \\ 0.2, & 15 \leq x < 16, \\ 0.3, & 16 \leq x < 18, \\ 0.4, & 18 \leq x < 20, \\ 0.5, & 20 \leq x < 25, \\ 0.6, & 25 \leq x < 26, \\ 0.7, & 26 \leq x < 28, \\ 0.9, & 28 \leq x < 30, \\ 1.0, & x \geq 30, \end{cases} \quad , \quad \widehat{F}_m(x) = \begin{cases} 0, & x < 15, \\ 0.1, & 15 \leq x < 16, \\ 0.2, & 16 \leq x < 17, \\ 0.3, & 17 \leq x < 19, \\ 0.4, & 19 \leq x < 21, \\ 0.5, & 21 \leq x < 26, \\ 0.6, & 26 \leq x < 27, \\ 0.7, & 27 \leq x < 29, \\ 0.9, & 29 \leq x < 31, \\ 1.0, & x \geq 31, \end{cases}
$$

$$\widehat{F}^L_m(x) = \begin{cases} 0, & x < 16, \\ 0.1, & 16 \le x < 17, \\ 0.3, & 17 \le x < 20, \\ 0.4, & 20 \le x < 22, \\ 0.5, & 22 \le x < 27, \\ 0.6, & 27 \le x < 28, \\ 0.7, & 28 \le x < 30, \\ 0.8, & 30 \le x < 32, \\ 1.0, & x \ge 32, \end{cases}, \quad \widehat{G}^U_n(x) = \begin{cases} 0, & x < 15, \\ 0.1, & 15 \le x < 17, \\ 0.2, & 17 \le x < 19, \\ 0.3, & 19 \le x < 20, \\ 0.4, & 20 \le x < 22, \\ 0.5, & 22 \le x < 24, \\ 0.7, & 24 \le x < 26, \\ 0.8, & 26 \le x < 27, \\ 0.9, & 27 \le x < 29, \\ 1.0, & x \ge 29, \end{cases}$$

$$\widehat{G}_n(x) = \begin{cases} 0.0, & x < 16, \\ 0.1, & 16 \le x < 18, \\ 0.2, & 18 \le x < 21, \\ 0.4, & 21 \le x < 24, \\ 0.5, & 24 \le x < 25, \\ 0.7, & 25 \le x < 27, \\ 0.8, & 27 \le x < 28, \\ 0.9, & 28 \le x < 30, \\ 1.0, & x \ge 30, \end{cases}, \quad \widehat{G}^L_n(x) = \begin{cases} 0.0, & x < 17, \\ 0.1, & 17 \le x < 19, \\ 0.2, & 19 \le x < 22, \\ 0.3, & 22 \le x < 23, \\ 0.4, & 23 \le x < 26, \\ 0.7, & 26 \le x < 28, \\ 0.9, & 28 \le x < 31, \\ 1.0, & x \ge 31. \end{cases}$$

Therefore, from Table 5.14, we have $\sqrt{\frac{mn}{m+n}}d^0_{mn} = \max\{0.2\sqrt{5}, 0.2\sqrt{5}, 0.2\sqrt{5}\}$ $= 0.4472$ *and thus* $p - value = P_{H_0}(\sqrt{\frac{mn}{m+n}}D_{mn} > 0.4472) = 1 - (1 - P(\sqrt{\frac{mn}{m+n}}D_{mn} > 0.4472))^3 = 1 - (1 - 0.988265)^3 = 0.999$. *It can be concluded that there is no evidence to reject* $\widetilde{H}_0$ *at significance level of* $\delta = 0.05$.

5.2.4 Kruskal-Wallis test

Assume we have k independent sets of observations $X_{11}, X_{12}, \ldots, X_{1n_1}, X_{21}, X_{22}, \ldots, X_{2n_2}, \ldots, X_{k1}, X_{k2}, \ldots, X_{kn_k}$ one from each of k populations with continuous cumulative distribution functions of F_1, F_2, ..., F_k. Let $N = n_1 + \ldots + n_k$ denote the total number of observations. The location model for the k sample problem is that the **CDF**s are $F_1(x) = F(x - \theta_1)$, $F_2(x) = F(x - \theta_2)$, $\ldots$, $F_k(x) = F(x - \theta_k)$, respectively, where θ_i denotes a location parameter of the ith population. Interest hypothesis is that all k samples are drawn from identical populations, i.e.

$$H_0 : F_1(x) = F_2(x) = \ldots = F_k(x), \ \textit{for all } x \textit{ or } \theta_1 = \theta_2 = \ldots = \theta_k,$$

and the alternative hypothesis is:

$$H_1 : \theta_i \neq \theta_j \ \textit{for at least one } i \neq j.$$

TABLE 5.14
Calculation the elements of **FECDF**s in Example 5.11.

i	Z^U	$\lvert\widehat{F}^L_m(z^U) - \widehat{G}^L_n(z^U)\rvert$	Z	$\lvert\widehat{F}_m(z) - \widehat{G}_n(z)\rvert$	Z^L	$\lvert\widehat{F}^U_m(z^L) - \widehat{G}^U_n(z^L)\rvert$
1	20	0.2	19	0.2	18	0.2
2	18	0.2	17	0.2	16	0.2
3	32	0	31	0	30	0
4	30	0.2	29	0.2	28	0
5	16	0.2	15	0.2	14	0.2
6	22	0.2	21	0.2	20	0
7	32	0	30	0.2	28	0
8	27	0.2	26	0.2	25	0.2
9	17	0.2	16	0.2	15	0.2
10	28	0.2	27	0.2	26	0
11	26	0.2	24	0	22	0.2
12	29	0.2	28	0.2	27	0.2
13	28	0.2	27	0.2	26	0
14	17	0.2	16	0.2	15	0.2
15	19	0.2	18	0.2	17	0.2
16	26	0.2	25	0.2	24	0.2
17	23	0.2	21	0.2	19	0
18	26	0.2	25	0.2	24	0.2
19	22	0.2	21	0.2	20	0
20	31	0.2	30	0.2	29	0

This test statistic is called Kruskal and Wallis:

$$h = \frac{12}{N(N+1)} \sum_{i=1}^{k} \frac{(r_i^2 - n_i(N+1)/2)^2}{n_i}, \tag{5.51}$$

where r_i stands for the sum of ranks assigned to the elements in the ith sample. At a significance level of δ, H_0 is rejected if $h \geq c_\delta$ where h and c_δ represent the observed value of H and critical region, respectively. Exact probabilities for H under H_0 are given in [54] for some specified k and n_i. Moreover, under H_0, if no n_i is very small, then H is distributed approximately as chi-square with $k-1$ degrees of freedom. When the null hypothesis is rejected, we can compare any two groups, say i and j (with $1 \leq i < j \leq k$), by a multiple comparisons procedure. In this regard, to test $H_0 : \theta_i = \theta_j$ versus $H_1 : \theta_i \neq \theta_j$ if:

$$|\overline{r}_i - \overline{r}_j| \in \left[Z_{\delta/k(k-1)} \sqrt{\frac{N(N+1)}{12}} (\frac{1}{n_i} + \frac{1}{n_j}), \infty\right), \tag{5.52}$$

where $\overline{r}_i = r_i/n_i$, $\overline{r}_j = r_j/n_j$ and $Z_{\delta/k(k-1)}$ is $[\delta/k(k-1)]^{th}$ the upper standard normal quantile.

Now, assume we have k independent sets of **FRS**s of $\widetilde{x}_{11}, \widetilde{x}_{12}, \ldots, \widetilde{x}_{1n_1}$, $\widetilde{x}_{21}, \widetilde{x}_{22}, \ldots, \widetilde{x}_{2n_2}$, ..., $\widetilde{x}_{k1}, \widetilde{x}_{k2}, \ldots, \widetilde{x}_{kn_k}$ one from each of k populations with continuous **FCDFs** $\widetilde{F}_1(x) = \widetilde{F}(x-\theta_1)$, $\widetilde{F}_2(x) = \widetilde{F}(x-\theta_2)$, ..., $\widetilde{F}_k(x) = \widetilde{F}(x-\theta_k)$. The combined **FRS** is denoted by $\widetilde{w}_1, \widetilde{w}_2, \ldots, \widetilde{w}_N$, $N = n_1+n_2+\ldots+n_k$. The hypothesis of interest is that all k samples are drawn from identical populations, i.e.,

$$H_0 : \theta_1 = \theta_2 = \ldots = \theta_k,$$

and the alternative is:

$$H_1 : \theta_i \neq \theta_j \ for\ at\ least\ one\ i \neq j.$$

The fuzzy test statistics is defined by a fuzzy set $\widetilde{h}$ with the following membership function:

$$\widetilde{h}(x) = \sup\{\alpha \in [0,1] : \widetilde{h}[\alpha] = [\widetilde{h}^L_\alpha, \widetilde{h}^U_\alpha]\}, \tag{5.53}$$

where

$$\widetilde{h}^L_\alpha = \inf_{\beta \in [\alpha/2, 1-\alpha/2]} \frac{12}{N(N+1)} \sum_{i=1}^{k} \frac{((\widetilde{r}_i)_\beta - n_i(N+1)/2)^2}{n_i},$$

$$\widetilde{h}^U_\alpha = \sup_{\beta \in [\alpha/2, 1-\alpha/2]} \frac{12}{N(N+1)} \sum_{i=1}^{k} \frac{((\widetilde{r}_i)_\beta - n_i(N+1)/2)^2}{n_i},$$

in which $(\widetilde{r}_i)_\beta = \sum_{l=1}^{n_i} \sum_{j=1}^{N} I(P_d(\widetilde{x}_{lj} \geq \widetilde{w}_j) \geq \beta)$. It is easy to check that $\widetilde{H}$ is a **FN**.

At the significance level of δ, consider the conventional rejection region $[c_\delta, \infty)$. Now, to accept or reject H_0, the credibility criterion [129] is employed to measure the degree to which $\widetilde{h}$ belongs to $[c_\delta, \infty)$. This index relies on possibility and necessity measures [43] defined as follows:

$$Cr(\widetilde{H} \in [c_\delta, \infty)) = \frac{\sup\{\widetilde{h}(x) : x \geq c_\delta\} + 1 - \sup\{\widetilde{h}(x) : x < c_\delta\}}{2}.$$

It is easy to check that $Cr(\widetilde{h} \in [c_\delta, \infty)) \in [0,1]$ and $Cr(\widetilde{h} \in [c_\delta, \infty)) = 1 - Cr(\widetilde{h} \in (0, c_\delta))$. In this regard, at a significance level of δ, we say the null hypothesis is accepted if $Cr(\widetilde{h} \in (0, c_\delta)) \geq Cr(\widetilde{h} \in [c_\delta, \infty))$ that is $Cr(\widetilde{h} \in (0, c_\delta)) \geq 0.5$.

Remark 5.2 *Let* $\{\widetilde{h}[\alpha]\}_{\alpha \in [0,1]}$ *denotes the* α*-cut of the fuzzy test statistics of* $\widetilde{H}$*. At the significance level of* δ*, the credibility that* $\widetilde{H}$ *belongs to the rejection region* $[c_\delta, \infty)$ *can be evaluated according to the following cases:*

(1) *If* $c_\delta \geq \sup \widetilde{h}[0]$ *then* $Cr(\widetilde{h} \in [c_\delta, \infty)) = 0$.

(2) *If* $\widetilde{h}[1] \leq c_\delta < \sup \widetilde{h}[0]$ *then*

$$Cr(\widetilde{h} \in [c_\delta, \infty)) = \frac{\sup\{\widetilde{h}^U_\alpha : c_\delta \in \widetilde{h}[\alpha]\}}{2}.$$

(3) *If* $\sup \widetilde{h}[0] < c_\delta < \widetilde{h}[1]$ *then*

$$Cr(\widetilde{h} \in [c_\delta, \infty)) = \frac{1 + (1 - \sup\{\widetilde{h}_\alpha^L : c_\delta \in \widetilde{h}[\alpha]\})}{2} = 1 - \frac{\sup\{\widetilde{h}_\alpha^L : c_\delta \in \widetilde{h}[\alpha]\}}{2}.$$

(4) *If* $c_\delta \leq \inf \widetilde{h}[0]$ *then* $Cr(\widetilde{h} \in [c_\delta, \infty)) = 1$.

Now consider the cases in which the null hypothesis $H_0 : \theta_1 = \theta_2 = \ldots = \theta_k$ is rejected. Therefore, multiple comparisons are required to find the differences. In this regard, to test $H_0 : \theta_i = \theta_j$ versus $H_1 : \theta_i \neq \theta_j$ with $i \neq j$, we reject H_0 if $D_R \geq D_A$ where

$$D_A = Cr\Big(\widetilde{d}(\widetilde{\overline{r}}_i, \widetilde{\overline{r}}_j) \in \big[Z_{[\delta/k(k-1)]}\sqrt{\frac{N(N+1)}{12}(\frac{1}{n_i} + \frac{1}{n_j})}, \infty)\Big),$$

and

$$D_R = Cr\Big(\widetilde{d}(\widetilde{\overline{r}}_i, \widetilde{\overline{r}}_j) \in \big[0, Z_{[\delta/k(k-1)]}\sqrt{\frac{N(N+1)}{12}(\frac{1}{n_i} + \frac{1}{n_j})})\Big),$$

in which

(1) $\widetilde{r}_i$ denotes the fuzzy total ranking of the ith group with the following α-cut:

$$(\widetilde{r}_i)[\alpha] = [\sum_{l=1}^{n_i}\sum_{j=1}^{N} I(P_d(\widetilde{x}_{lj} \succeq \widetilde{w}_j) \geq 1 - \alpha/2), \sum_{l=1}^{n_i}\sum_{j=1}^{N} I(P_d(\widetilde{x}_{lj} \succeq \widetilde{w}_j) \geq \alpha/2)],$$

(2) $\widetilde{\overline{r}}_i = (\frac{1}{n_i}) \otimes \widetilde{r}_i$, and

(3)
$$\widetilde{d}(\widetilde{\overline{r}}_i, \widetilde{\overline{r}}_j)(d) = \sup_{x,y:d=|x-y|} \min\{\widetilde{\overline{r}}_i(x), \widetilde{\overline{r}}_j(y)\},$$

is the fuzzy distance between two fuzzy sets of $\widetilde{\overline{r}}_i$ and $\widetilde{\overline{r}}_j$ as discussed in chapter 1.

Example 5.12 *Experts have long claimed that speakers who use some sort of audiovisual aids in their presentations are much more effective in communicating with their audience. A consulting agency would like to test this claim; however, it is very difficult to find many speakers who can be regarded as virtually identical in their speaking capabilities. The agency managed to find only 15 such speakers from a nationwide search. The speakers were randomly assigned to one of three groups. The first group of speakers was not allowed to use any aids while the second group was allowed to use a regular overhead projector and a microphone, and the third group could use a 35 mm color slide*

TABLE 5.15
The judges ($\widetilde{x}_{ij}$) for each presentation in each group in Example 5.12.

Group 1	Group 2	Group 3
$(61, 63, 68)_T$	$(69, 74, 76)_T$	$(60, 62, 67)_T$
$(66, 68, 73)_T$	$(67, 72, 74)_T$	$(63, 67, 70)_T$
$(65, 70, 72)_T$	$(70, 73, 78)_T$	$(68, 71, 72)_T$
$(67, 69, 73)_T$	$(71, 76, 78)_T$	$(72, 74, 79)_T$
$(63, 66, 71)_T$	$(72, 77, 79)_T$	$(76, 78, 83)_T$

TABLE 5.16
The values of$P_d(\widetilde{x}_{1l} \succ \widetilde{w_j})$ in Example 5.12.

j	1	2	3	4	5	6	7	8	9	10	11	12	13	14	15
$P_d(\widetilde{x}_{11} \succ \widetilde{w_j})$	0.5	0	0	0	0	0	0	0	0	0	1	0	0	0	0
$P_d(\widetilde{x}_{12} \succ \widetilde{w_j})$	1	0.5	0.2	0	1	0	0	0	0	0	1	1	0.033	0	0
$P_d(\widetilde{x}_{13} \succ \widetilde{w_j})$	1	0.8	0.5	0.31	1	0	0	0	0	0	1	1	0	0	0
$P_d(\widetilde{x}_{14} \succ \widetilde{w_j})$	1	1	0.68	0.5	1	0	0	0	0	0	1	1	0.071	0	0
$P_d(\widetilde{x}_{15} \succ \widetilde{w_j})$	1	0	0	0	0.5	0	0	0	0	0	1	0.25	0	0	0

projector together with a microphone and a tape recorder (which played prerecorded audio messages). After a certain period of time, each speaker made a presentation in an auditorium, on a certain issue, in front of a live audience and a selected panel of judges. The contents of their presentations were virtually the same so that any differences in effectiveness could be only attributed to the audiovisual aids used by the speakers. The judges scored each presentation on a fuzzy scale of $30 - 100$*, depending on their judgments and the reaction of the audience, with larger scores denoting greater effectiveness; the scores are given in Table 5.15. At significance level* $\delta = 0.05$*, the hypothesis of interest is that three samples are drawn from identical populations, i.e.,*

$$H_0 : \theta_1 = \theta_2 = \theta_3,$$

and the alternative hypothesis is:

$$H_1 : \theta_i \neq \theta_j \text{ for at least one } i \neq j.$$

To compute the fuzzy statistics of $\widetilde{H}$*, first, there is a need to evaluate the fuzzy sum ranks of the first, second, and third samples. For this purpose, the values of* $P_d(\widetilde{x}_i \succ \widetilde{w}_j)$ *are calculated in Tables 5.16, 5.17, and 5.18. Now, the* α*-cuts*

TABLE 5.17
The values of $P_d(\widetilde{x}_{2l} \succ \widetilde{w_j})$ in Example 5.12.

j	1	2	3	4	5	6	7	8	9	10	11	12	13	14	15
$P_d(\widetilde{x}_{21} \succ \widetilde{w_j})$	1	1	1	1	1	0.5	1	0.31	0	0	1	1	1	0	0
$P_d(\widetilde{x}_{22} \succ \widetilde{w_j})$	1	1	1	1	1	0	0.5	0	0	0	1	1	0.87	0	0
$P_d(\widetilde{x}_{23} \succ \widetilde{w_j})$	1	1	1	1	1	0.68	1	0.5	0	0	1	1	1	0	0
$P_d(\widetilde{x}_{24} \succ \widetilde{w_j})$	1	1	1	1	1	1	1	1	0.5	0	1	1	1	0.8	0
$P_d(\widetilde{x}_{25} \succ \widetilde{w_j})$	1	1	1	1	1	1	1	1	1	0.5	1	1	1	1	0

TABLE 5.18
The values of $P_d(\widetilde{x}_{3l} \succ \widetilde{w_j})$ in Example 5.12.

j	1	2	3	4	5	6	7	8	9	10	11	12	13	14	15
$P_d(\widetilde{x}_{31} \succ \widetilde{wj})$	0	0	0	0	0	0	0	0	0	0	0.5	0	0	0	0
$P_d(\widetilde{x}_{32} \succ \widetilde{w_j})$	1	0	0	0	0.75	0	0	0	0	0	1	0.5	0	0	0
$P_d(\widetilde{x}_{33} \succ \widetilde{w_j})$	1	0.97	1	0.93	1	0	0.125	0	0	0	1	1	0.5	0	0
$P_d(\widetilde{x}_{34} \succ \widetilde{w_j})$	1	1	1	1	1	1	1	1	0.2	0	1	1	1	0.5	0
$P_d(\widetilde{x}_{35} \succ \widetilde{w_j})$	1	1	1	1	1	1	1	1	1	1	1	1	1	1	0.5

of $\widetilde{r}_i$ *can be evaluated as follows:*

$$(\widetilde{r}_i)^L_\alpha = \sum_{l=1}^{n_i}\sum_{j=1}^{15} I(P_d(\widetilde{x}_{lj} \geq \widetilde{w}_j) \geq 1 - \alpha/2),$$

$$(\widetilde{r}_i)^U_\alpha = \sum_{l=1}^{n_i}\sum_{j=1}^{15} I(P_d(\widetilde{x}_{lj} \geq \widetilde{w}_j) \geq \alpha/2).$$

Accordingly, one can find that

$$\widetilde{r}_1[\alpha] = \begin{cases} \{16, 17, ..., 28\}, & 0 < \alpha \leq 0.06, \\ \{16, 17, ..., 27\}, & 0.06 < \alpha \leq 0.14, \\ \{16, 17, ..., 26\}, & 0.14 < \alpha \leq 0.40, \\ \{17, ..., 25\}, & 0.40 < \alpha \leq 0.50, \\ \{17, ..., 24\}, & 0.50 < \alpha \leq 0.62, \\ \{18, ..., 23\}, & 0.62 < \alpha \leq 0.99, \\ \{23\}, & \alpha = 1, \end{cases} \tag{5.54}$$

$$\widetilde{r}_2[\alpha] = \begin{cases} \{49, 50, ..., 58\}, & 0 < \alpha \leq 0.24, \\ \{50, 51, ..., 58\}, & 0.24 < \alpha \leq 0.40, \\ \{51, ..., 58\}, & 0.40 < \alpha \leq 0.62, \\ \{52, ..., 57\}, & 0.62 < \alpha \leq 0.99, \\ \{57\}, & \alpha = 1, \end{cases} \tag{5.55}$$

$$\widetilde{r}_3[\alpha] = \begin{cases} \{32, 33, ..., 42\}, & 0 < \alpha \leq 0.06, \\ \{33, 34, ..., 42\}, & 0.06 < \alpha \leq 0.14, \\ \{34, 35, ..., 42\}, & 0.14 < \alpha \leq 0.26, \\ \{34, 35, ..., 41\}, & 0.26 < \alpha \leq 0.40, \\ \{34, 35, ..., 40\}, & 0.40 < \alpha \leq 0.49, \\ \{35, ..., 40\}, & 0.49 < \alpha \leq 0.99, \\ \{40\}, & \alpha = 1, \end{cases} \tag{5.56}$$

Therefore, the lower and upper bounds of the fuzzy test statistics can be evaluated as follows:

$$\widetilde{h}_\alpha^L = \inf_{\beta \in [\alpha/2, 1-\alpha/2]} \frac{12}{15(16)} \sum_{i=1}^{3} \frac{((\widetilde{r}_i)_\beta - 5(16)/2)^2}{n_i},$$

$$\widetilde{h}_\alpha^U = \sup_{\beta \in [\alpha/2, 1-\alpha/2]} \frac{12}{15(16)} \sum_{i=1}^{3} \frac{((\widetilde{r}_i)_\beta - 5(16)/2)^2}{n_i}.$$

The calculations show that:

$$\widetilde{h}[\alpha] = \begin{cases} [4.72, 7.21], & 0 < \alpha \leq 0.06, \\ [4.97, 7.12], & 0.06 < \alpha \leq 0.14, \\ [5.21, 7.12], & 0.14 < \alpha \leq 0.40, \\ [5.49, 6.88], & 0.40 < \alpha \leq 0.49, \\ [5.49, 6.75], & 0.49 < \alpha \leq 0.50, \\ [5.78, 6.53], & 0.50 < \alpha \leq 0.99, \\ \{5.78\}, & \alpha = 1. \end{cases} \tag{5.57}$$

At the significance level of $\delta = 0.05$*, one can check that* $R_{0.05} = [5.6, \infty)$*. This concludes to:*

$$\begin{aligned} Cr(\widetilde{h} \in [5.6, \infty)) &= \frac{1 + (1 - \sup\{\widetilde{h}_\alpha^L : c_\delta \in \widetilde{h}[\alpha]\})}{2} \\ &= 1 - \frac{\sup\{\widetilde{h}_\alpha^L : c_\delta \in \widetilde{h}[\alpha]\}}{2} \\ &= 1 - 0.4/2 = 0.80. \end{aligned}$$

Therefore, we should reject the null hypothesis. This means there is a significant difference between the three populations. The proposed fuzzy multiple comparisons procedure is applied to check those populations. For this purpose, first, note that the membership functions of the sum of the ranks of each group are:

$$\widetilde{r}_1 = \{\frac{0.4}{16}, \frac{0.62}{17}, \frac{0.99}{\{18, 19, ..., 22\}}, \frac{1}{23}, \frac{0.62}{24}, \frac{0.5}{25}, \frac{0.4}{26}, \frac{0.14}{27}, \frac{0.06}{28}\},$$

$$\widetilde{r}_2 = \{\frac{0.24}{49}, \frac{0.40}{50}, \frac{0.62}{51}, \frac{0.99}{\{52, 53, , ..., 56\}}, \frac{1}{57}, \frac{0.62}{58}\},$$

$$\widetilde{r}_3 = \{\frac{0.06}{32}, \frac{0.14}{33}, \frac{0.49}{34}, \frac{0.99}{\{35, 36, ..., 39\}}, \frac{1}{40}, \frac{0.40}{41}, \frac{0.26}{42}\}.$$

The fuzzy distances between fuzzy sets of $\widetilde{\overline{r}}_i$ *and* $\widetilde{\overline{r}}_j$ *can be evaluated as follows:*

$$\widetilde{D}(\widetilde{\overline{r}}_1, \widetilde{\overline{r}}_2) = \{\frac{0.5}{5.2}, \frac{0.62}{\{5.4, 5.6\}}, \frac{0.99}{\{5.8, 6, 6.2, 6.4, 6.6\}}, \frac{1}{6.8}, \frac{0.99}{\{7, 7.2, 7.4\}}, \frac{0.64}{\{8, 8.2\}}\}.$$

$$\widetilde{D}(\widetilde{\overline{r}}_1, \widetilde{\overline{r}}_3) = \{\frac{0.6}{1}, \frac{0.14}{\{1.2, 1.4\}}, \frac{0.4}{\{1.6\}}, \frac{0.49}{1.8}, \frac{0.5}{2}, \frac{0.62}{\{2.2\}},$$

$$\frac{0.99}{\{2.4, 2.6, 2.8, 3, 3.2\}}, \frac{1}{3.4}, \frac{0.99}{\{3.6, 3.8, 4, 4.2, 4.4\}}, \frac{0.64}{\{4.6, 4.8\}}, \frac{0.4}{5}, \frac{0.26}{5.2}\},$$

$$\widetilde{D}(\widetilde{\overline{r}}_2, \widetilde{\overline{r}}_3) = \{\frac{0.26}{1.6}, \frac{0.40}{\{1.8, 2\}}, \frac{0.62}{2.2}, \frac{0.99}{\{2.4, 2.6, 2.8, 3, 3.2\}},$$

$$\frac{1}{3.4}, \frac{0.99}{\{3.6, 3.8, 4, 4.2, 4.4\}}, \frac{0.62}{4.6}, \frac{0.49}{4.8}, \frac{0.14}{5}\}.$$

At significance level of $\delta = 0.05$*, we have* $Z_{0.05/6} = Z_{[0.0083333]} = 2.394$ *and thus* $Z_{[0.0083333]}\sqrt{\frac{15\times16\times2}{12\times5}} = 6.7713$*. Therefore, the degrees of rejections to test* $H_0 : \theta_i = \theta_j$ *versus* $H_1 : \theta_i \neq \theta_j$, $i, j = 1, 2, 3$ *can be evaluated as follows:*

$$Cr\Big(\widetilde{d}(\widetilde{\overline{r}}_1, \widetilde{\overline{r}}_2) \in (6.7713, \infty)\Big) = \frac{\sum_{d\geq6.7713} \widetilde{d}(\widetilde{\overline{r}}_1, \widetilde{\overline{r}}_2)(d)}{\sum_d \widetilde{d}(\widetilde{\overline{r}}_1, \widetilde{\overline{r}}_2)(d)} = \frac{7.23}{13.29} = 0.5194.$$

Additionally, it is readily seen that:

$$Cr\Big(\widetilde{d}(\widetilde{\overline{r}}_1, \widetilde{\overline{r}}_3) \in (6.7713, \infty)\Big) = 0,$$

and also

$$Cr\Big(\widetilde{d}(\widetilde{\overline{r}}_2, \widetilde{\overline{r}}_3) \in (6.7713, \infty)\Big) = 0.$$

The above results reveal that, at a significance level of $\delta = 0.05$*, there is a significant difference between the first and the second populations. Further, it is significantly evident that the first and third as well as the second and third populations are the same.*

5.3 One-Way ANOVA Based on FRVs

Analysis of variance (ANOVA) is a statistical technique enabling us to test the null hypothesis that the three or more population means are equal against the non-equal alternative hypothesis [103, 175]. The simplest type of ANOVA is one-way ANOVA which can be used to compare the means of several populations. Analysis of Variance (ANOVA) is a hypothesis-testing technique to test the equality of the means of two or more populations (or treatments) by examining the variances of their corresponding samples. ANOVA is based on comparing the variance (or variation) between the data samples to variation within each particular sample. If the between variation is much larger than the within variation, the means of different samples will not be equal. If the between and within variations are approximately the same, then there will be no significant difference between samples. Assumptions of ANOVA are: (i) all involved populations follow a normal distribution, (ii) all populations have the same variance (or standard deviation), and (iii) the samples are randomly selected and independent of one another. Consider, r denotes the number of levels of the factor under study, any of these levels is denoted by the index i, $i = 1, 2, \ldots, r$. The number of cases for the ith factor level is denoted by n_i, and the total number of cases in the study is denoted by N, where $N = \sum_{i=1}^{r} n_i$. Assume that X_{ij} denotes the jth observation on the dependent (response) variable for the ith factor level. The ANOVA model can be stated as $X_{ij} = \mu_i + \epsilon_{ij}$ where X_{ij} are the value of the response variables in the jth trial for the ith factor level, μ_i is the factor level mean, and ϵ_{ij} are independent random variables with normal distribution $N(0, \sigma^2)$. So X_{ij}s are independent random variables having the normal distribution $N(\mu_i, \sigma^2)$ for any $i = 1, 2, \ldots, r$. The null hypothesis for an one-way ANOVA always assumes the population means are equal. Hence, we may write the null hypothesis as $H_0 : \mu_1 = \mu_2 = \ldots = \mu_r$ versus $H_1 : \mu_i \neq \mu_j$ for at least on pair (i, j). The test statistic in ANOVA is the ratio of the between and within variations of the data:

$$F = \frac{MSR}{MSE} = \frac{SSR/r-1}{SSE/N-r}, \tag{5.58}$$

in which

$$SSR = \sum_{i=1}^{r} n_i(\overline{X}_{i.} - \overline{X}_{..})^2, \quad SSE = \sum_{i=1}^{r}\sum_{j=1}^{n_i}(X_{ij} - \overline{X}_{i.})^2,$$

where $\overline{X}_{i.} = \frac{1}{n_i}\sum_{j=1}^{n_i} X_{ij}$ and $\overline{X}_{..} = \frac{1}{n}\sum_{j=1}^{r}\sum_{j=1}^{n_i} X_{ij}$. SSR and SSE stand for between and within variations, respectively. When the null hypothesis of H_0 holds true, F is distributed as f-distribution with $r-1$ and $N-r$ degrees of freedom, i.e., $F_{r-1,N-r}$. Finally, at a given significance level of δ, the hypothesis of H_0 is rejected if and only if $f > c_\delta$, where $P(F_{r-1,N-r} > c_\delta) = \delta$

and f is the observed test statistics of F. When an ANOVA gives a significant result to reject H_0, this indicates that at least one group differs from the other groups. Moreover the test does not indicate which group differs. To analyze the pattern of difference between means, the ANOVA is often followed by specific comparisons, and the most commonly comparing two means. One of the most commonly used technique is called the least significant difference test. This procedure uses F^* statistics for testing $H_0 : \mu_i = \mu_j$ versus $H_1 : \mu_i \neq \mu_j$ as follows:

$$F^* = \frac{(\overline{X}_{i.} - \overline{X}_{j.})^2}{MSE(\frac{1}{n_i} + \frac{1}{n_j})}. \tag{5.59}$$

At a given significance level of δ, H_0 is rejected if $f^* > c_\delta$ where $P(F_{1,N-r} > c_\delta) = \delta$ and f^* is the observed test statistics.

As noted, the main assumption is the homogeneity of variances. That is, in an ANOVA we assume that treatment variances are equal, i.e., $H_0 : \sigma_1^2 = \sigma_2^2 = \ldots = \sigma_r^2$ against $H_1 : \sigma_i^2 \neq \sigma_j^2$ for some i, j. Bartlett's test is often used in such cases. The test statistics is $\chi_0^2 = 2.3026\frac{q}{c}$ where

$$q = (N - r)\log_{10} S_p^2 - \sum_{i=1}^{r}(n_i - 1)\log_{10} S_i^2, \tag{5.60}$$

in which

$$c = 1 + \frac{1}{3(r-1)}\left(\sum_{i=1}^{r}\frac{1}{n_i - 1} - \frac{1}{N-r}\right),$$

$$S_p^2 = \frac{1}{N-r}\sum_{i=1}^{r}(n_i - 1)S_i^2,$$

and S_i^2 is the sample variance of the ith population. At a given significance level of δ, we reject H_0 when $\S_0^2 > c_\delta$ where $P(\chi_{r-1}^2 > c_\delta) = \delta$ and $\mathcal{X}_0^2$ is the observed value of χ_0^2. For other analysis of variance methods see for example [174].

Now assume that $\widetilde{X}_{11}, \widetilde{X}_{12}, \ldots, \widetilde{X}_{1n_1}, \widetilde{X}_{21}, \widetilde{X}_{21}, \ldots, \widetilde{X}_{2n_2}$ and $\widetilde{X}_{r1}, \widetilde{X}_{r2}, \ldots, \widetilde{X}_{rn_r}$ are sets of r independent **FRSs** from $N(\widetilde{\mu}_i, \sigma^2)$, where $\widetilde{X}_{ij}$ is **TFRV** as $\widetilde{X}_{ij} = (X_{ij}^L, X_{ij}, X_{ij}^U)_T$. At a significance level of $\delta \in (0,1)$, we wish to examine the following fuzzy hypotheses:

$$\begin{cases} \widetilde{H}_0 : & \widetilde{\mu}_1 = \widetilde{\mu}_2 = \ldots = \widetilde{\mu}_k, \\ \widetilde{H}_1 : & \widetilde{\mu}_i \neq \widetilde{\mu}_j \ \ for\ at\ least\ one\ i\ and\ j. \end{cases}$$

The total variation is defined as follows:

$$SST = \sum_{i=1}^{r}\sum_{j=1}^{n_i} d_2^2(\widetilde{X}_{ij}, \widetilde{X}_{..}),$$

where

$$d_2(\widetilde{A}, \widetilde{B}) = \sqrt{\frac{(a^L - b^L)^2 + (a - b)^2 + (a^U - b^U)^2}{3}}.$$

In addition, the between and within variations are also defined as:

$$SSR = \sum_{i=1}^{r} n_i d_2^2(\widetilde{X}_{i.}, \widetilde{\overline{X}}_{..}),$$

and

$$SSE = \sum_{i=1}^{r} \sum_{j=1}^{n_i} d_2^2(\widetilde{X}_{ij}, \widetilde{\overline{X}}_{i.}),$$

in which $\widetilde{X}_{i.} = (1/n_i) \bigoplus_{j=1}^{n_i} \widetilde{X}_{ij}$.

Lemma 5.17 *Assume that $\widetilde{X}_{11}, \widetilde{X}_{12}, \ldots, \widetilde{X}_{1n_1}$, $\widetilde{X}_{21}, \widetilde{X}_{21}, \ldots, \widetilde{X}_{2n_2}$ and $\widetilde{X}_{r1}$, $\widetilde{X}_{r2}, \ldots, \widetilde{X}_{rn_r}$ are independent **FRSs** from $N(\widetilde{\mu}_i, \sigma^2)$. Then:*

(1) *$SST = SSR + SSE$.*

(2) *Under the null hypothesis $\widetilde{\mu}_1 = \widetilde{\mu}_2 = \ldots = \widetilde{\mu}_k$,*

$$F = \frac{MSR}{MSE} = \frac{SSR/3(r-1)}{SSE/3(N-r)} \sim F_{3(r-1),3(N-r)}. \quad (5.61)$$

Proof *The direct calculations show that*

$$SST = \sum_{i=1}^{r} \sum_{j=1}^{n_i} \frac{(X_{ij}^L \pm \overline{X}_{i.}^L - \overline{X}_{..}^L)^2 + (X_{ij} \pm \overline{X}_{i.} - \overline{X}_{..})^2 + (X_{ij}^U \pm \overline{X}_{i.}^U - \overline{X}_{..}^U)^2}{3}$$

$$= \frac{1}{3} \sum_{i=1}^{r} \sum_{j=1}^{n_i} \Big([(X_{ij}^L - \overline{X}_{i.}^L)^2 + (X_{i.}^L - \overline{X}_{..}^L)^2] + [(X_{ij} - \overline{X}_{i.})^2 + (X_{i.} - \overline{X}_{..})^2]$$

$$+ [(X_{ij}^L - \overline{X}_{i.}^U)^2 + (X_{i.}^U - \overline{X}_{..}^U)^2] \Big).$$

*This simply concludes that $SST = SSR + SSE$ which verifies (1). Furthermore, since $\widetilde{X}_{11}, \widetilde{X}_{12}, \ldots, \widetilde{X}_{1n_1}$, $\widetilde{X}_{21}, \widetilde{X}_{22}, \ldots, \widetilde{X}_{2n_2}$ and $\widetilde{X}_{r1}, \widetilde{X}_{r2}, \ldots, \widetilde{X}_{rn_r}$ are independent **FRSs** from $N(\widetilde{\mu}_i, \sigma^2)$, it is readily seen that $3SSR/\sigma^2 \sim \chi^2_{3(r-1)}$ and $3SSE/\sigma^2 \sim \chi^2_{3(N-r)}$, and SSR and SSE are independent and hence (2) is verified. The observed values of SST, SSE, and SSR are represented by sst, sse, and ssr, respectively.*

At a given significance level of δ, therefore $\widetilde{H}_0$ is rejected if $f > F_{3(r-1),3(N-r),\alpha}$ where $f = (ssr/r-1)/(sse/N-r)$ is the observed test statistics. In cases that the null hypothesis is rejected, for testing $H_0 : \mu_i = \mu_j$ versus $H_1 : \mu_i \neq \mu_j$ $(i \neq j)$ at a significance level of δ, the least significant difference test can be extended as:

$$F^* = \frac{d_2^2(\widetilde{\overline{X}}_{i.}, \widetilde{\overline{X}}_{j.})}{MSE(\frac{1}{n_i} + \frac{1}{n_j})}, \quad (5.62)$$

where

$$d_2^2(\widetilde{A}, \widetilde{B}) = \frac{|a^L - b^L|^2 + |a - b|^2 + |a^U - b^U|^2}{3}.$$

It can be shown that $F^* \sim F_{3,3(N-r)}$. At a given significance level of δ, so H_0 is rejected if $f^* > c_\delta$ where $P(F^* > c_\delta) = \delta/2$ and f^* is the observed value of F^*. To test the homogeneity of variances assumption, i.e., $H_0 : \sigma_1^2 = \sigma_2^2 = \ldots = \sigma_r^2$ against $H_1 : \sigma_i^2 \neq \sigma_j^2$ for some $i \neq j$, Bartlett's test can be developed as follows. The test statistic is defined by $\chi_0^2 = 2.3026\frac{q}{c}$ where:

$$q = (N - r)\log_{10} S_p^2 - \sum_{i=1}^{r}(n_i - 1)\log_{10} S_i^2, \tag{5.63}$$

in which

$$c = 1 + \frac{1}{3(r-1)}(\sum_{i=1}^{r}\frac{1}{n_i - 1} - \frac{1}{N - r}), \tag{5.64}$$

$$S_p^2 = \frac{1}{N - r}\sum_{i=1}^{r}(n_i - 1)S_i^2,$$

and $S_i^2 = (1/(n_i - 1))\sum_{j=1}^{n_j} d_2^2(\widetilde{X}_{ij}, \overline{\widetilde{X}}_{i.})$ is the sample variance of the ith population. At a given significance level of δ, we reject H_0 if $\mathrm{x}_0^2 > c_\delta$ where $P(\chi^2_{3(r-1)} > c_\delta) = \delta$ and x_0^2 represents the observed test statistics of χ_0^2.

Example 5.13 *Suppose that the development engineer is intended to determine if varying the cotton content in a synthetic fiber affects its tensile strength. He ran a completely randomized experiment with five levels of cotton percentage in five replicates. The data are shown in Table 5.19. Because of some impreciseness in the experimental environment, the data are reported as* **TFNs***. Based on these observations, a test is taken to see whether or not the means of five tensile strengths are the same. In other words, at a significance level of* $\delta = 0.05$, *we wish to test* $H_0 : \widetilde{\mu}_1 = \widetilde{\mu}_2 = \widetilde{\mu}_3 = \widetilde{\mu}_4 = \widetilde{\mu}_5$ *v.s.* $H_1 : \widetilde{\mu}_i \neq \widetilde{\mu}_j$ *for at least one pair* (i, j). *So, we need to evaluate the observed test statistics of* F. *For this, it is easy to check that* $\overline{\widetilde{x}}_{1.} = (6.36, 7.10, 8.14)_T$, $\overline{\widetilde{x}}_{2.} = (14, 15.4, 16.72)_T$, $\overline{\widetilde{x}}_{3.} = (16.3, 17.64, 19.24)_T$, $\overline{\widetilde{x}}_{4.} = (20.1, 21.60, 23.22)_T$, $\overline{\widetilde{x}}_{5.} = (9.66, 10.8, 12.38)_T$, *and* $\widetilde{x}_{..} = (13.286, 14.51, 15.942)_T$. *Therefore, according to Table 5.20, we get:*

$$\begin{aligned} sst = \sum_{i=1}^{r}\sum_{j=1}^{5} d_2^2(\widetilde{x}_{ij}, \widetilde{x}_{..}) &= \frac{1}{3}(\sum_{i=1}^{3}\sum_{j=1}^{5}(x_{ij}^L - 13.286)^2 \\ &= \sum_{i=1}^{r}\sum_{j=1}^{5}(x_{ij} - 14.51)^2 + \sum_{i=1}^{r}\sum_{j=1}^{5}(x_{ij}^U - 15.942)^2) \\ &= \frac{740.51 + 814.84 + 860.12}{3} = 805.156. \end{aligned}$$

TABLE 5.19
Fuzzy values of tensile strength in Example 5.13.

Cotton percentage	$\widetilde{x}_{ij} = (x_{ij}^L, x_{ij}, x_{ij}^U)_T$
15	$(6.3, 7, 8.4)_T$
15	$(6.4, 7, 8.1)_T$
15	$(0.7, 1.5, 2.4)_T$
15	$(10.3, 11, 11.8)_T$
15	$(8.1, 9, 10)_T$
20	$(10.8, 12, 13.3)_T$
20	$(15.6, 17, 18.4)_T$
20	$(10.5, 12, 14)_T$
20	$(16.1, 18, 19.3)_T$
20	$(17, 18, 18.6)_T$
25	$(12.6, 14, 15, 7)_T$
25	$(16.4, 18, 19.3)_T$
25	$(17.2, 18, 19.8)_T$
25	$(17.2, 19, 21)_T$
25	$(18.1, 19.20.4)_T$
30	$(18.2, 19.20.5)_T$
30	$(23, 25, 26.6)_T$
30	$(19.8, 22, 24.5)_T$
30	$(18, 19, 20)_T$
30	$(21.5, 23, 24.5)_T$
35	$(6, 7, 8.3)_T$
35	$(9, 10, 12)_T$
35	$(10, 11, 12)_T$
35	$(13.5, 15, 17)_T$
35	$(9.8, 11, 12.6)_T$

Moreover, from Table 5.21, $ssr = 5\sum_{j=1}^{5} d_2^2(\widetilde{\overline{x}}_{i.}, \widetilde{\overline{x}}_{..}) = 641.137$. This yields to $sse = sst - ssr = 805.156 - 641.137 = 164.02$ and thus $f = \frac{641.137/4}{164.02/20} = 19.5445$. Since $f > F_{12,60,0.05} = 1.92$, at the significance level of $\delta = 0.05$, we then reject the null hypothesis of $\widetilde{H}_0 : \widetilde{\mu}_1 = \widetilde{\mu}_2 = ... = \widetilde{\mu}_5$. Therefore, it is requred to conduct multiple comparisons to determine which fuzzy mean(s) is/are different. For this purpose, the developed least significance difference test was applied. According to Table 5.22, the values of test statistic F^* are summarized in Table 5.23. Therefore, there is significant evidence for concluding that $\widetilde{\mu}_1 \neq \widetilde{\mu}_4$, $\widetilde{\mu}_3 \neq \widetilde{\mu}_5$, and $\widetilde{\mu}_i = \widetilde{\mu}_j$ for other cases. Now, the null hypothesis of interest is the homogeneity of variances for five stores at the significance level of $\delta = 0.05$, i.e.,

$$\begin{cases} H_0: & \sigma_1^2 = \sigma_2^2 = \sigma_1^3 = \sigma_1^4 = \sigma_5^5 = \sigma^2, \\ H_1: & \sigma_i^2 \neq \sigma_j^2 \text{ for at least one pair } (i,j). \end{cases}$$

For this purpose, note that $c = 1.1$, $s_1^2 = 37.666$, $s_2^2 = 27.442$, $s_3^2 = 13.411$, $s_4^2 = 19.537$, and $s_5^2 = 24.960$. This concludes that $s_p^2 = 24.6032$,

TABLE 5.20
Computing the elements of SST in Example 5.13.

j	$(x_{1j}^L - 13.286)$	$(x_{1j} - 14.51)$	$(x_{1j}^U - 15.942)$
1	1.6	0.8	0
2	−6.4	−5.2	−4
3	4.6	4.8	5
4	−2.4	−3.2	−4
5	2.6	2.8	3
j	$(x_{2j}^L - 13.286)$	$(x_{2j} - 14.51)$	$(x_{2j}^U - 15.942)$
1	−2	−3.6	−5.2
2	−2	−1.6	−1.2
3	0	1.4	2.8
4	6	6.4	6.8
5	−2	−2.6	−3.2
j	$(x_{3j}^L - 13.286)$	$(x_{3j} - 14.51)$	$(x_{3j}^U - 15.942)$
1	−4.2	−5.2	−6.2
2	−0.2	−0.2	−0.2
3	−0.2	0.8	1.8
4	5.8	6.8	7.8
5	−1.2	−2.2	−3.2
j	$(x_{4j}^L - 13.286)$	$(x_{4j} - 14.51)$	$(x_{4j}^U - 15.942)$
1	−1.9	−2.6	−2.72
2	2.9	3.4	3.38
3	−0.3	0.4	1.28
4	−2.1	−2.6	−3.22
5	1.4	1.4	1.28
j	$(x_{5j}^L - 13.286)$	$(x_{5j} - 14.51)$	$(x_{5j}^U - 15.942)$
1	−3.66	−3.8	−4.08
2	−0.66	−0.8	−0.38
3	0.34	0.2	−0.38
4	3.84	4.2	4.62
5	0.14	0.2	0.22

and $q = 20\log_{10}(24.6032) - 4\sum_{i=1}^{5}\log_{10} s_i^2 = 0.500116$. *Thus* $\chi_0^2 = 2.3026(0.500116/1.1) = 1.0469$. *At a given significance level of* $\delta = 0.05$, *since* $\chi_0^2 = 1.0469 < \chi^2_{12,0.05} = 21.03$, *the null hypothesis* $H_0 : \sigma_1^2 = \sigma_2^2 = \sigma_3^2 = \sigma_4^2 = \sigma_5^2$ *is accepted.*

5.4 Exercise

Exercise 5.1 *The data in Table 5.24 gives the efficiency of a chemical process using three different catalysts (A, B, and C) on each of four days. At the significance level of* $\delta = 0.05$, *is there evidence that the different catalysts*

TABLE 5.21
Computing the elements of SSR in Example 5.13.

i	$(\overline{x}^L_{i.} - 13.286)^2$	$(\overline{x}_{i.} - 14.51)^2$	$(\overline{x}^U_{i.} - 15.942)^2$
1	47.9695	54.9081	60.8712
2	0.509796	0.7921	0.605284
3	9.0842	9.7969	10.8768
4	46.4306	50.2681	52.9693
5	13.1479	13.7641	12.6878

TABLE 5.22
Values of $d_2^2(\overline{x}_{i.}, \overline{x}_{j.})$ in Example 5.13.

j	1	2	3	4	5
$d_2^2(\overline{x}_{1.}, \overline{x}_{j.})$	-	8.173	10.527	14.15	3.757
$d_2^2(\overline{x}_{2.}, \overline{x}_{j.})$	-	-	2.278	6.267	4.427
$d_2^2(\overline{x}_{3.}, \overline{x}_{j.})$	-	-	-	3.913	6.78
$d_2^2(\overline{x}_{4.}, \overline{x}_{j.})$	-	-	-	-	10.693

TABLE 5.23
Multiple comparison of fuzzy means in Example 5.13.

No.	Hypothesis	f^*	$F_{3,60,0.05}$	Conclusion
1	$\widetilde{H}_0 : \widetilde{\mu}_1 = \widetilde{\mu}_2$	$\frac{8.173}{4.1005} = 1.994$	2.77	$\widetilde{H}_0$ is accepted
2	$\widetilde{H}_0 : \widetilde{\mu}_1 = \widetilde{\mu}_3$	$\frac{10.577}{4.1005} = 2.567$	2.77	$\widetilde{H}_0$ is accepted
3	$\widetilde{H}_0 : \widetilde{\mu}_1 = \widetilde{\mu}_4$	$\frac{14.44}{4.1005} = 3.522$	2.77	$\widetilde{H}_0$ is rejected
4	$\widetilde{H}_0 : \widetilde{\mu}_1 = \widetilde{\mu}_5$	$\frac{3.757}{4.1005} = 1.212$	2.77	$\widetilde{H}_0$ is accepted
5	$\widetilde{H}_0 : \widetilde{\mu}_2 = \widetilde{\mu}_3$	$\frac{2.287}{4.1005} = 0.558$	2.77	$\widetilde{H}_0$ is accepted
6	$\widetilde{H}_0 : \widetilde{\mu}_2 = \widetilde{\mu}_4$	$\frac{6.267}{4.1005} = 1.528$	2.77	$\widetilde{H}_0$ is accepted
7	$\widetilde{H}_0 : \widetilde{\mu}_2 = \widetilde{\mu}_5$	$\frac{4.427}{4.1005} = 1.08$	2.77	$\widetilde{H}_0$ is accepted
8	$\widetilde{H}_0 : \widetilde{\mu}_3 = \widetilde{\mu}_4$	$\frac{3.913}{4.1005} = 0.954$	2.77	$\widetilde{H}_0$ is accepted
9	$\widetilde{H}_0 : \widetilde{\mu}_3 = \widetilde{\mu}_5$	$\frac{6.78}{4.1005} = 9.0916$	2.77	$\widetilde{H}_0$ is rejected
10	$\widetilde{H}_0 : \widetilde{\mu}_4 = \widetilde{\mu}_5$	$\frac{10.693}{4.1005} = 2.608$	2.77	$\widetilde{H}_0$ is accepted

result in different efficiencies? Assume in this example, that the data may not be normally distributed.

Exercise 5.2 *Assume a **FRS** of size $n = 20$ from a population of $N(\widetilde{\mu}, \sigma^2)$ whose fuzzy data are listed in Table 5.25. At the significance level of $\delta = 0.01$, conduct a hypothesis for testing the following hypotheses:*

TABLE 5.24
Data set in Exercise 5.1.

Catalyst	Day 1	Day 2	Day 3	Day 4
A	$(82, 84, 86)_T$	$(80, 82, 83)_T$	$(78, 79, 80)_T$	$(75, 77, 80)_T$
B	$(77, 78, 79)_T$	$(75, 76, 78)_T$	$(78, 80, 82)_T$	$(82, 83, 85)_T$
C	$(81, 83, 85)_T$	$(77, 78, 79)_T$	$(75, 76, 77)_T$	$(78, 80, 81)_T$

TABLE 5.25
Data set in Exercise 5.2.

$\widetilde{x}_1 = (1.19, 1.29, 1.79)_T$	$\widetilde{x}_2 = (1.01, 1.51, 1.61)_T$	$\widetilde{x}_3 = (2.58, 2.63, 2.70)_T$
$\widetilde{x}_4 = (3.49, 3.53.3.58)_T$	$\widetilde{x}_5 = (0.18, 0.23, 0.27)_T$	$\widetilde{x}_6 = (2.68, 2.722.80)_T$
$\widetilde{x}_7 = (1.31, 1.37, 1.42)_T$	$\widetilde{x}_8 = (3, 3.08, 3.12)_T$	$\widetilde{x}_9 = (0.85, 0.9, 0.94)_T$
$\widetilde{x}_{10} = (6.73, 6.8, 6.95)_T$	$\widetilde{x}_{11} = (1.80, 1.9, 1.98)_T$	$\widetilde{x}_{12} = (0.73, 0.8, 0.87)_T$
$\widetilde{x}_{13} = (0.66, 0.70, 0.80)_T$	$\widetilde{x}_{14} = (2.01, 2.11, 2.20)_T$	$\widetilde{x}_{15} = (1.55, 1.62, 170)_T$
$\widetilde{x}_{16} = (1.78, 3.05, 3.12)_T$	$\widetilde{x}_{17} = (2.18, 2.25, 2.31)_T$	$\widetilde{x}_{18} = (0.48, 0.54, 0.62)_T$
$\widetilde{x}_{19} = (2.92, 3.0, 3.08)_T$	$\widetilde{x}_{20} = (1.54, 1.61, 1.69)_T$	–

1.

$$\widetilde{H}_0 : \widetilde{\mu} \simeq_M \widetilde{\mu}_0 = (1.75, 2, 2.25)_T \quad \textit{versus } \widetilde{H}_1 : \widetilde{\mu} \neq_M \widetilde{\mu}_0 = (1.75, 2, 2.25)_T,$$

2.

$$H_0 : \sigma^2 = 1.5 \qquad \textit{versus } H_1 : \sigma^2 \neq 1.5.$$

Exercise 5.3 *Table 5.26 shows the lifetime (in 1000 km) of front disk brake pads on a randomly selected set of 40 cars (same model) that were monitored by a dealer network. In practice, however, measuring the lifetime of a disk may not yield an exact result. A disk may work perfectly over a certain period but partially operate for some time, and finally, be unusable at a certain time. So, such data can be reported as imprecise quantities. Assume that the lifetimes of front disk brake pads are normally distributed with a fuzzy mean of $\widetilde{\mu}$ and unknown variance of σ^2. At the significance level of $\delta = 0.05$, do the fuzzy data provide enough evidence to show that the fuzzy lifetime mean of front disk brake pads is equal to $(42, 45, 48)_T$?*

Exercise 5.4 *Table 5.27 lists the lifetime of the batteries (hours) produced by a company. Note that measuring the lifetime of a battery may not yield an exact number since a battery may work perfectly for some specific time (for instance 700 (h)) before it begins to lose power during the next time (for instance 500 (h)). Based on the aforementioned data set, perform a one-tailed*

TABLE 5.26
Data set in Exercise 5.3.

$\widetilde{x}_i$	$\widetilde{x}_i$	$\widetilde{x}_i$	$\widetilde{x}_i$
$(82.6, 86.2, 89.5)_T$	$(42, 45.1, 47.6)_T$	$(50.3, 52.1, 54, 6)_T$	$(52.4, 54.2, 60, 3)_T$
$(37.4, 38.4, 42.6)$	$(39.1, 41.0, 42.5)_T$	$(53.2, 56.4, 58.5)_T$	$(77.8, 81.3, 84.2)_T$
$(43.7, 45.5, 48.6)$	$(35.4, 36.7, 38.4)_T$	$(40.3, 42.2, 43.8)_T$	$(49.2, 51.6, 55.3)_T$
$(21.4, 22.7, 23.6)$	$(21.3, 22.6, 25.6)_T$	$(38.5, 40.0, 42.8)_T$	$(37.1, 38.8, 39.5)_T$
$(46.3, 48.8, 51.4)$	$(78, 6, 81.7, 83.4)_T$	$(58.2, 61.5, 62.7)_T$	$(51.3, 53.6, 55.6)_T$
$(40.2, 42.8, 43.8)$	$(67.6, 72.5, 75.5)_T$	$(39.8, 42.7, 43.6)_T$	$(68.6, 80.6, 83.7)_T$
$(70.5, 73.1, 75.6)$	$(25.6, 28.4, 31.6)_T$	$(44.3, 46.9, 49.8)_T$	$(43.1, 45.9, 48.9)_T$
$(56.2, 59.8, 62.5)$	$(30.1, 31.7, 32.5)_T$	$(32.1, 33.9, 34.5)_T$	$(46.8, 50.6, 52.6)_T$
$(56.2, 59.3, 62.1)_T$	$(59.2, 62.4, 63.5)_T$	$(33.2, 34.4, 37.4)_T$	$(28.2, 50.2, 55.1)_T$
$(47.2, 50.7, 55.3)_T$	$(61.2, 64.5, 66.5)_T$	$(29.6, 33.8, 37.4)_T$	$(53.8, 56.7, 60.2)_T$

TABLE 5.27
Data set in Exercise 5.4

No.	$\widetilde{x}_i$	No.	$\widetilde{x}_i$
1	$(500, 700, 800)_T$	7	$(500, 600, 700)_T$
2	$(950, 1050, 1100)_T$	8	$(800, 900, 1150)_T$
3	$0(700, 900, 1000)_T$	9	$(550, 650, 850)_T$
4	$(450, 550, 600)_T$	10	$(600, 750, 800)_T$
5	$(1000, 1100, 1200)_T$	11	$(1100, 1150, 1200)_T$
6	$(850, 900, 1000)_T)_T$	12	$(600, 700, 750)_T$

hypothesis test at the significance level of $\delta = 0.05$*:*

$$\begin{cases} \widetilde{H}_0 : \widetilde{M} \succeq_{P_d} \widetilde{M}_0 = (600, 700, 800)_T, \\ \widetilde{H}_1 : \widetilde{M} \prec_{P_d} \widetilde{M}_0 = (600, 700, 800)_T. \end{cases} \tag{5.65}$$

Exercise 5.5 *A product development engineer is interested in investigating the tensile strength of a new synthetic fiber that will be used to make man shirts. It is known from previous experiences that the strength is affected by the weight percent of cotton used in the blend of fiber materials. Also, it is known that cotton content should range in* $10 - 40\%$ *to obtain the final product with the desired quality (such as the ability to take a permanent-press finishing treatment). The engineer decided to test the specimen at five levels of cotton weight percent: 15, 20, 25, 30, and 35% in five replicates. Due to the ambiguity of technical conditions of measurements, the response variable cannot be measured exactly and data cannot be recorded clearly with precise numbers and only linguistic terms can justify the required tolerance of the errors in measurements. So, the evaluator decides to record the observations of*

TABLE 5.28
Data set in Exercise 5.5.

Cotton	x_{ij}				
Weight percentage	1	2	3	4	5
15	7	10	12	10	9
20	12	17	12	18	16
25	14	18	17	13	15
30	19	25	22	18	23
35	8	11	13	9	14

the tensile strength by ***STFNs*** $\widetilde{x}_{ij} = (0.99x_{ij}, x_{ij}, 1.01x_{ij})_T$ *(see Table 5.28). At a significance level of* $\alpha = 0.05$*:*

1. *Check the homogeneity of variances assumption.*

2. *Is there a significant difference between groups? that is*

$$\begin{cases} \widetilde{H}_0 : & \widetilde{\mu}_1 = \widetilde{\mu}_2 = ... = \widetilde{\mu}_5, \\ \widetilde{H}_1 : & \widetilde{\mu}_i \neq \widetilde{\mu}_j \;\; for \; at \; least \; one \; i \; and \; j. \end{cases}$$

3. *If there is evidence to reject* $\widetilde{H}_0$, *analyze the pattern of difference between means.*

Exercise 5.6 *An organization has three different branches. Turnover level differs across the three branches and management wants to know whether this may be explained by the extent to which employees are satisfied with their working environment across the branches. Thirty employees were randomly selected at each branch and their job satisfaction was reported by some* ***STFNs*** $\widetilde{x}_{ij} = (0.9x_{ij}, x_{ij}, 1.1x_{ij})_T$ *where* x_{ij} *are listed in Table 5.29. Using fuzzy Kruskal-Walis test:*

1. *Perform a hypothesis test at the* 5% *level of significance for*

$$\begin{cases} H_0 : & \theta_1 = \theta_2 = \theta_3, \\ H_1 : & \theta_i \neq \theta_j \;\; for \; at \; least \; one \; i \neq j. \end{cases}$$

2. *If there is evidence to reject* H_0, *what is the direction of that difference?*

Exercise 5.7 *Two brands of flashlight batteries were compared by testing their products with five flashlights. Ten flashlights were randomly selected and divided into two groups comprising five flashlights. Then, each group of flashlights uses a different brand of battery. The lifetimes of the batteries (in hour) were reported in* ***TFNs*** *as listed in Table 5.30. Preliminary data analyses indicated that the independent samples come from normal populations. At the significance level of* 5%,

TABLE 5.29
Data set in Exercise 5.6.

	x_{ij}									
Branch	1	2	3	4	5	6	7	8	9	10
1	48	62	55	65	58	53	66	54	60	63
2	65	70	85	73	77	68	82	85	76	63
3	45	68	52	60	58	52	67	56	64	67

TABLE 5.30
Data set in Exercise 5.7.

No.	Brand A	No.	Brand B
1	$(41, 42, 42.5)_T$	6	$(27.5, 28, 28.5)_T$
2	$(29, 30, 30.5)_T$	7	$(34.5, 36, 36.5)_T$
3	$0(38, 39, 40)_T$	8	$(30.5, 31, 32)_T$
4	$(27, 28, 28.5)_T$	9	$(31, 32, 33)_T$
5	$(28, 29, 30.5)_T$	10	$(26.5, 27, 27.5)_T$

1. *Conduct a hypothesis test for* $\sigma_1^2 = \sigma_2^2$.
2. *Is there any difference in the fuzzy mean lifetime of the two brands?*

Exercise 5.8 *A dining hall manager planned to introduce new boxed lunch services and took a survey to investigate what price for a boxed lunch would be acceptable to male and female customers. 20 customers (10 males and 10 females) were randomly selected who resided around this dining hall in the city of Taipei. The investigator asked them, how many dollars they would be willing to spend (can answer with interval) for a boxed lunch in a Japanese dining hall. The answers are presented in Table 5.31. At the significance level of 5%, do males and females express different acceptable prices for a boxed lunch? that is*

$$\begin{cases} \widetilde{H}_0: & \widetilde{F}(x) = \widetilde{G}(x), \quad \text{for all } x \in \mathbb{R}, \\ \widetilde{H}_1: & \widetilde{F}(x) \neq \widetilde{G}(x), \quad \text{for some } x \in \mathbb{R}. \end{cases}$$

Exercise 5.9 *A car manufacturing company is producing low-consuming cars. It claimed that the consumption rate of its new car generationis less than about 5 liter in per 100 kilometers. But it seems that the customers of these cars did not accept this claim. To investigate this claim, we selected 10 car owners as* ***TFN****s (Table 5.32). Suppose the observed fuzzy observations are distributed according to a normal* ***FRV*** *with a known variance of one. At the significance level of 5%, examine the null hypothesis of* $\widetilde{H}_0 : \widetilde{\mu} \preceq_M (4.5, 5, 5.5)_T$ *versus the alternative hypothesis of* $\widetilde{H}_1 : \widetilde{\mu} \succ_M (4.5, 5, 5.5)_T$.

TABLE 5.31
Data set in Exercise 5.8.

No.	Males	Females
1	$(60,,65,70)_T$	$(50,55,60)_T$
2	$(70,80,90)_T$	$(60,65,70)_T$
3	$(50,70,80)_T$	$(80,90,100)_T$
4	$(50,55,60)_T$	$(90,100,110)_T$
5	$(80,90,100)_T$	$(90,100)_T$
6	$(70,80,90)_T$	$(55,60,75)_T$
7	$(50,80)_T$	$(70,85,90)_T$
8	$(50,60,70)_T$	$(100,110,120)_T$
9	$(65,75,95)_T$	$(80,95,110)_T$
10	$(50,70,100)_T$	$(90,110,120)_T$

TABLE 5.32
Data set in Exercise 5.9.

No.	$\widetilde{x}_i$	No.	$\widetilde{x}_i$
1	$(4.3,4.5,4.8)_T$	6	$(5.2,5.5,5.8)_T$
2	$(4.6,5,5.5)_T$	7	$(4.1,4.6,5)_T$
3	$(4.7,5.1,5.5)$	8	$(5,5.3,5.6)_T$
4	$(4.6,5,5.3)_T$	9	$(4.2,4.7,5.2)_T$
5	$(4.4,4.7,4.9)_T$	10	$(4.7,5,5.3)_T$

Exercise 5.10 *Three organic wastes were added at three rates of 0%, 1.25%, and 2.5% to air-dried contaminated soil (w/w) and mixed well to get a homogenized soil-organic fertilizer. They were filled in plastic pots of 3 (kg) and 5 fenugreek seeds were sown in each pot and all pots were irrigated with distilled water, during the growing season. Sixty days after sowing, the above plant parts were harvested, dried, powdered by an electric grinder, and kept in an electric oven at 5500 C° followed by acid digestion to obtain the plant extracts. The Pb and Cd concentrations were determined. The data set includes the effect of three levels (0%, 1.25%, and 2.5%) of municipal waste compost, manure sheep, and sewage sludge on Cd concentration in the root of the corn plant. Some unavoidable aspects of the experiment resulted in the vagueness of the Atomic Absorption Spectrophotometer data. Here, such data can be regarded as a fuzzy random sample of a fuzzy normal random variable with a fuzzy mean of* $\widetilde{\mu}$ *and exact variance of* σ^2. *The observed fuzzy data can be represented as* $\widetilde{x}_{ij} = (0.8x_{ij}, x_{ij}, 1.2x_{ij})_T$ *where* x_{ij} *are listed in Table 5.33. Test whether* $\widetilde{\mu}_i$*s are equal for each organic fertilizer at the significance level of* 0.05, *where* $i = 1, 2, 3$. *In other words, we want to decide to accept only one*

of the following hypotheses:

$$\begin{cases} \widetilde{H}_0: & \widetilde{\mu}_1 = \widetilde{\mu}_2 = \widetilde{\mu}_3, \\ \widetilde{H}_1: & \widetilde{\mu}_i \neq \widetilde{\mu}_j \text{ for at least one } i \text{ and } j. \end{cases}$$

for all three studied rates (0%, 1.25%, and 2.5%) at the significance level of $\delta = 0.05$*. If there is evidence to reject* $\widetilde{H}_0$*, analyze the pattern of difference between the mean values. Also, check the homogeneity of variances assumption.*

Exercise 5.11 *Prove Lemma 5.3.*

Exercise 5.12 *Prove Lemma 5.4.*

Exercise 5.13 *Prove Lemma 5.5.*

Exercise 5.14 *Prove Lemma 5.6.*

Exercise 5.15 *Prove Lemma 5.7.*

Exercise 5.16 *Prove Lemma 5.8.*

Exercise 5.17 *Prove Lemma 5.9.*

Exercise 5.18 *Prove Lemma 5.10.*

Exercise 5.19 *Prove Lemma 5.11.*

Exercise 5.20 *Prove Lemma 5.12.*

Exercise 5.21 *Prove Lemma 5.13.*

Exercise 5.22 *Prove Lemma 5.14.*

Exercise 5.23 *Prove Lemma 5.15.*

5.5 Glossary

Fuzzy ANOVA: An extension of the ANOVA that allows for comparing the fuzzy means of normal fuzzy random variables and homogeneity of variances.

Fuzzy hypothesis: A fuzzy mathematical form to construct the null and alternative hypotheses.

TABLE 5.33
Fuzzy data in Exercise 5.10.

Municipal waste compost	Cadmium concentration
0%	3.88 4.43 3.88 3.33
1.25%	6.4 6.4 6.64 6.17
2.50%	0.77 0.77 0.7 0.85
Manure Sheep	Cadmium concentration
0%	2.88 3.43 5.88 4.33
1.25%	1.82 2.16 2.16 2.50
2.50%	4.37 4.37 5.1 3.65
sewage sludge	Cadmium concentration
0%	3.11 2.23 6.88 1.33
1.25%	2.67 2.67 2.64 2.7
2.50%	5.86 5.86 6.1 5.63

Fuzzy hypothesis testing: A generalization for the procedure of assessing whether fuzzy sample data is consistent or otherwise with fuzzy hypotheses made about the population.

Fuzzy test: An extension of the classical binary decision to accept or reject the null hypothesis in fuzzy domain.

Normal fuzzy random variable: A fuzzy random variable induced by normal random variable with fuzzy mean and exact variance.

6

*Fuzzy Quality Control and Reliability Systems for **FRVs***

The random variable was adopted to describing the objective randomness in the conventional quality control charts and reliability analysis. As one of the objective facts, fuzziness exists in every problem related to real-life. Therefore, fuzzy quality control and fuzzy reliability can help to combine the classical quality control process and reliability theory with fuzzy mathematics. Fuzzy control charts have been well reviewed in the last decades by Zavvar Sabegha et al. [221] (before 2014) and Hesamian et al. [73] (2014 – 2018). For the recent works on control charts based on fuzzy information, see [4, 9, 70, 108, 109, 135, 158, 159, 173]. Assessment of system reliability based on fuzzy data has been also extensively investigated by several authors during the last decades (see, for instance, [5, 45, 66, 96, 98, 120, 119, 150, 148, 177]).

In this chapter, the problem of constructing the most common control processes and evaluating system reliability will be considered in which the observed data are described by **FRV**s.

6.1 Fuzzy Control Charts and Capability Indices Based on NFRVs

Quality control refers to the set of procedures adopted to ensure the compliance of a product or service with a defined set of quality criteria or some specific requirements set by the client or customer. Two major aspects of statistical quality control processes include the establishment of well-defined controls and process capability. These criteria help to standardize both the production and reactions to potential quality issues [139]. The most common quality control technique is the statistical control chart, which has become a vital tool for professionals who seek to improve the quality of their products. In reality, however, some elements of the control chart processes are often observed or defined imprecisely. Specifically, fuzzy data may occur during a control process. For instance, electronic circuit thickness is one of the most important quality characteristics in a process that produces electronic boards for vacuum cleaners. However, due to environmental conditions and the limited

DOI: 10.1201/9781003248644-6

resolution of the measuring instrument, the expert can express each measured item in the form of a fuzzy quantities. For such cases, the theory of fuzzy sets can help to construct a statistical control chart where observed data are fuzzy quantities.

In this section, a fuzzy quality control process is presented by extending (1) some common control charts, (2) the most commonly used process capability, and (3) statistical hypothesis testing of the proposed fuzzy process capability in cases where the observed data are outcomes of **FRVs**. Accordingly, a method will be proposed to construct the most interesting fuzzy control charts, namely $\overline{x}$-chart, s-chart, and exponentially-weighted moving average. A degree of violence is also suggested to verify the degree to which a subgroup fuzzy sample mean is excluded from specific fuzzy control limits.

6.1.1 Fuzzy $\overline{x}$-chart and s-chart for NFRVs

Recall that $\widetilde{X} = (X^L, X, X^U)_T$ is said to be a **NFRV** with a fuzzy mean of $\widetilde{\mu} = (\mu; l_\mu, r_\mu)_T$ and exact mean of σ^2 (denoted by $\widetilde{X} \sim N(\widetilde{\mu}, \sigma^2)$ if:

1) $\widetilde{\mu} = (\mu; l_\mu, r_\mu)_T$,

2) $X^L = Y - l_\mu + \epsilon$, $X = Y + \epsilon$ and $X^U = Y + r_\mu + \epsilon$,

3) $Y \sim N(\mu, \sigma^2/2)$ and $\epsilon \sim N(0, \sigma^2/2)$ are independent random variables,

This section constructs a fuzzy quality control chart based on m **FRS** subgroups from $N(\widetilde{\mu}, \sigma^2)$.

Definition 6.1 *Assume $\widetilde{X}_{i1}, \cdots, \widetilde{X}_{im}$, $i = 1, 2, \ldots, n$, as a subgroup, each of size m, from $N(\widetilde{\mu}, \sigma^2)$ with unknown σ^2. The fuzzy control limits for $\widetilde{\mu}$ (say $\widetilde{\overline{x}}$-chart) are defined to be exact values as:*

1) Fuzzy lower control limit:

$$\widetilde{LCL} = \widetilde{\overline{\overline{x}}} \ominus A_3 \overline{s}, \tag{6.1}$$

2) Fuzzy Central Control Limit:

$$\widetilde{CL} = \widetilde{\overline{\overline{x}}}, \tag{6.2}$$

3) Fuzzy upper control limit:

$$\widetilde{UCL} = \widetilde{\overline{\overline{x}}} \oplus A_3 \overline{s}, \tag{6.3}$$

where

1) $\widetilde{\overline{\overline{x}}} = \frac{1}{nm} \otimes (\bigoplus_{i=1}^{n} \bigoplus_{j=1}^{m} \widetilde{x}_{ij})$ is the average of all the fuzzy observations,

2) $\bar{s} = \frac{1}{m}\sum_{i=1}^{m} s_i$ *is the average of* m *standard deviations where*

$$s_i = \sqrt{\frac{1}{m-1}\sum_{j=1}^{m} d_2^2(\widetilde{x}_{ij}, \widetilde{\bar{\bar{x}}})}, \tag{6.4}$$

in which

$$d_2^2(\widetilde{A}, \widetilde{B}) = \frac{(a^L - b^L)^2 + (a - b)^2 + (a^U - b^U)^2}{3}, \tag{6.5}$$

*is the square error distance between to **TFNs** of* $\widetilde{A} = (a^L, a, a^U)_T$ *and* $\widetilde{B} = (b^L, b, b^U)_T$.

3) $A_3 = \frac{3}{c_4\sqrt{n}}$ *is a constant which depends on* n *[139].*

Remark 6.1 *Let* σ^2 *be an unknown quantity. We know that* $S_i^2 = \frac{1}{m-1}\sum_{j=1}^{m} d_2^2(\widetilde{X}_{ij}, \widetilde{\bar{X}}_i)$ *is an unbiased (and consistent) estimator of* σ^2 *for any* $i = 1, 2, \ldots, m$. *Therefore* $\widehat{\sigma}^2 = \bar{s}$ *can be used to estimate* σ^2 *in* $\widetilde{\bar{x}}$*-chart.*

A classical control chart is a graph used to study the changes in a process over time [139]. In some cases, some points may be beyond the control limits due to some specific reasons. Such a special reason should be found and permanently removed from the process. However, based on an **FRS**, the concepts of 'under control' or 'out of control' are vague. Hence, a criterion is required to describe the extent to which a subgroup of fuzzy sample means (or a fuzzy sample variance) is under control or out of control. Here, a criterion is utilized to describe the degree of exclusion from the specific fuzzy control limits using the preference ranking criterion, P_d, discussed in chapter 1.

Definition 6.2 *[73] Consider a **FRS** from a **NFRV**. Let* $[\widetilde{LCL}, \widetilde{UCL}]$ *be a fuzzy control interval for* $\widetilde{\mu}$. *Then, the degree that a fuzzy sample mean* $\widetilde{\bar{x}}^*$ *of a subgroup is excluded from the specific fuzzy control interval* $[\widetilde{LCL}, \widetilde{UCL}]$*, called degree of violence, can be defined by:*

$$D_v = \max\left\{P_d\{\widetilde{\bar{x}}^* \succ \widetilde{UCL}\}, P_d\{\widetilde{\bar{x}}^* \prec \widetilde{LCL}\}\right\}. \tag{6.6}$$

Remark 6.2 *From Definition 6.2, we can say that* $\widetilde{\bar{x}}^*$ *is out of control if* $D_v \geq 0.5$. *However, the linguistic interpretation of* D_v *can be derived in ten cases as shown in Table 6.1. Moreover, for an observed fuzzy mean of* $\widetilde{\bar{x}}^*$*, the degree of violence* D_v *has the following properties:*

1. $D_v \in [0, 1]$,

2. $1 - D_v = \min\left\{P_d(\widetilde{\bar{x}}^* \prec \widetilde{UCL}), P_d(\widetilde{\bar{x}}^* \succ \widetilde{LCL})\right\}$ *which can be interpreted as the degree to which* $\widetilde{\bar{x}}^*$ *is under control.*

TABLE 6.1
A linguistic interpretation of degree of violence D_v.

No.	Range of D_1	Interpretation
1	$D_v \in [0.0, 0.05)$	$\widetilde{\overline{x}}^*$ is complectly under control
2	$D_v \in [0.05, 0.15)$	$\widetilde{\overline{x}}^*$ is absolutely under control
3	$D_v \in [0.15, 0.25)$	$\widetilde{\overline{x}}^*$ is strongly under control
4	$D_v \in [0.25, 0.35)$	$\widetilde{\overline{x}}^*$ is more or less under control
5	$D_v \in [0.35, 0.5)$	$\widetilde{\overline{x}}^*$ is weakly under control
6	$D_v \in (0.50, 0.65]$	$\widetilde{\overline{x}}^*$ is weakly out of control
7	$D_v \in (0.65, 0.75]$	$\widetilde{\overline{x}}^*$ is more or less out of control
8	$D_v \in (0.75, 0.85]$	$\widetilde{\overline{x}}^*$ is strongly out of control
9	$D_v \in (0.85, 0.95]$	$\widetilde{\overline{x}}^*$ is absolutely out of control
10	$D_v \in (0.95, 1]$	$\widetilde{\overline{x}}^*$ is completely out of control

3. $D_v = 1$ *if and only if* $P_d\{\widetilde{\overline{x}}^* \succ \widetilde{UCL}\} = 1$ *or* $P_d\{\widetilde{\overline{x}}^* \prec \widetilde{LCL}\} = 1$.

4. $D_v = 0$ *if and only if* $P_d\{\widetilde{\overline{x}}^* \prec \widetilde{UCL}\} = 1$ *and* $P_d\{\widetilde{\overline{x}}^* \succeq \widetilde{LCL}\} = 1$.

Definition 6.3 *Let* $\widetilde{X}_{i1}, \cdots, \widetilde{X}_{im}$, $i = 1, 2, \ldots, n$ *be a subgroup, with a size of* m, *from* $N(\widetilde{\mu}, \sigma^2)$ *with unknown* σ^2. *The control limits of* σ^2 *(s-chart) are defined as follows:*

1) Lower control limit:

$$LCL = B_3\overline{s}, \tag{6.7}$$

2) Central control limit:

$$CL = \overline{s}, \tag{6.8}$$

3) Upper control limit:

$$UCL = B_4\overline{s}, \tag{6.9}$$

where $B_3 = 1 - \frac{3}{c_4}\sqrt{1 - c_4^2}$ *and* $B_4 = 1 + \frac{3}{c_4}\sqrt{1 - c_4^2}$ *[139].*

The possible applications of the proposed fuzzy control limits $\widetilde{\overline{x}}$-chart is examined in the quality control process by the following numerical evaluation.

Example 6.1 *Table 6.2 shows a random sample of electronic circuit thickness in the production of electronic boards for vacuum cleaners in a factory. The data are measured as* ***TFNs****. Here, it is assumed that the circuit thickness is an* ***NFRV*** *with an unknown fuzzy mean,* $\widetilde{\mu}$, *and exact variance* σ^2. *The values of* $\widetilde{\overline{x}}_i$ *and* s_i^2 *are listed in Table 6.3 for each sample. Accordingly, we get* $\widetilde{\overline{\overline{x}}} = (69.57, 73.07, 75.63)_T$ *and* $\overline{s}^2 = 5.238$. *Since* $A_3 = 0.148$, *it can be shown*

TABLE 6.2
Data set in Example 6.1.

Sample No.	Subgroups reported by **TFNs**		
	$\widetilde{x}_{i1}$	$\widetilde{x}_{i2}$	$\widetilde{x}_{i3}$
1	$(70.33, 71.27, 75.10)_T$	$(70.19, 71.40, 74.15)_T$	$(64.13, 68.67, 71.13)_T$
2	$(67.34, 70.16, 75.93)_T$	$(68.34, 73.72, 78.71)_T$	$(69.47, 71.79, 75.16)_T$
3	$(72.97, 77.53, 80.89)_T$	$(67.12, 74.44, 77.77)_T$	$(73.73, 75.35, 77.71)_T$
4	$(71.85, 74.15, 77.94)_T$	$(75.17, 75.85, 78.84)_T$	$(71.78, 72.12, 74.75)_T$
5	$(67.30, 70.12, 71.18)_T$	$(71.19, 72.1174.79)_T$	$(73.06, 74.71, 76.00)_T$
6	$(76.64, 78.67, 79.83)_T$	$(75.43, 78.56, 83.56)_T$	$(74.45, 79.00, 82.68)_T$
7	$(73.11, 76.00, 77.99)_T$	$(74.18, 77.06, 85.67)_T$	$(75.21, 76.47, 82.04)_T$
8	$(60.65, 62.85, 67.57)_T$	$(53.86, 57.13, 67.14)_T$	$(61.36, 64.13, 69.11)_T$
9	$(70.96, 74.31, 83.16)_T$	$(71.23, 72.92, 79.14)_T$	$(69.93, 73.16, 82.15)_T$
10	$(73.74, 75.18, 75.49)_T$	$(76.35, 77.70, 81.63)_T$	$(75.16, 75.89, 79.16)_T$
11	$(67.12, 72.25, 75.31)_T$	$(63.37, 66.58, 77.31)_T$	$(64.13, 68.15, 71.88)_T$
12	$(73.39, 75.90, 80.46)_T$	$(67.33, 74.03, 83.53)_T$	$(71.14, 71.98, 81.87)_T$
13	$(70.61, 70.65, 72.34)_T$	$(66.32, 68.77, 79.53)_T$	$(68.59, 71.49, 74.45)_T$
14	$(77.93, 80.03, 82.25)_T$	$(75.81, 76.57, 81.81)_T$	$(75.95, 78.22, 80.02)_T$
15	$(74.49, 76.32, 78.29)_T$	$(66.67, 71.89, 74.48)_T$	$(69.56, 73.02, 76.13)_T$

that the membership function of the fuzzy control limits ($\widetilde{LCL}$ and $\widetilde{UCL}$) of $\widetilde{\overline{x}}$-chart are:

$$\widetilde{LCL} = (69.57, 73.3, 75.63)_T \ominus \frac{\sqrt{5.238}}{0.148\sqrt{15}} = (65.587, 69.317, 71.647)_T,$$

$$\widetilde{CL} = (69.57, 73.3, 75.63)_T,$$

$$\widetilde{UCL} = (69.57, 73.3, 75.63)_T \oplus \frac{\sqrt{5.238}}{0.148\sqrt{15}} = (73.553, 77.283, 79.613)_T.$$

*The plots of fuzzy control limits for each fuzzy sample mean is shown in Fig. 6.1. Along the degree of violence introduced in Eq. (6.6), we can evaluate the degree to which a subgroup fuzzy mean $\widetilde{\overline{X}}_i$ lies between (or out of) fuzzy control limits of $\widetilde{LCL}$ and $\widetilde{UCL}$. The results are shown in Table 6.3 which reveal that (1) subgroup 8 is completely out of control, (2) subgroup 11 is weakly under control, while subgroups 7 and 12 are absolutely under control, and (3) the others are completely under control. Further, the control limits of **s**-chart are $LCL = 0$, $CL = \overline{s} = 2.2888$, and $UCL = B_4\overline{s} = 2.568 \times 2.888 = 5.8755$. This shows that the **s**-chart is in control that is the sample standard deviations of circuit thickness are similar.*

6.1.2 Fuzzy EWMA chart based on NFRVs

An exponentially-weighted moving average (**EWMA**) chart is an alternative to Shewhart control charts and can serve as an effective tool for the detection

TABLE 6.3
$\widetilde{\overline{x}}_i$, s_i^2 and $(D_v)_i$ in Example 6.1.

Sample No.	$\widetilde{\overline{x}}_i$	s_i^2	$(D_v)_i$
1	$(68.220, 70.447, 73.460)_T$	6.40	0
2	$(68.383, 71.890, 76.600)_T$	2.602	0
3	$(71.270, 75.773, 78.790)_T$	6.30	0
4	$(72.930, 74.040, 77.180)_T$	3.953	0
5	$(70.520, 72.310, 73.960)_T$	6.717	0
6	$(75.507, 78.743, 82.020)_T$	1.680	0
7	$(74.167, 76.510, 81.900)_T$	5.380	0.09
8	$(58.620, 61.370, 66.440)_T$	10.700	1
9	$(70.707, 73.463, 81.480)_T$	1.800	0.0008
10	$(75.083, 76.257, 78.760)_T$	4.310	0
11	$(64.870, 68.990, 74.830)_T$	6.680	0.49
12	$(70.620, 73.970, 81.953)_T$	5.19	0.057
13	$(68.510, 70.303, 75.440)_T$	6.74	0
14	$(76.563, 78.273, 81.360)_T$	1.932	0.006
15	$(70.240, 73.740, 76.300)_T$	5.238	0

of shift in the small persistent process. The statistical **EWMA** charts, introduced by Roberts [167], have received considerable attention for detecting small changes in the process mean or its variability [35, 144]. It assigns weight to the observations in geometrically decreasing order so that the most recent observations have higher contribution while the contribution of the oldest ones is very little.

In this section, the **EWMA** control limits are developed to detect small shifts in the process based on a random sample of **NFRVs**. For this purpose, a notion of fuzzy **EWMA** statistic is first introduced inspired by Hesamian et al. [70].

Definition 6.4 *Assume that $\widetilde{X} \sim N(\widetilde{\mu}, \sigma^2)$ and $\widetilde{X}_{i1}, \cdots, \widetilde{X}_{im}$, $i = 1, 2, ..., n$ are n **FRS**. The fuzzy **EWMA** statistic is defined as follows:*

$$\widetilde{Z}_i = (\gamma \otimes \widetilde{\overline{X}}_i) \oplus \big((1-\gamma) \otimes \widetilde{Z}_{i-1}\big), \tag{6.10}$$

where $i \in \{1, 2, \ldots, n\}$ is the sample number, γ denotes the smoothing constant such that $0 \leq \gamma \leq 1$, $\widetilde{\overline{X}}_i$ shows the fuzzy average of the ith subgroup and $\widetilde{Z}_0 = \widetilde{\mu}$.

Remark 6.3 *It should be mentioned that the $\widetilde{Z}_{i-1}$ quantity can be regarded as the fuzzy past information and its initial value (i.e., $\widetilde{Z}_0$) is taken equal to the fuzzy target mean of $\widetilde{\mu}$ or the fuzzy average of the preliminary **FRSs** ($\widetilde{\overline{\overline{X}}} = \frac{1}{mn} \otimes (\bigoplus_{i=1}^{n} \bigoplus_{j=1}^{m} \widetilde{X}_{ij})$). The parameter γ determines the rate at which the previous information comes into the calculation of the fuzzy **EWMA** statistic. A large value of γ gives more weight to the current information and less weight to the past information while a small value of γ gives more weight to the past*

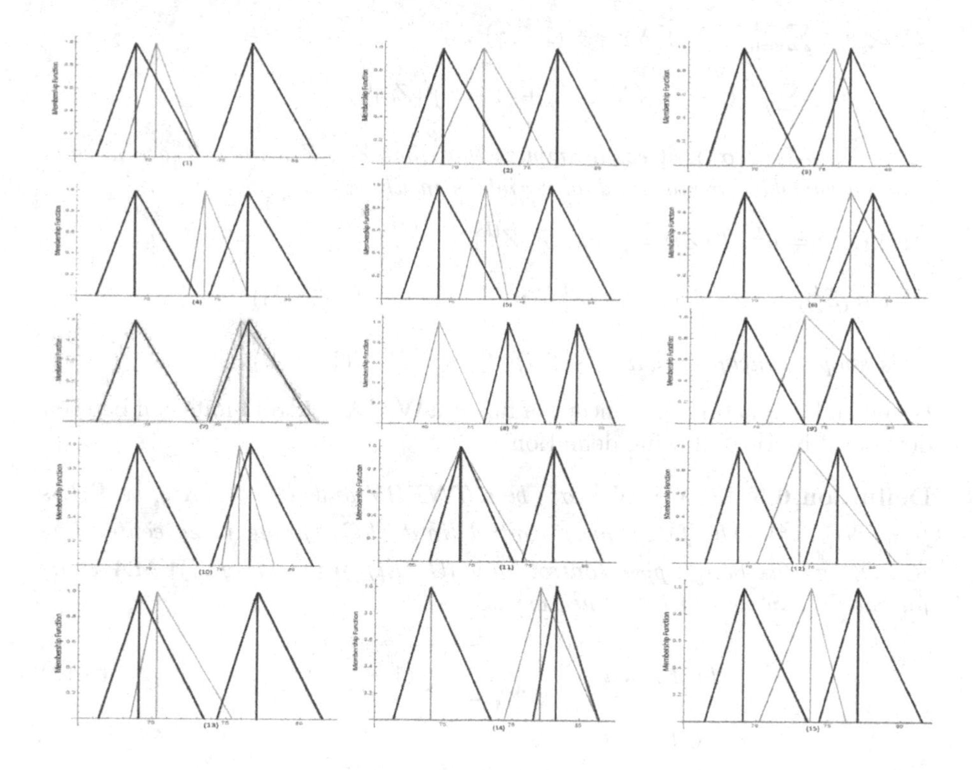

FIGURE 6.1
Fuzzy control limits for each fuzzy sample means in Example 6.1.

*information and less weight to the current information. It should be noted that the fuzzy **EWMA** will be reduced to $\widetilde{\overline{x}}$-chart if $\gamma = 1$.*

Lemma 6.1 *Let $\widetilde{X} \sim N(\widetilde{\mu}, \sigma^2)$ be a **TNFRV** and $\widetilde{X}_{i1}, \cdots, \widetilde{X}_{im}$ be a **FRS** from $\widetilde{X}$. Then, $\widetilde{Z}_i$ introduced in Eq.* (6.10) *is a **TNFRV** as $\widetilde{Z}_i \sim N(\widetilde{\mu}, \frac{\sigma^2}{n}(\frac{\gamma}{2-\gamma}(1-(1-\gamma)^i)))$ for each sample number.*

Proof *By solving the recurrence relation of Eq.* (6.10), *note that $\widetilde{Z}_i$ can be evaluated as follows:*

$$\widetilde{Z}_i = \left(\gamma \otimes \bigoplus_{j=0}^{i-1} ((1-\gamma)^j \otimes \widetilde{\overline{X}}_{i-j})\right) \oplus \left((1-\gamma)^i \otimes \widetilde{Z}_0\right) \tag{6.11}$$
$$= (Z_i^L, Z_i, Z_i^U)_T.$$

where

1) $Z_i^L = \gamma \sum_{j=0}^{i-1} (1-\gamma)^j (\overline{X}_{i-j})^L + (1-\gamma)^i (Z_0)^L,$

2) $Z_i = \gamma \sum_{j=0}^{i-1}(1-\gamma)^j \overline{X}_{i-j} + (1-\gamma)^i Z_0,$

3) $Z_i^U = \gamma \sum_{j=0}^{i-1}(1-\gamma)^j (\overline{X}_{i-j})^U + (1-\gamma)^i (Z_0)^U.$

Since $\widetilde{X}_i \sim N(\widetilde{\mu}, \sigma^2)$, *it easily implies that that* Z_i^L, Z_i, *and* Z_i^U *are normal random variable normal random variables in which*

1) $E(Z_i^L) = \mu^L$, $E(Z_i) = \mu$ *and* $E(Z_i^U) = \mu^U$,

2) $var(Z_i^L) = var(Z_i) = var(Z_i^U) = \frac{\sigma^2}{n}(\frac{\gamma}{2-\gamma}(1-(1-\gamma)^i)),$

This simply concludes that $\widetilde{Z}_i \sim N(\widetilde{\mu}, \frac{\sigma^2}{n}(\frac{\gamma}{2-\gamma}(1-(1-\gamma)^i)))$.

Based on Lemma 6.1, the concept of fuzzy **EWMA** control limits can be then developed by the following definition.

Definition 6.5 *Let* $\widetilde{X} \sim N(\widetilde{\mu}, \sigma^2)$ *be a* ***TNFRV*** *and* $\widetilde{X}_{i1}, \cdots, \widetilde{X}_{im}$ *be* ***FRSs*** *from* $N(\widetilde{\mu}, \sigma^2)$. *The fuzzy lower control limit* ($\widetilde{LCL}_i$), *the fuzzy center line* ($\widetilde{CL}_i$), *and the fuzzy upper control limit* ($\widetilde{UCL}_i$) *of the fuzzy* ***EWMA*** *chart for the* i^{th} *sample can be defined as:*

$$\widetilde{LCL}_i = \widetilde{\mu} \ominus L\frac{\sigma}{\sqrt{n}}\sqrt{\frac{\gamma}{2-\gamma}(1-(1-\gamma)^i)}, \tag{6.12}$$

$$\widetilde{CL}_i = \widetilde{\mu},$$

$$\widetilde{UCL}_i = \widetilde{\mu} \oplus L\frac{\sigma}{\sqrt{n}}\sqrt{\frac{\gamma}{2-\gamma}(1-(1-\gamma)^i)},$$

where L is the width of the fuzzy control limits of the fuzzy ***EWMA*** *chart.*

Remark 6.4 *It is worth noting that a combination of selecting* $L = 3$ *and a larger value of* $\gamma \in [0.05, 0.25]$ *works reasonably well for a* ***EWMA*** *chart [144]. However, there is an advantage in reducing the width of the control limits in cases where* $L \in [2.6, 2, 8]$ *and* $\gamma \leq 0.1$ *[35, 144].*

Remark 6.5 *Assume that* $\widetilde{X} \sim N(\widetilde{\mu}, \sigma^2)$ *and* $\widetilde{x}_1, \cdots, \widetilde{x}_n$ *is a* ***FRS*** *from* $\widetilde{X}$. *If the* i^{th} *sample number is moderately large, the proposed fuzzy* ***EWMA*** *control limits are reduced as follows:*

$$\widetilde{LCL}_i = \widetilde{\mu} \ominus L\frac{\sigma}{\sqrt{n}}\sqrt{\frac{\gamma}{2-\gamma}}, \tag{6.13}$$

$$\widetilde{CL}_i = \widetilde{\mu},$$

$$\widetilde{UCL}_i = \widetilde{\mu} \oplus L\frac{\sigma}{\sqrt{n}}\sqrt{\frac{\gamma}{2-\gamma}}.$$

Remark 6.6 *If* σ^2 *is unknown then* $\overline{s} = \frac{1}{n}\sum_{i=1}^{n} s_i^2$ *can be used to estimate* σ^2 *in this situation. Thus, the fuzzy lower control limit* ($\widetilde{LCL}$), *the fuzzy center* *line* $(\widetilde{CL})$, *and the fuzzy upper control limit* $(\widetilde{UCL})$ *of the fuzzy* **EWMA** *control limits can be evaluated as follows:*

$$\widetilde{LCL}_i = \widetilde{\mu} \ominus L\frac{\overline{s}}{\sqrt{n}}\sqrt{\frac{\gamma}{2-\gamma}(1-(1-\gamma)^i)}, \tag{6.14}$$

$$\widetilde{CL}_i = \widetilde{\mu},$$

$$\widetilde{UCL}_i = \widetilde{\mu} \oplus L\frac{\overline{s}}{\sqrt{n}}\sqrt{\frac{\gamma}{2-\gamma}(1-(1-\gamma)^i)}.$$

Moreover, if $\widetilde{\mu}$ *is an unknown parameter, it is enough to replace* $\widetilde{\mu}$ *with* $\widetilde{\overline{\overline{x}}} = \frac{1}{nm} \otimes (\bigoplus_{i=1}^{n} \bigoplus_{j=1}^{m} \widetilde{x}_{ij})$ *in the fuzzy* **EWMA** *limits. Therefore, the proposed fuzzy* **EWMA** *control limits can be evaluated as follows:*

$$\widetilde{LCL}_i = \widetilde{\overline{\overline{x}}} \ominus L\frac{\overline{s}}{\sqrt{n}}\sqrt{\frac{\gamma}{2-\gamma}(1-(1-\gamma)^i)}, \tag{6.15}$$

$$\widetilde{CL}_i = \widetilde{\overline{\overline{x}}},$$

$$\widetilde{UCL}_i = \widetilde{\overline{\overline{x}}} \oplus L\frac{\overline{s}}{\sqrt{n}}\sqrt{\frac{\gamma}{2-\gamma}(1-(1-\gamma)^i)}.$$

Example 6.2 *An engine manufacturer uses a special kind of molds to make piston rings. The quality engineers want to assess the process capability. The data consists of 25 samples, each with 5 observations, where the quality characteristic is the internal diameter of the ring. The data set was reported by* **TFNs** *as* $\widetilde{x}_{ij} = (x_{ij}, l_{x_{ij}}, r_{x_{ij}})_T$ *(Table 6.4). However, using molds for production led to some deviations in the production process which motivated us to apply the proposed fuzzy* **EWMA** *control chart to monitor the small deviations of each fuzzy sample for this data set. Since both* $\widetilde{\mu}$ *and* σ^2 *are unknown quantities (Eq. (6.15)) can be utilized to evaluate lower and upper bounds of the proposed fuzzy* **EWMA** *control chart with* $L = 2.7$, $\gamma = 0.3$. *For this purpose, note that* $\widehat{\widetilde{\mu}} = \widetilde{\overline{\overline{x}}} = \frac{1}{125} \otimes (\bigoplus_{i=1}^{25} \bigoplus_{j=1}^{5} \widetilde{x}_{ij}) = (73.9796, 74.001, 74.0215)_T$ *and* $\widehat{\sigma}^2 = \frac{1}{25}\sum_{i=1}^{25} s_i^2 = 0.0048$ *where* $s_i^2 = \frac{1}{4}\sum_{j=1}^{5} d_2^2(\widetilde{x}_{ij}, \widetilde{\overline{x}}_i)$. *Therefore, the fuzzy control limits of the fuzzy* **EWMA** *charts for each sample can be examined as follows:*

$$\begin{aligned}\widetilde{LCL}_i &= (73.9796, 74.001, 74.0215)_T \ominus 2.7\frac{0.0048}{5}\sqrt{\frac{0.3}{1.7}(1-(0.7)^i)}\\ &= (73.9785\sqrt{1-(0.7)^i}, 73.9999\sqrt{1-(0.7)^i}, 74.0226\sqrt{1-(0.7)^i})_T,\\ \widetilde{CL}_i &= (73.9796, 74.001, 74.0215)_T,\\ \widetilde{UCL}_i &= (73.9796, 74.001, 74.0215)_T \oplus 2.7\frac{0.0048}{5}\sqrt{\frac{0.3}{1.7}(1-(0.7)^i)}\\ &= (73.9807\sqrt{1-(0.7)^i}, 74.0021\sqrt{1-(0.7)^i}, 74.0226\sqrt{1-(0.7)^i})_T.\end{aligned}$$

The fuzzy control limits of the fuzzy **EWMA** *charts for each sample are summarized in Table 6.5. To monitor the process conditions for each fuzzy sample* *mean, the degrees of violence are summarized in Table 6.4 for all 25 fuzzy sample means. Since* $\min_{i\in\{1,2,\ldots,25\}} D_{vi} \geq 0.5$, *it can be concluded that the process is under control.*

Remark 6.7 *It should be noted that Erginel and Senturk [44] also generalized the conventional* ***EWMA*** *in a fuzzy environment. Their work relied on fuzzy control limits based on fuzzy data and fuzzy variance. For modeling fuzzy* ***EWMA*** *control charts,* ***TFN****s and fuzzy control limits were also transformed by the median transformation technique to investigate the process condition. Shu et al. [182] proposed a fuzzy multivariate exponentially by combining multivariate statistical quality control and the fuzzy set theory.*

6.1.3 Process capability indices based on NFRVs

Process capability is a measure of the producer's capability to produce a product in compliance with the customer's level of tolerances. These measures ensure that the measuring equipment and measurement processes are appropriate for their intended use and can fulfill the quality objectives. Vännman [196] defined a class of capability indices based on two non-negative parameters of u and v as:

$$\mathcal{C}_p(u,v) = \frac{d - u|\mu - m|}{3\sqrt{\sigma^2 + v(\mu - T)^2}},$$

where μ is the process mean, σ shows the process standard deviation, $d = (USL - LSL)/2$, where USL and LSL are the upper and lower specification limits, $m = (USL + LSL)/2$, is the specification center, and T denotes the target value, respectively. The four basic indices of $\mathcal{C}_p$, $\mathcal{C}_{pm}$, $\mathcal{C}_{pk}$, and $\mathcal{C}_{pmk}$ are special cases of $\mathcal{C}_p(u,v)$ by considering $u = 0$ or 1 and $v = 0$ or 1. $\mathcal{C}_p$ only takes into account the process variance. It fails to detect the departure of the process mean from the specification center, and therefore cannot be used to fully evaluate the process capability. The six quality conditions and their corresponding $\mathcal{C}_p$ values are summarized in Table 6.6 [156]. $\mathcal{C}_{pk}$ takes into account both the process mean and the process variance. However, $\mathcal{C}_{pk}$ is not able to detect departure from the target value. Although $\mathcal{C}_{pm}$ overcomes this drawback by taking into account $(\mu - T)^2$, it fails in detecting the location of the process mean in the interval (LSL, USL). $\mathcal{C}_{pmk}$ considers the process variability, a departure from the target value and location of the process mean in (LSL, USL). The relation between process capabilities of $\mathcal{C}_p$, $\mathcal{C}_{pk}$, $\mathcal{C}_{pm}$, and $\mathcal{C}_{pmk}$ is listed as follows:

1) $\mathcal{C}_{pk} = \mathcal{C}_p - \frac{|\mu - m|}{3\sigma}$,

2) $\mathcal{C}_{pm} = (\frac{1}{\sqrt{1+\frac{|\mu-T|^2}{\sigma^2}}})\mathcal{C}_p$,

3) $\mathcal{C}_{pmk} = (\frac{1}{\sqrt{1+\frac{|\mu-T|^2}{\sigma^2}}})\mathcal{C}_{pk}$.

TABLE 6.4
Data set in Example 6.2.

No.	Fuzzy subgroup samples				
1	$(74.030; 0.017, 0.012)_T$	$(74.002; 0.008, 0.037)_T$	$(74.019; 0.040, 0.036)_T$	$(73.992; 0.002, 0.039)_T$	$(74.008; 0.0140.039)_T$
2	$(73.995; 0.021, 0.040)_T$	$(73.992; 0.024, 0.016)_T$	$(74.001; 0.024, 0.024)_T$	$(74.011; 0.028, 0.030)_T$	$(74.004; 0.031, 0.012)_T$
3	$(73.988; 0.001, 0.027)_T$	$(74.024; 0.017, 0.013)_T$	$(74.021; 0.000, 0.035)_T$	$(74.005; 0.038, 0.012)_T$	$(74.002; 0.028, 0.029)_T$
4	$(74.002; 0.022, 0.011)_T$	$(73.996; 0.026, 0.002)_T$	$(73.993; 0.004, 0.004)_T$	$(74.015; 0.010, 0.016)_T$	$(74.009; 0.017, 0.013)_T$
5	$(73.992; 0.002, 0.005)_T$	$(74.007; 0.010, 0.022)_T$	$(74.015; 0.027, 0.007)_T$	$(73.989; 0.010, 0.023)_T$	$(74.014; 0.020, 0.037)_T$
6	$(74.009; 0.036, 0.035)_T$	$(73.994; 0.025, 0.019)_T$	$(73.997; 0.027, 0.005)_T$	$(73.985; 0.036, 0.040)_T$	$(73.993; 0.005, 0.001)_T$
7	$(73.995; 0.002, 0.016)_T$	$(74.006; 0.022, 0.031)_T$	$(73.994; 0.016, 0.006)_T$	$(74.000; 0.009, 0.031)_T$	$(74.005; 0.006, 0.029)_T$
8	$(73.985; 0.028, 0.001)_T$	$(74.003; 0.006, 0.015)_T$	$(73.993; 0.022, 0.027)_T$	$(74.015; 0.026, 0.004)_T$	$(73.988; 0.037, 0.001)_T$
9	$(74.008; 0.019, 0.035)_T$	$(73.995; 0.026, 0.007)_T$	$(74.009; 0.008, 0.013)_T$	$(74.005; 0.039, 0.019)_T$	$(74.004; 0.038, 0.021)_T$
10	$(73.998; 0.039, 0.033)_T$	$(74.000; 0.034, 0.022)_T$	$(73.990; 0.010, 0.021)_T$	$(74.007; 0.003, 0.034)_T$	$(73.995; 0.010, 0.037)_T$
11	$(73.994; 0.031, 0.039)_T$	$(73.998; 0.034, 0.011)_T$	$(73.994; 0.018, 0.032)_T$	$(73.995; 0.038, 0.003)_T$	$(73.990; 0.008, 0.039)_T$
12	$(74.004; 0.016, 0.022)_T$	$(74.000; 0.029, 0.037)_T$	$(74.007; 0.032, 0.019)_T$	$(74.000; 0.018, 0.021)_T$	$(73.996; 0.038, 0.003)_T$
13	$(73.983; 0.005, 0.006)_T$	$(74.002; 0.025, 0.021)_T$	$(73.998; 0.018, 0.004)_T$	$(73.997; 0.012, 0.021)_T$	$(74.012; 0.038, 0.033)_T$
14	$(74.006; 0.008, 0.010)_T$	$(73.967; 0.028, 0.023)_T$	$(73.994; 0.026, 0.031)_T$	$(74.000; 0.025, 0.013)_T$	$(73.984; 0.015, 0.015)_T$
15	$(74.012; 0.001, 0.036)_T$	$(74.014; 0.039, 0.010)_T$	$(73.998; 0.035, 0.032)_T$	$(73.999; 0.017, 0.034)_T$	$(74.007; 0.028, 0.014)_T$
16	$(74.000; 0.026, 0.022)_T$	$(73.984; 0.008, 0.031)_T$	$(74.005; 0.038, 0.028)_T$	$(73.998; 0.032, 0.012)_T$	$(73.996; 0.022, 0.028)_T$
17	$(73.994; 0.004, 0.038)_T$	$(74.012; 0.035, 0.009)_T$	$(73.986; 0.022, 0.028)_T$	$(74.005; 0.018, 0.031)_T$	$(74.007; 0.028, 0.024)_T$
18	$(74.006; 0.034, 0.005)_T$	$(74.010; 0.035, 0.033)_T$	$(74.018; 0.027, 0.037)_T$	$(74.003; 0.035, 0.015)_T$	$(74.000; 0.034, 0.024)_T$
19	$(73.984; 0.029, 0.000)_T$	$(74.002; 0.006, 0.040)_T$	$(74.003; 0.040, 0.035)_T$	$(74.005; 0.006, 0.029)_T$	$(73.997; 0.017, 0.009)_T$
20	$(74.000; 0.027, 0.001)_T$	$(74.010; 0.000, 0.039)_T$	$(74.013; 0.027, 0.013)_T$	$(74.020; 0.036, 0.025)_T$	$(74.003; 0.007, 0.027)_T$
21	$(73.982; 0.004, 0.026)_T$	$(74.001; 0.030, 0.014)_T$	$(74.015; 0.025, 0.009)_T$	$(74.005; 0.027, 0.010)_T$	$(73.996; 0.018, 0.024)_T$
22	$(74.004; 0.022, 0.014)_T$	$(73.999; 0.030, 0.039)_T$	$(73.990; 0.013, 0.039)_T$	$(74.006; 0.028, 0.008)_T$	$(74.009; 0.024, 0.008)_T$
23	$(74.010; 0.029, 0.035)_T$	$(73.989; 0.000, 0.034)_T$	$(73.990; 0.018, 0.005)_T$	$(74.009; 0.038, 0.018)_T$	$(74.014; 0.032, 0.002)_T$
24	$(74.015; 0.023, 0.027)_T$	$(74.008; 0.024, 0.025)_T$	$(73.993; 0.038, 0.011)_T$	$(74.000; 0.007, 0.003)_T$	$(74.010; 0.024, 0.025)_T$
25	$(73.982; 0.013, 0.006)_T$	$(73.984; 0.007, 0.008)_T$	$(73.995; 0.020, 0.007)_T$	$(74.017; 0.016, 0.003)_T$	$(74.013; 0.021, 0.009)_T$

TABLE 6.5
Degrees of violence in Example 6.2.

No.	$\widetilde{LCL}_i$	$\widetilde{UCL}_i$	$\widetilde{\widetilde{x}}_i$	D_{vi}
1	$(40.5200, 40.5315, 40.5438)_T$	$(40.5209, 40.532640.5438)_T$	$(73.994, 74.010, 74.0426)_T$	0.96031
2	$(52.8316, 52.8466, 52.8627)_T$	$(52.8328, 52.8481, 52.8627)_T$	$(73.9753, 74.001, 74.0253)_T$	0.91214
3	$(59.9641, 59.9811, 59.9994)_T$	$(59.9655, 59.9828, 59.9994)_T$	$(73.9909, 74.008, 74.0312)_T$	0.84286
4	$(64.4892, 64.5075, 64.5272)_T$	$(64.4906, 64.5093, 64.5272)_T$	$(73.9872, 74.003, 74.0123)_T$	1.00000
5	$(67.4764, 67.4955, 67.5162)_T$	$(67.4779, 67.4975, 67.5162)_T$	$(73.9892, 74.003, 74.0218)_T$	0.92718
6	$(69.4911, 69.5108, 69.5320)_T$	$(69.4927, 69.5128, 69.5320)_T$	$(73.9704, 73.996, 74.0161)_T$	1.00000
7	$(70.8673, 70.8874, 70.9091)_T$	$(70.8689, 70.8894, 70.9091)_T$	$(73.9888, 74., 74.0228)_T$	0.98313
8	$(71.8150, 71.8354, 71.8573)_T$	$(71.8166, 71.8374, 71.8573)_T$	$(73.9731, 73.997, 74.0066)_T$	1.00000
9	$(72.4710, 72.4915, 72.5137)_T$	$(72.4726, 72.4936, 72.5137)_T$	$(73.9778, 74.004, 74.0229)_T$	0.95542
10	$(72.9267, 72.9474, 72.9696)_T$	$(72.9283, 72.9494, 72.9696)_T$	$(73.9791, 73.998, 74.0273)_T$	0.88402
11	$(73.2439, 73.2647, 73.2871)_T$	$(73.2456, 73.2668, 73.2871)_T$	$(73.968, 73.994, 74.0188)_T$	0.99800
12	$(73.4652, 73.4861, 73.5085)_T$	$(73.4669, 73.4882, 73.5085)_T$	$(74.001; 0.02671, 0.02048)_T$	0.85945
13	$(73.6197, 73.6406, 73.6631)_T$	$(73.6214, 73.6427, 73.6631)_T$	$(73.9782, 73.998, 74.0152)_T$	1.00000
14	$(73.7277, 73.7486, 73.7712)_T$	$(73.7294, 73.7507, 73.7712)_T$	$(73.9746, 73.9951, 74.0134)_T$	1.00000
15	$(73.8032, 73.8241, 73.8467)_T$	$(73.8049, 73.8262, 73.8467)_T$	$(73.982, 74.006, 74.031)_T$	0.99300
16	$(73.8560, 73.8769, 73.8995)_T$	$(73.8577, 73.8790, 73.8995)_T$	$(73.9717, 73.997, 74.0215)_T$	0.98574
17	$(73.8929, 73.9139, 73.9365)_T$	$(73.8946, 73.9160, 73.9365)_T$	$(73.9794, 74.001, 74.0271)_T$	0.99592
18	$(73.9187, 73.9397, 73.9623)_T$	$(73.9204, 73.9418, 73.9623)_T$	$(73.9741, 74.007, 74.0298)_T$	0.89285
19	$(73.9368, 73.9578, 73.9804)_T$	$(73.9385, 73.9599, 73.9804)_T$	$(73.9786, 73.998, 74.0207)_T$	0.95648
20	$(73.9495, 73.9705, 73.9931)_T$	$(73.9512, 73.9726, 73.9931)_T$	$(73.9897, 74.009, 74.03)_T$	0.98306
21	$(73.9583, 73.9793, 74.0019)_T$	$(73.9600, 73.9814, 74.0019)_T$	$(73.979, 74., 74.0166)_T$	1.00000
22	$(73.9645, 73.9855, 74.0081)_T$	$(73.9662, 73.9876, 74.0081)_T$	$(73.9787, 74.002, 74.0235)_T$	0.89506
23	$(73.9689, 73.9899, 74.0125)_T$	$(73.9706, 73.9920, 74.0125)_T$	$(73.9786, 74.002, 74.0209)_T$	0.89134
24	$(73.9719, 73.9929, 74.0155)_T$	$(73.9736, 73.9950, 74.0155)_T$	$(73.9818, 74.005, 74.0233)_T$	0.86065
25	$(73.9740, 73.9950, 74.0176)_T$	$(73.9757, 73.9971, 74.0176)_T$	$(73.9826, 73.998, 74.0049)_T$	1.00000

TABLE 6.6
Quality conditions for $\mathcal{C}_p$.

Quality condition	$\mathcal{C}_p$ value
Super excellent	$\mathcal{C}_p \in I_6 = (2.00, \infty)$
Excellent	$\mathcal{C}_p \in I_5 = (1.67, 2.00]$
Satisfactory	$\mathcal{C}_p \in I_4 = (1.33, 1.67]$
Capable	$\mathcal{C}_p \in I_3 = (1.00, 1.33]$
Inadequate	$\mathcal{C}_p \in I_2 = (0.67, 1.00]$
Poor	$\mathcal{C}_p \in I_1 = (0.00, 0.67]$

It can be observed that the evaluation of the process capability relies on three issues: (1) variability in process, (2) degree of departure of the process mean from the target value, and (3) location of the process mean in the $[LSL, USL]$ interval. Note that $\mathcal{C}_p$ only takes into account (1), while $\mathcal{C}_{pk}$considers (1) and (3), $\mathcal{C}_{pm}$ includes (1) and (2), whereas $\mathcal{C}_{pmk}$ considers all (i.e., (1), (2) and (3)). The larger a capability index value of a process, the more capable the process will be.

The classical context of the process capability indices, however, suffers from some limitations preventing a deep and flexible analysis when exact data is used. The observed data could be imprecise quantities due to the inherent uncertainty of real applications. In such cases, both lower and upper specification limits can be also considered as fuzzy quantities instead of exact values. The reason is that there are often more than two alternates than defective and non-defective cases. A specific product may be belonging to the fuzzy set of non-defective products with a specified degree. Therefore, the conventional process capability indices are not suitable for such cases. In order to overcome this problem, fuzzy set theory can be applied to improve the fuzzy situation. In this regard, several fuzzy process capability indices have been proposed [122, 153, 154, 206, 75].

In the following, the classical process capability index $\mathcal{C}_p(u, v)$ for **NFRS** is first extended based on fuzzy data, specific limits and fuzzy targets by utilizing a fuzzy distance measure. Then, some of the main relations between proposed fuzzy process capability indices are verified.

Definition 6.6 *Let $\widetilde{X}$ be a* ***TNFRV*** *with known fuzzy mean of $\widetilde{\mu}$ and exact variance of σ^2. Also, let $\widetilde{USL}$ and $\widetilde{LSL}$ (with* $\sup \widetilde{LSL}[0] < \inf \widetilde{USL}[0]$*) be the fuzzy upper and lower specification limits of the quality characteristic, respectively. The fuzzy process capability, $\widetilde{C}(u, v)$, is defined to be a* ***TFN*** *as follows:*

$$\widetilde{\mathcal{C}}(u, v) = \left(\frac{1}{3\sqrt{\sigma^2 + v d_2^2(\widetilde{\mu}, \widetilde{T})}}\right) \otimes (\widetilde{d} \ominus (u \otimes \widetilde{d}_a(\widetilde{\mu}, \widetilde{m})), \tag{6.16}$$

where

1) $\widetilde{USL} = ((USL)^L, USL, (USL)^U)_T$, $\widetilde{LSL} = ((LSL)^L, LSL, (LSL)^U)_T$,

2) $\widetilde{d} = \frac{1}{2} \otimes (\widetilde{USL} \ominus \widetilde{LSL})$,

3) $\widetilde{m} = \frac{1}{2} \otimes (\widetilde{USL} \oplus \widetilde{LSL})$,

4) $\widetilde{T} = (T^L, T, T^U)_T$ *is a fuzzy target,*

5) $\widetilde{d}_a(\widetilde{A}, \widetilde{B})$ *is the fuzzy absolute distance between two* ***TFNs*** *of* $\widetilde{A}$ *and* $\widetilde{B}$ *defined as:*

$$\widetilde{d}_a(\widetilde{A}, \widetilde{B}) = (d_a^L(\widetilde{A}, \widetilde{B}), d_a(\widetilde{A}, \widetilde{B}), d_a^U(\widetilde{A}, \widetilde{B}))_T,$$

where

1) $d_a^L(\widetilde{A}, \widetilde{B}) = \frac{|a^L - b^L| + |a-b| + |a^U - b^U|}{3}$,

2) $d_a(\widetilde{A}, \widetilde{B}) = \sqrt{\frac{|a^L - b^L|^2 + |a-b|^2 + |a^U - b^U|^2}{3}}$,

3) $d_a^U(\widetilde{A}, \widetilde{B}) = \max\{|a^L - b^L|, |a-b|, |a^U - b^U|\}$, *and*

6) $d_2^2(\widetilde{A}, \widetilde{B}) = \frac{|a^L - b^L|^2 + |a-b|^2 + |a^U - b^U|^2}{3}$.

It should be noted that $\widetilde{C}_p(u, v)$ can be also represented as follows:

$$\widetilde{C}_p(u, v) = \left(\frac{d^L - u d_a^U(\widetilde{\mu}, \widetilde{m})}{3\sqrt{\sigma^2 + v d_2^2(\widetilde{\mu}, \widetilde{T})}}, \frac{d - u d_a(\widetilde{\mu}, \widetilde{m})}{3\sqrt{\sigma^2 + v d_2^2(\widetilde{\mu}, \widetilde{T})}}, \frac{d^U - u d_a^L(\widetilde{\mu}, \widetilde{m})}{3\sqrt{\sigma^2 + v d_2^2(\widetilde{\mu}, \widetilde{T})}}\right)_T. \tag{6.17}$$

Definition 6.7 *Recall all assumptions in Definition 6.6. It is easy to show that the fuzzy process capabilities* $\widetilde{C}_p$, $\widetilde{C}_{pk}$, $\widetilde{C}_{pm}$, *and* $\widetilde{C}_{pmk}$ *can be expressed as:*

1) $\widetilde{C}_p = \widetilde{C}(0, 0) = (\frac{d^L}{3\sigma}, \frac{d}{3\sigma}, \frac{d^U}{3\sigma})_T$.

2) $\widetilde{C}_{pk} = \widetilde{C}(1, 0) = (\frac{d^L - d_a^U(\widetilde{\mu}, \widetilde{m})}{3\sigma}. \frac{d - d_a(\widetilde{\mu}, \widetilde{m})}{3\sigma}. \frac{d^U - d_a^L(\widetilde{\mu}, \widetilde{m})}{3\sigma})_T$.

3) $\widetilde{C}_{pm} = \widetilde{C}(0, 1) = (\frac{d^L}{3\sqrt{\sigma^2 + d_2^2(\widetilde{\mu}, \widetilde{T})}}. \frac{d}{3\sqrt{\sigma^2 + d_2^2(\widetilde{\mu}, \widetilde{T})}}. \frac{d^U}{3\sqrt{\sigma^2 + d_2^2(\widetilde{\mu}, \widetilde{T})}})_T$.

4) $\widetilde{C}_{pmk} = \widetilde{C}(1, 1) = (\frac{d^L - d_a^U(\widetilde{\mu}, \widetilde{m})}{3\sqrt{\sigma^2 + d_2^2(\widetilde{\mu}, \widetilde{T})}}. \frac{d - d_a(\widetilde{\mu}, \widetilde{m})}{3\sqrt{\sigma^2 + d_2^2(\widetilde{\mu}, \widetilde{T})}}. \frac{d^U - d_a^L(\widetilde{\mu}, \widetilde{m})}{3\sqrt{\sigma^2 + d_2^2(\widetilde{\mu}, \widetilde{T})}})_T$.

By the following lemma, some relationships governing the proposed fuzzy process capability indices are investigated.

Lemma 6.2 *Recall all assumptions in Definition 6.6. Then, the following relations are valid between* $\widetilde{C}_p$, $\widetilde{C}_{pm}$, $\widetilde{C}_{pk}$, *and* $\widetilde{C}_{pmk}$:

1) $\widetilde{C}_{pk} = \widetilde{C}_p \ominus ((\frac{1}{3\sigma}) \otimes \widetilde{d}_a(\widetilde{\mu}, \widetilde{m}))$.

2) $\widetilde{C}_{pm} = (\frac{1}{\sqrt{1+\frac{d_2^2(\widetilde{\mu},\widetilde{T})}{\sigma^2}}}) \otimes \widetilde{C}_p$.

3) $\widetilde{C}_{pmk} = (\frac{1}{\sqrt{1+\frac{d_2^2(\widetilde{\mu},\widetilde{T})}{\sigma^2}}}) \otimes \widetilde{C}_{pk}$.

4) $\widetilde{C}_p = (\frac{1}{3\sigma}) \otimes \widetilde{d}$.

Proof *The proof is easily verified according to the arithmetic operations of* ***TFN****s.*

Lemma 6.3 *Recall the assumptions in Definition 6.6. Then:*

1) $\widetilde{C}_p \succeq_{P_d} \widetilde{C}_{pk} \succeq_{P_d} \widetilde{C}_{pmk}$.

2) $\widetilde{C}_p \succeq_{P_d} \widetilde{C}_{pm} \succeq_{P_d} \widetilde{C}_{pmk}$.

Proof *The first assertion will be only verified. For this purpose,* $(C_p)^L \geq (C_{pk})^L \geq (C_{pmk})^L$, $C_p \geq C_{pk} \geq C_{pmk}$ *and* $(C_p)^U \geq (C_{pk})^U \geq (C_{pmk})^U$ *simply concludes that* $P_d(\widetilde{C}_p \succeq \widetilde{C}_{pk}) = 1$ *and* $P_d(\widetilde{C}_{pk} \succeq \widetilde{C}_{pmk}) = 1$. *These complete the proof.*

Remark 6.8 *Let* $\widetilde{X}$ *be a* ***TNFRV*** *with unknown fuzzy mean* $(\widetilde{\mu})$ *and variance of* σ^2. *Recall the assumptions in Definition 6.6. Then, the fuzzy estimated value of* $\widetilde{C}(u, v)$ *is defined to be a* ***TFN*** *as follows:*

$$\widehat{\widetilde{C}}_p(u, v) = (\frac{1}{3\sqrt{s_n^2 + vd_2^2(\widetilde{\overline{x}}, \widetilde{T})}}) \otimes (\widetilde{d} \ominus (u \otimes \widetilde{d}_a(\widetilde{\overline{X}}, \widetilde{m})), \tag{6.18}$$

where $s_n^2 = \frac{1}{n-1}\sum_{i=1}^n d_2^2(\widetilde{x}_i, \widetilde{\overline{x}})$ *and* $\widetilde{\overline{x}} = (\frac{1}{n}) \otimes (\bigoplus_{i=1}^n \widetilde{x}_i)$.

It should be noted that $\widehat{\widetilde{C}}_p(u, v)$ can be represented by a **TFN** as follows:

$$\widehat{\widetilde{C}}_p(u, v) = (\frac{d^L - ud_a^U(\widetilde{\overline{X}}, \widetilde{m})}{3\sqrt{s_n^2 + vd_2^2(\widetilde{\overline{x}}, \widetilde{T})}}, \frac{d - ud_a(\widetilde{\overline{X}}, \widetilde{m})}{3\sqrt{s_n^2 + vd_2^2(\widetilde{\overline{x}}, \widetilde{T})}}, \tag{6.19}$$

Remark 6.9 *Having a fuzzy process capability index, a criterion for interpreting the process conditions is required. Here, the degree to which* $\widetilde{C}_p(u, v)$ *belongs to a quality condition* I_j, $j = 1, 2, ..., k$ *is defined according to the principle of the maximum degree of belonging :*

$$d(\widetilde{A} \in I_j) = \frac{\int_{I_j} \widetilde{C}_p(u, v)(x)dx}{\int_{\mathbb{R}} \widetilde{C}_p(u, v)(x)dx}.$$

TABLE 6.7
Fuzzy data in Example 6.3.

$\widetilde{x}_1 = (5.85, 6.15, 6.35)_T$	$\widetilde{x}_2 = (5.79, 5.90, 5.98)_T$	$\widetilde{x}_3 = (5.71, 5.83, 5.99)_T$
$\widetilde{x}_4 = (6.05, 6.18, 6.32)_T$	$\widetilde{x}_5 = (5.89, 6.06, 6.23)_T$	$\widetilde{x}_6 = (6.01, 6.10, 6.25)_T$
$\widetilde{x}_7 = (6.15, 6.20, 6.30)_T$	$\widetilde{x}_8 = (5.64, 5.81, 6.05)_T$	$\widetilde{x}_9 = (5.80, 5.90, 5.98)_T$
$\widetilde{x}_{10} = (6.01, 6.12, 6.24)_T$	$\widetilde{x}_{11} = (5.86, 6.04, 6.25)_T$	$\widetilde{x}_{12} = (6.13, 6.23, 6.33)_T$
$\widetilde{x}_{13} = (5.95, 6.05, 6.19)_T$	$\widetilde{x}_{14} = (5.60, 5.65, 5.70)_T$	$\widetilde{x}_{15} = (5.65, 5.74, 5.84)_T$
$\widetilde{x}_{16} = (5.70, 5.77, 5.83)_T$	$\widetilde{x}_{17} = (6.23, 6.32, 6.40)_T$	$\widetilde{x}_{18} = (5.60, 5.70, 5.80)_T$
$\widetilde{x}_{19} = (5.85, 5.95, 6.05)_T$	$\widetilde{x}_{20} = (5.90, 6.00, 6.10)_T$	$\widetilde{x}_{21} = (5.50, 5.81, 5.99)_T$
$\widetilde{x}_{22} = (5.60, 5.92, 6.05)_T$	$\widetilde{x}_{23} = (5.50, 5.75, 5.95)_T$	$\widetilde{x}_{24} = (5.84, 6.03, 6.15)_T$
$\widetilde{x}_{25} = (6.05, 6.30, 6.50)_T$	$\widetilde{x}_{26} = (6.25, 6.35, 6.45)_T$	$\widetilde{x}_{27} = (5.65, 5.86, 6.05)_T$
$\widetilde{x}_{28} = (5.70, 5.87, 5.95)_T$	$\widetilde{x}_{29} = (5.75, 5.95, 6.15)_T$	$\widetilde{x}_{30} = (6.10, 6.23, 6.46)_T$

Example 6.3 *A total of 30 fuzzy data are collected from a factory as shown in Table 6.7. It is assumed that such data are observed* ***FRS*** *from a normal distribution with an unknown fuzzy mean of* $\widetilde{\mu}$ *and variance of* σ^2. *The specifications of a particular product are given as* $\widetilde{LSL} = (5.15, 5.50, 5.87)_T$, $\widetilde{USL} = (6.12, 6.40, 6.76)_T$, *and* $\widetilde{T} = (5.85, 6.00, 6.25)_T$. *According to the observed fuzzy data and the product specifications, one may check that* $\widetilde{\overline{\overline{x}}} = (5.8477, 6.0067, 6.1218)_T$, $s_{30}^2 = 0.0464$ *and*

$$\widetilde{d} = \frac{1}{2} \otimes (\widetilde{USL} \ominus \widetilde{LSL}) = (0.125, 0.45, 0.805)_T,$$

$$\widetilde{m} = \frac{1}{2} \otimes (\widetilde{USL} \oplus \widetilde{LSL}) = (5.635, 5.95, 6.315)_T.$$

Therefore, we get $\widetilde{d}_a(\widetilde{\overline{\overline{x}}}, \widetilde{m}) = (0.1533, 0.21, 0.291)_T$ *and* $d_2^2(\widetilde{\overline{\overline{x}}}, \widetilde{T}) = 0.0165$. *Thus, the membership function of* $\widetilde{\widetilde{C}}_p(u, v)$ *can be constructed as follows:*

$$\widetilde{\widetilde{C}}_p(u, v) = (\frac{0.125 - 0.291u}{3\sqrt{0.0464 + 0.0165v}}, \frac{0.45 - 0.21u}{3\sqrt{0.0464 + 0.0165v}}, \frac{0.805 - 0.21u}{3\sqrt{0.0464 + 0.0165v}})_T,$$

where $u, v \in \{0, 1\}$. *Especially, we obtain:*

$$\widetilde{\widetilde{C}}_p = (0.193, 0.696, 1.246)_T, \quad (6.20)$$

$$\begin{aligned}
\widetilde{\widetilde{C}}_{pm} &= (\frac{0.125}{3\sqrt{0.0464 + 0.0165}}, \frac{0.45}{3\sqrt{0.0464 + 0.0165}}, \frac{0.805}{3\sqrt{0.0464 + 0.0165}})_T \\
&= (0.01045, 0.03762, 0.06729)_T, \\
\widetilde{\widetilde{C}}_{pk} &= (\frac{0.125 - 0.291}{3\sqrt{0.0464}}, \frac{0.45 - 0.21}{3\sqrt{0.0464}}, \frac{0.805 - 0.21}{3\sqrt{0.0464}})_T \\
&= (-0.0119, 0.01142, 0.03691)_T, \\
\widetilde{\widetilde{C}}_{pmk} &= (\frac{0.125 - 0.291}{3\sqrt{0.0464 + 0.0165}}, \frac{0.45 - 0.21}{3\sqrt{0.0464 + 0.0165}}, \frac{0.805 - 0.21}{3\sqrt{0.0464 + 0.0165}})_T \\
&= (-0.03143, 0.02006, 0.04297)_T.
\end{aligned}$$

TABLE 6.8
Degrees to which $\widetilde{\widehat{\mathcal{C}}}_p$ belongs to I_i, $i = 1, 2, \ldots, 6$ in Example 6.3.

	$j=1$	$j=2$	$j=3$	$j=4$	$j=5$	$j=6$
$d(\widetilde{\widehat{\mathcal{C}}}_p \in I_j)$	0	0.7168	0.2897	0	0	0

To evaluate the process conditions based on $\widetilde{\widehat{\mathcal{C}}}_p$, there is a need to calculate $d(\widetilde{\widehat{\mathcal{C}}}_p \in I_j)$ for any $j = 1, 2, \ldots, 6$ where I_j are given in Table 6.6. The results are shown in Table 6.8. In this regards, calculations show that:

$$d(\widetilde{\widehat{\mathcal{C}}}_p \in I_2) = \frac{\int_{0.67}^{1} (\frac{1.246-x}{0.55}) dx}{\int_{0.193}^{0.696} (\frac{1.246-x}{0.55}) dx} = \frac{0.3774}{0.5265} = 0.7168,$$

$$d(\widetilde{\widehat{\mathcal{C}}}_p \in I_3) = \frac{\int_{1.246}^{1} (\frac{1.246-x}{0.55}) dx}{\int_{0.193}^{0.696} (\frac{1.246-x}{0.55}) dx} = \frac{0.1525}{0.5265} = 0.2897.$$

Accordingly, $d(\widetilde{\widehat{\mathcal{C}}}_p \in I_2) = \max_{j=1}^{6} d(\widetilde{\widehat{\mathcal{C}}}_p \in I_j) = 0.7168$ that is $\widetilde{\widehat{\mathcal{C}}}_p \in I_2$. Thus, the information shows that this factory has a quality of level of 'inadequate' based on the principle of the maximum degree of belonging.

6.1.4 Testing hypotheses for fuzzy process capability $\widetilde{C}_p$

In fuzzy quality control scenarios such as fuzzy process capability analysis, it might be required to decide either accepting or rejecting the capability of the process. That is, we have to test the hypothesis as H_0: the process is not capable, which is against H_1 : the process is capable. In this regard, it should be noted that Wu [206] proposed an approach for testing the process performance of C_{pk} based on a notion of **TNFRVs** with exact parameters.

Let $\widetilde{X}_1, \cdots, \widetilde{X}_n$ be a **FRS** from a **TNFRV** with fuzzy mean of $\widetilde{\mu}$ and exact variance of σ^2. A procedure was proposed for statistical hypothesis testing of the fuzzy capability process of $\widetilde{C}_p$ based on the aforementioned **TNFRVs**. For this purpose, the methods of Chapter 5 was employed here.

Definition 6.8 *Let $\widetilde{USL}$ and $\widetilde{LSL}$ be the fuzzy upper and lower specification limits of the quality characteristic and $\widetilde{X}_1, \cdots, \widetilde{X}_n$ be a **FRS** from a **TNFRV** with a fuzzy mean $\widetilde{\mu}$ and exact variance of $\widetilde{\sigma}^2$. At a given significance level of $\delta \in (0, 1]$, the null and alternative hypothesis for fuzzy capability process is defined as:*

$$\begin{cases} H_0: & \widetilde{C}_p \preceq_M \widetilde{C}_0, \equiv M_{\widetilde{C}_p} \leq M_{\widetilde{C}_0}, \\ H_1: & \widetilde{C}_p \succ_M \widetilde{C}_0, \equiv M_{\widetilde{C}_p} > M_{\widetilde{C}_0}, \end{cases}$$

*where $\widetilde{C}_0$ is a specific **TFN**.*

Lemma 6.4 *Recall all assumptions given in Definition 6.8. Then, under the null hypothesis of* $\widetilde{H}_0$, *we have*

$$3(n-1)(\frac{M_{\widetilde{C}_0}}{M_{\widetilde{\widehat{C}}_p}})^2 \sim \chi^2_{3(n-1)},$$

where $\chi^2_{n-1,\delta}$ *denotes the* δ^{th} *upper of the chi-square distribution with* $n-1$ *degrees of freedom.*

Proof *The proof is immediately followed since* $3(n-1)(\frac{M_{\widetilde{C}_0}}{M_{\widetilde{\widehat{C}}_p}})^2 = 3(n-1)\frac{S_n^2}{\sigma^2}$ *which is distributed according to Chi-square distribution with* $3(n-1)$ *degrees of freedom.*

At a given significance level of δ, *we therefore reject* $\widetilde{H}_0$ *if* $3(n-1)(\frac{M_{\widetilde{C}_0}}{M_{\widetilde{\widehat{C}}_p}})^2 \geq \chi^2_{3(n-1),\delta}$, *otherwise, we accept it.*

Example 6.4 *Assume a stable process with the respective upper and lower specifications of* $\widetilde{USL} = (60, 62, 65)_T$ *and* $\widetilde{LSL} = (35, 38, 41)_T$. *Furthermore, with a sample of size* $n = 20$, *suppose that the mean and standard deviation are reported by* ***TFNs*** *of* $\widetilde{\overline{x}} = (40, 47, 52)_T$ *and* $s_{20} = 1.75$, *respectively. To qualify for business with his company, a customer has told his supplier that the supplier must demonstrate that his process capability exceeds* $\widetilde{C}_0 = (1.25, 1.33, 1.44)_T$. *Therefore, we wish to test the following fuzzy hypotheses:*

$$\begin{cases} H_0: & \widetilde{C}_p \preceq \widetilde{C}_0, \equiv M_{\widetilde{C}_p} \leq M_{\widetilde{C}_0}, \\ H_1: & \widetilde{C}_p \succ \widetilde{C}_0, \equiv M_{\widetilde{C}_p} > M_{\widetilde{C}_0}. \end{cases}$$

For this purpose, according to the given assumptions, note that:

$$\widetilde{\widehat{C}}_p = (\frac{60-41}{6\times 1.75}, \frac{62-38}{6\times 1.75}, \frac{65-35}{6\times 1.75})_T = (1.81, 2.286, 2.857)_T.$$

This concludes that $M_{\widetilde{\widehat{C}}_p} = 2.318$ *and thus* $3(n-1)(\frac{M_{\widetilde{C}_0}}{M_{\widetilde{\widehat{C}}_p}})^2 = 3 \times 19(\frac{1.34}{2.318})^2 = 19.048$. *At the level of* $\delta = 0.05$, *since* $19.048 > \chi^2_{57,0.05} = 75.623$, *we should accept* $\widetilde{H}_0$.

6.2 Reliability Evaluation Using FRVs

Reliability analysis is an important research topic in engineering and applied science. Classical reliability assessment is mainly based on [58] exactly. These methods are inadequate for some types of imprecision due to due to the uncertainties in observed data. Therefore, imprecise data is often inevitable in a reliability analysis. For instance, a ball bearing may work perfectly over a certain period. But it may break for some time and finally it may be unusable at a certain time. So, the observed failure times of the ball bearings can be reported by fuzzy quantities. In this section, some common reliability systems and their life time components are extended based on **FRV**s. For this purpose, a notion of fuzzy system reliability is constructed based on **TNFRV**s. Such notions are taken from Hesamian et al. [74].

Definition 6.9 *A life time **FRV** (**LTFRV**) is defined as a **SFRV*** $\widetilde{X} = X \otimes (1 - U_1, 1, 1 + U_2)_T$ *where*

1) U_1 *and* U_2 *are independent random variables uniformly distributed on* $(0, 1)$,

2) X *is an ordinary life time random variable.*

Definition 6.10 *Let* $\widetilde{X} = X \otimes (1 - U_1, 1, 1 + U_2)_T$ *be a **LTFRV**. Then, the fuzzy system reliability function (**FSRF**) at the time of t is defined to be a **TFN** as* $\widetilde{R}(t) = (R^L(t), R(t), R^U(t))_T$ *where*

1) $R^L(t) = P(X(1 - U_1) > t) = \int_0^1 (1 - F_X(t/(1 - u_1)))du$,

2) $R(t) = P(X > t) = 1 - F_X(t)$,

3) $R^U(t) = P(X(1 + U_2) > t) = \int_0^1 (1 - F_X(t/(1 + u_2)))du$.

It should be pointed out that reliability estimation is an essential part of engineering systems. Consider a system consisting of several components with two states of operation or failure. The status of the system can be determined when the set of operating and failure components are specified. The problem is to compute the probability of the system operation, i.e., reliability of the system. Let $c_1, c_2, \ldots, c_n$ denote n components in a reliability systems. Assume that n components operate independently, and $P(c_j$ works until time $t) = R_j(t)$. In general, the reliability of a system can be written as $R_S(t) = g(R_1(t), R_2(t), \ldots, R_k(t))$. Certain types of systems frequently arise in practice, usually in k-out-of-n system. The reliability of the k-out-of-n system is given by:

$$R_S(t) = \sum_{j=k}^{n} \binom{n}{j} (R_j(t))^j (1 - R_j(t))^{n-j}.$$

If $k = 1$, then the system is reduced to a parallel system and if $k = n$, then the system is reduced to a system in series. Now, having a **LTFRV**, we can develop $R_S(t)$ of a system into space of **TFN**s as follows.

Definition 6.11 *Let $\widetilde{X}$ be a* ***LTFRV*** *where t shows the time to failure X is distributed according to* ***CDF*** *of F_X. The fuzzy reliability of a specified system $R_S(t) = g(R_1(t), R_2(t), \ldots, R_k(t))$ is defined to be a* ***TFN*** *$\widetilde{R}_S(t) = g(\widetilde{R}_1(t), \widetilde{R}_2(t), \ldots, \widetilde{R}_k(t))$ by extension principle.*

Here, a common reliability system is extended for **LTFRV**s.

Definition 6.12 *Let $\widetilde{X}$ be a* ***LTFRV****. The fuzzy reliability of the k-out-of-n system is given by a* ***TFN*** *as follows:*

$$\widetilde{R}_S(t) = \bigoplus_{j=k}^{n} \left(\binom{n}{j} \otimes (\widetilde{R}_j(t))^j \otimes (1 \ominus \widetilde{R}_j(t))^{n-j} \right)$$

$$= \Big(\prod_{j=k}^{n} \left(\binom{n}{j} (R_j^L(t))^j (1 - R_j^U(t))^{n-j} \right),$$

$$\prod_{j=k}^{n} \left(\binom{n}{j} (R_j(t))^j (1 - R_j(t))^{n-j} \right), \prod_{j=k}^{n} \left(\binom{n}{j} (R_j^U(t))^j (1 - R_j^L(t))^{n-j} \right) \Big)_T,$$

If $k = 1$, then the system is reduced to the parallel system; while if $k = n$ then the system is reduced to the system in series. It is noticeable that other forms of system reliability functions can be similarly defined using the arithmetic operations introduced in this paper.

6.2.1 Components of a reliability system based on LTFRVs

To track and improve the reliability of products, manufacturing organizations often utilize an accurate and concise method to specify and measure that reliability [217]. Here, some components of a reliability system are extended into the space of **TFN**s using the following definition.

Definition 6.13 *Let $\widetilde{X}$ be a* ***LTFRV****. Then, some common fuzzy components of a* ***LTFRV*** *at time $t > 0$ are defined as follows:*

1. *The fuzzy reliability function (**FRF**): $\widetilde{S}(t) = 1 \ominus \widetilde{R}_S(t)$.*

2. *The fuzzy mean time to failure (**IMTTF**):*

$$\widetilde{E}_S = \int_0^\infty \widetilde{R}_S(t)dt = (\int_0^\infty R_S^L(t)dt, \int_0^\infty R_S(t)dt, \int_0^\infty R_S^U(t)dt)_T.$$

3. *The fuzzy mean residual life function (**FMRLF**):*

$$\widetilde{M}_S(t) = \int_t^\infty \widetilde{R}_S(x)dx \oslash \widetilde{R}_S(t) = (\frac{\int_x^\infty R_S^L(x)dx}{R_S^U(t)}, \frac{\int_x^\infty R_S(x)dx}{R_S(t)}, \frac{\int_x^\infty R_S^U(x)dx}{R_S^L(t)})_T.$$

4. *The fuzzy conditional reliability (**FCR**) is defined as:*

$$\widetilde{C}_S(T,t) = \widetilde{R}_S(T+t) \oslash \widetilde{R}_S(t) = (\frac{R_S^L(T+t)}{R_S^U(t)}, \frac{R_S(T+t)}{R_S(t)}, \frac{R_S^U(T+t)}{R_S^L(t)})_T.$$

Remark 6.10 *A typical interpretation of fuzzy components induced by a **LT-FRV** at t can be regarded as follows:*

1. $\widetilde{R}_S(t)$ *can be interpreted as a membership degree to which the time to failure of a system will be consistent with the notion of 'system is reliable at t'.*

2. $\widetilde{M}_S(t)$ *is interpreted as a membership degree to which a system engendering, after a fixed time t, will be consistent with the notion of 'remaining life'.*

3. $\widetilde{E}_S$ *is interpreted as a membership degree to which the mean life-time of a system engineering will be consistent with the notion of 'remaining life'.*

4. $\widetilde{C}_S(T,t)$ *can be interpreted as a membership degree to which the chance of system reliability will be consistent with the notion of 'successfully completing another mission following the successful completion of a previous mission'.*

Example 6.5 *Consider a system with five synchronous computers that analyze all other systems and compare their results. For launching, four out of five computers must agree on system parameters. If all agree, the launch takes place. If one computer fails and four agree, the fifth computer is ignored and the launch occurs. If two computers fail to agree, the launch is scrubbed. Assume that **LTFRVs** of all components have the same distribution function with **CDF** of $F_{X_i}(t) = 0.2\exp(-0.2t)$. In this case, the fuzzy value that 'system is reliable at t' can be evaluated:*

$$\widetilde{R}_S(t) = \bigoplus_{j=4}^{5}\left(\binom{5}{j} \otimes (\widetilde{R}(t))^j \otimes (1 \ominus \widetilde{R}(t))^{5-j}\right)$$

$$= \Big(\sum_{j=4}^{5}\Big(\binom{5}{j}(R^L(t))^j(1-R_S^U(t))^{5-j}\Big),$$

$$\sum_{j=4}^{5}\Big(\binom{5}{j}(R(t))^j(1-R(t))^{5-j}\Big), \sum_{j=4}^{5}\Big(\binom{5}{j}(R^U(t))^j(1-R_S^L(t))^{5-j}\Big)\Big)_T,$$

where $\widetilde{R}(t)$ is evaluated by Definition 6.13 as $\widetilde{R}(t) = (R^L(t), R(t), R^U(t))_T$ in which

1) $R^L(t) = P(X(1-U) > t) = \int_0^1 (1 - F_X(t/(1-u)))du = \int_0^1 \exp(-0.2t/(1-u))du$,

2) $R(t) = P(X > t) = \exp(-0.2t)$,

3) $R^U(t) = P(X(1+U) > t) = \int_1^2 (\exp(-0.2t/(1+u))du$.

Therefore, the ***FMRLF*** *and* ***FCR*** *of* $\widetilde{X}$ *can be evaluated by:*

$$\widetilde{M}_S(t) = (\frac{\int_t^\infty R_S^L(x)dx}{R_S^U(t)}, \frac{\int_t^\infty R_S(x)dx}{R_S(t)}, \frac{\int_t^\infty R_S^U(x)dx}{R_S^L(t)})_T,$$

and

$$\widetilde{C}_S(T,t) = (\frac{R_S^L(T+t)}{R_S^U(t)}, \frac{R_S(T+t)}{R_S(t)}, \frac{R_S^U(T+t)}{R_S^L(t)})_T.$$

To simplify the calculation, one can employ the law of large numbers [46] to evaluate $R^L(t)$ *and* $R^U(t)$ *as follows:*

1) $R^L(t) \simeq \frac{1}{n}\sum_{i=1}^n e^{-0.2t/(1-u_i)}$ *where* $u_1, u_2, ..., u_n$ *is a random sample of uniform distribution on* $(0,1)$.

2) $R^U(t) \simeq \frac{1}{n}\sum_{i=1}^n e^{-0.2t/(1+u_i)}$ *where* $u_1, u_2, ..., u_n$ *is a random sample of uniform distribution on* $(0,1)$,

where n *is a large integer (such as 1000). Therefore, one can finds:*

1)

$$R_S^L(t) \simeq (\frac{1}{(1000)^5})\sum_{j=4}^{5}\binom{5}{j}(\sum_{i=1}^{1000} e^{-0.2t/(1-u_i)})^j(1000 - \sum_{i=1}^{1000} e^{-0.2t/(1+u_i)})^{5-j}.$$

2) $R(t) = \sum_{j=4}^5 \binom{5}{4}(e^{-0.2t})^j(1-e^{-0.2t})^{n-j} = 5e^{-0.8t} - 4e^{-t}$,

3)

$$R_S^U(t) \simeq (\frac{1}{(1000)^5})\sum_{j=4}^{5}\binom{5}{j}(\sum_{i=1}^{1000} e^{-0.2t/(1+u_i)})^j(1000 - \sum_{i=1}^{1000} e^{-0.2t/(1-u_i)})^{5-j}.$$

Therefore, the ***IMTTF*** *can be evaluated as:*

$$\widetilde{E}_S = \int_0^\infty \widetilde{R}(t)dt = (\int_0^\infty R_S^L(t)dt, \int_0^\infty R_S(t)dt, \int_0^\infty R_S^U(t)dt)_T.$$

where

1) $\int_0^\infty R_S^L(t)dt \simeq$

$$\int_0^\infty (\frac{1}{(1000)^5})\sum_{j=4}^{5}\binom{5}{j}(\sum_{i=1}^{1000} e^{-0.2t/(1-u_i)})^j(1000 - \sum_{i=1}^{1000} e^{-0.2t/(1+u_i)})^{5-j}dt$$

$$= (\frac{1}{1000})^5\Big((5\times 1000)\sum_{i=1}^{1000}\frac{u_i}{0.8} - 1000\sum_{i=1}^{1000}\sum_{j=1}^{1000}\frac{u_i(1+u_i)}{0.2((1-u_i)+4(1+u_i))}$$

$$+\sum_{i=1}^{1000}(1-u_i)\Big).$$

2) $\int_0^\infty R_S(t)dt = \int_0^\infty (5e^{-0.8t} - 4e^{-t})dt = 9/4,$

3)

$$\int_0^\infty R_S^U(t)dt \simeq$$

$$\int_0^\infty \Big((\frac{1}{(1000)^5})\sum_{j=4}^{5}\binom{5}{j}(\sum_{i=1}^{1000} e^{-0.2t/(1+u_i)})^j(1000 - \sum_{i=1}^{1000} e^{-0.2t/(1-u_i)})^{5-j}\Big)dt$$

$$= (\frac{1}{1000})^5\Big((5\times 1000)\sum_{i=1}^{1000}\frac{(1+u_i)}{0.8} - 1000\sum_{i=1}^{1000}\sum_{j=1}^{1000}\frac{(1-u_i)(1+u_i)}{0.2(4(1-u_i)+(1+u_i))}$$

$$+\sum_{i=1}^{1000}(1+u_i)\Big).$$

Now, having a simulated uniform distribution on $(0,1)$ *with size of* $n = 1000$*, the* ***FMRLF*** *can be evaluated as:*

$$\widetilde{M}_S(t) = (\frac{\int_t^\infty R_S^L(x)dx}{R_S^U(t)}, \frac{\int_t^\infty R_S(x)dx}{R_S(t)}, \frac{\int_t^\infty R_S^U(x)dx}{R_S^L(t)})_T,$$

where

1) $\int_t^\infty R_S^L(x)dx \simeq$

$$\int_t^\infty (\frac{1}{(1000)^5})\sum_{j=4}^{5}\binom{5}{j}(\sum_{i=1}^{1000} e^{-0.2x/(1-u_i)})^j(1000 - \sum_{i=1}^{1000} e^{-0.2x/(1+u_i)})^{5-j}dx$$

$$= (5\times 1000)\sum_{i=1}^{1000} e^{-\frac{0.8}{(1-u_i)}}\frac{(1-u_i)}{0.8}$$

$$-5\sum_{i=1}^{1000}\sum_{j=1}^{1000} e^{-0.2t(\frac{4}{(1-u_i)}+\frac{1}{(1+u_j)})}\frac{(1-u_i)(1+u_i)}{0.2((1-u_i)+4(1+u_j))}$$

$$+\sum_{i=1}^{1000} e^{-\frac{0.8t}{(1-u_i)}}(1-u_i),$$

2) $R_S^U(t) = (\frac{1}{(1000)^5})\sum_{j=4}^5\binom{5}{j}(\sum_{i=1}^{1000} e^{-0.2t/(1+u_i)})^j(1000- \sum_{i=1}^{1000} e^{-0.2t/(1-u_i)})^{5-j},$

3) $\frac{\int_t^\infty R_S(x)dx}{R_S(t)} = \frac{\int_t^\infty (5e^{-0.8x}-4e^{-x})dx}{5e^{-0.8t}-4e^{-t}} = \frac{\frac{25}{4}e^{-0.8t}-4e^{-t}}{5e^{-0.8t}-4e^{-t}},$

4) $\int_t^\infty R_S^U(x)dx \simeq$

$$\int_t^\infty (\frac{1}{(1000)^5})\sum_{j=4}^{5}\binom{5}{j}(\sum_{i=1}^{1000} e^{-0.2x/(1+u_i)})^j(1000-\sum_{i=1}^{1000} e^{-0.2x/(1-u_i)})^{5-j}dx$$

$$= (5\times 1000)\sum_{i=1}^{1000} e^{-\frac{0.8}{(1+u_i)}}\frac{(1-u_i)}{0.8} - 5\sum_{i=1}^{1000}\sum_{j=1}^{1000} e^{-0.2t(\frac{4}{(1+u_i)}+\frac{1}{(1-u_j)})}$$

$$\times \frac{(1-u_i)(1+u_i)}{0.2((1-u_i)+4(1+u_j))} + \sum_{i=1}^{1000} e^{-\frac{0.8t}{(1+u_i)}}(1+u_i),$$

5) $R_S^L(t) = \sum_{j=4}^{5}\binom{5}{j}(\sum_{i=1}^{1000} e^{-0.2t/(1-u_i)})^j(1000-\sum_{i=1}^{1000} e^{-0.2t/(1+u_i)})^{5-j}$.

Furthermore, it can be shown that the ***FCR*** *is:*

$$\widetilde{C}_S(T,t) = \widetilde{R}_S(T+t)\oslash\widetilde{R}_S(t) = (\frac{R_S^L(T+t)}{R_S^U(t)}, \frac{R_S(T+t)}{R_S(t)}, \frac{R_S^U(T+t)}{R_S^L(t)})_T,$$

where

1) $\frac{R_S^L(T+t)}{R_S^U(t)} =$

$$\frac{\sum_{j=4}^{5}\binom{5}{j}(\sum_{i=1}^{1000} e^{-0.2(t+T)/(1-u_i)})^j(1000-\sum_{i=1}^{1000} e^{-0.2(t+T)/(1+u_i)})^{5-j}}{\sum_{j=4}^{5}\binom{5}{j}(\sum_{i=1}^{1000} e^{-0.2t/(1+u_i)})^j(1000-\sum_{i=1}^{1000} e^{-0.2t/(1-u_i)})^{5-j}},$$

2) $\frac{R_S(T+t)}{R_S(t)} = \frac{5e^{-0.8(t+T)}-4e^{-(t+T)}}{5e^{-0.8t}-4e^{-t}}$,

3) $\frac{R_S^U(T+t)}{R_S^L(t)} =$

$$\frac{\sum_{j=4}^{5}\binom{5}{j}(\sum_{i=1}^{1000} e^{-0.2(t+T)/(1+u_i)})^j(1000-\sum_{i=1}^{1000} e^{-0.2(t+T)/(1-u_i)})^{5-j}}{\sum_{j=4}^{5}\binom{5}{j}(\sum_{i=1}^{1000} e^{-0.2t/(1-u_i)})^j(1000-\sum_{i=1}^{1000} e^{-0.2t/(1+u_i)})^{5-j}}.$$

Now, based on a simulated random sample $u_1, u_2, ..., u_{1000}$ *of uniform distribution on* $(0,1)$*, one can get* $\int_0^\infty R_S^L(t)dt \simeq 4/7$ *and* $\int_0^\infty R_S^U(t)dt \simeq 16/3$ *and hence* $\widetilde{E}_S = (4/7, 9/4, 16/3)_T$*. Therefore, one may conclude that the mean of the system is consistent with 'remaining life is 9/4' with a degree of one. Further, some specific values of* $\widetilde{R}_S(t)$*,* $\widetilde{M}_S(t)$*, and* $\widetilde{C}_S(2t,t)$ *for* $t \in \{1,2,...,6\}$ *are listed in Table 6.9. The possible interpretation of each reliability criterion reliability criterion can be achieved by Remark 6.10. For instance, based on Table 6.9, it can be said that the system is consistent with 'system is reliable at* $t = 1$*' is 'about 0.693'.*

TABLE 6.9
Some specified values of $\widetilde{R}_S(t)$ and $\widetilde{M}_S(t)$ in Example 6.5.

t	$\widetilde{R}_S(t)$	$\widetilde{M}_S(t)$	$\widetilde{C}_S(2t,t)$
1	$(0.655, 0.693, 0.766)_T$	$(1.698, 1.724, 1.796)_T$	$(0.325, 0.328, 0.329)_T$
2	$(0.347, 0.401, 0.439)_T$	$(1.221, 1.539, 2.495)_T$	$(0.062, 0.067, 0.070)_T$
3	$(0.176, 0.213, 0.282)_T$	$(0.654, 1.445, 1.734)_T$	$(0.011, 0.012, 0.017)_T$
4	$(0.069, 0.108, 0.190)_T$	$(1.092, 1.390, 2.075)_T$	$(0.0002, 0.0003, 0.0007)_T$
5	$(0.017, 0.053, 0.087)_T$	$(0.566, 1.354, 1.512)_T$	$(0.0001, 0.0004, 0.001)_T$
6	$(0.018, 0.025, 0.112)_T$	$(1.062, 1.329, 1.889)_T$	$(0.00002, 0.00008, 0.004)_T$

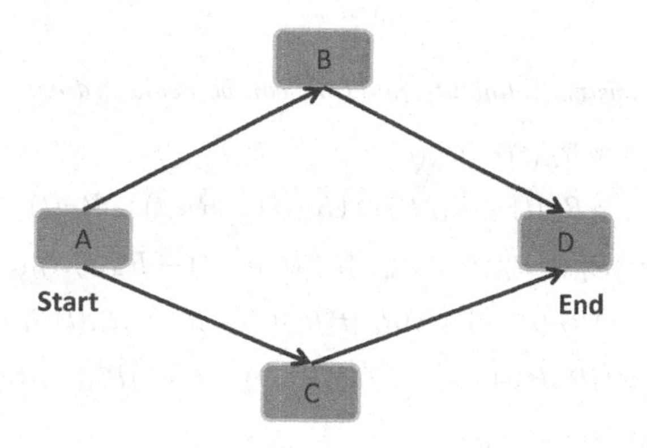

FIGURE 6.2
The system reliability in Example 6.6.

Example 6.6 *Consider a complex system as shown in Fig. 6.2 where the starting and ending blocks can't fail. Assume that the* ***LTFRV*** *of components A through D are* $\widetilde{X} = (X(1-U_1), X, X(1+U_2))_T$ *where*

1) $F(x) = 1/(1+x)$, $x > 0$,

2) $X(1-U_1) \sim^d X/U_1'$ *and* $X(1+U_2) \sim^d X/U_1''$ *in which* U_1' *and* U_2'' *are two independent random variables with* $U_1' \sim U(1, 3/2)$ *and* $U_2'' \sim U(2/3, 1)$, *and*

3) (U_1', U_2'') *is independent of* X.

First, note that

$$\widetilde{R}_i(t) = (R_i^L(t), R_i(t), R_i^U(t))_T,$$

where

$$R^L(t) = P(X/U_1' \geq t) = 2\int_1^{3/2} (1 - F_X(tu_1'))du_1$$
$$= 2\int_1^{3/2} (\frac{1}{1+tu_1'})du_1' = \frac{2}{t}\ln(\frac{1+1.5t}{1+t}),$$
$$R_i(t) = \frac{1}{1+t},$$
$$R^U(t) = P(X/U_2'' \geq t) = 3\int_{2/3}^{1} (1 - F_X(tu_2''))du_2''$$
$$= \frac{3}{t}\ln(\frac{1+t}{1+1.5t}).$$

Therefore, the system reliability function can be evaluated as:

$$\widetilde{R}_S(t) = (\widetilde{R}_A(t) \otimes \widetilde{R}_B(t) \otimes \widetilde{R}_D(t))$$
$$\oplus (\widetilde{R}_A(t) \otimes \widetilde{R}_C(\widetilde{t}) \otimes \widetilde{R}_D(t)) \ominus (\widetilde{R}_A(t) \otimes \widetilde{R}_B(t) \otimes \widetilde{R}_C(t) \otimes \widetilde{R}_D(t))$$
$$= \Big(R_A^L(t)R_B^L(t)R_D^L(t) + R_A^L(t)R_C^L(t)R_D^L(t) - R_A^U(t)R_B^U(t)R_C^U(t)R_D^U(t),$$
$$R_A(t)R_B(t)R_D(t) + R_A(t)R_C(t)R_D(t) - R_A(t)R_B(t)R_C(t)R_D(t),$$
$$R_A^U(t)R_B^U(t)R_D^U(t) + R_A^U(t)R_C^U(t)R_D^U(t) - R_A^L(t)R_B^L(t)R_C^L(t)R_D^L(t)\Big)_T$$
$$= (g_L(t), g(t), g_U(t))_T,$$

in which

1) $g_L(t) = 2(\frac{2}{t}\ln(\frac{(1+(3/2)t)}{(1+t)}))^3 - (\frac{3}{t}\ln(\frac{1+t}{1+(2/3)t}))^4,$

2) $g(t) = 2(\frac{1}{1+t})^3 - (\frac{1}{1+t})^4,$

3) $g_U(t) = 2(\frac{3}{t}\ln(\frac{(1+t)}{(1+(2/3)t)}))^3 - (\frac{2}{t}\ln(\frac{1+(3/2)t}{1+t}))^4,$

Let $t_1, t_2, ..., t_{1000}$ *be a observed random sample of* $exp(1)$. *Based on the strong law of large numbers, the* ***IMTTF*** *can be evaluated as:*

$$\widetilde{E}_S = \int_0^{\infty} \widetilde{R}(t)dt = (\int_0^{\infty} R_S^L(t)dt, \int_0^{\infty} R_S(t)dt, \int_0^{\infty} R_S^U(t)dt)_T.$$

where

1) $\int_0^{\infty} R_S^L(t)dt \simeq$

$$(\frac{1}{1000})\sum_{i=1}^{1000}(2(\frac{2}{t_i}\ln(\frac{(1+(3/2)t)}{(1+t_i)}))^3 - (\frac{3}{t_i}\ln(\frac{1+t_i}{1+(2/3)t_i}))^4)e^{t_i} = 0.4117,$$

2) $\int_0^{\infty} R_S(t)dt \simeq (\frac{1}{1000})\sum_{i=1}^{1000}(2(\frac{1}{1+t_i})^3 - (\frac{1}{1+t_i})^4)e^{t_i} = 0.6192,$

3) $\int_0^\infty R_S^U(t)dt \simeq$

$$(\frac{1}{1000})\sum_{i=1}^{1000}(2(\frac{3}{t_i}\ln(\frac{(1+t_i)}{(1+(2/3)t_i)}))^3-(\frac{2}{t_i}\ln(\frac{1+(3/2)t_i}{1+t_i}))^4)e^{t_i}=0.9293.$$

Now, having a simulated double exponential distribution of $f_X(x) = e^{-(x-t)}I(x>t)$ *with the size of* $n = 1000$*, the* ***FMRLF*** *can be evaluated as:*

$$\widetilde{M}_S(t)=(\frac{\int_t^\infty R_S^L(x)dx}{R_S^U(t)},\frac{\int_t^\infty R_S(x)dx}{R_S(t)},\frac{\int_t^\infty R_S^U(x)dx}{R_S^L(t)})_T,$$

where

1) $\frac{\int_t^\infty R_S^L(x)dx}{R_S^U(t)}$

$$=\frac{\int_t^\infty(2(\frac{2}{x}\ln(\frac{(1+(3/2)x)}{(1+x)}))^3-(\frac{3}{x}\ln(\frac{1+x}{1+(2/3)x}))^4)dx}{2(\frac{3}{t}\ln(\frac{(1+t)}{(1+(2/3)t)}))^3-(\frac{2}{t}\ln(\frac{1+(3/2)t}{1+t}))^4}$$

$$\simeq\frac{(\frac{1}{1000})\sum_{i=1}^{1000}(2(\frac{2}{x_i}\ln(\frac{(1+(3/2)x_i)}{(1+x_i)}))^3-(\frac{3}{x_i}\ln(\frac{1+x_i}{1+(2/3)x_i}))^4)e^{x_i-t}}{2(\frac{3}{t}\ln(\frac{(1+t)}{(1+(2/3)t)}))^3-(\frac{2}{t}\ln(\frac{1+(3/2)t}{1+t}))^4},$$

2) $\frac{\int_t^\infty R_S(x)dx}{R_S(t)}=$

$$\frac{\int_t^\infty(2(\frac{1}{1+x})^3-(\frac{1}{1+x})^4)dx}{2(\frac{1}{1+t})^3-(\frac{1}{1+t})^4}\simeq\frac{(\frac{1}{1000})\sum_{i=1}^{1000}(2(\frac{1}{1+x_i})^3-(\frac{1}{1+x_i})^4)e^{x_i-t}dx}{2(\frac{1}{1+t})^3-(\frac{1}{1+t})^4},$$

3) $\frac{\int_t^\infty R_S^U(x)dx}{R_S^L(t)}=$

$$\frac{\int_t^\infty(2(\frac{3}{x}\ln(\frac{(1+x)}{(1+(2/3)x)}))^3-(\frac{2}{x}\ln(\frac{1+(3/2)x}{1+x}))^4)dx}{2(\frac{2}{t}\ln(\frac{(1+(3/2)t)}{(1+t)}))^3-(\frac{3}{t}\ln(\frac{1+t}{1+(2/3)t}))^4}$$

$$\simeq\frac{(\frac{1}{1000})\sum_{i=1}^{1000}(2(\frac{3}{t}\ln(\frac{(1+x_i)}{(1+(2/3)x_i)}))^3-(\frac{2}{x_i}\ln(\frac{1+(3/2)x_i}{1+x_i}))^4)e^{x_i-t}dx}{2(\frac{2}{t}\ln(\frac{(1+(3/2)t)}{(1+t)}))^3-(\frac{3}{t}\ln(\frac{1+t}{1+(2/3)t}))^4}.$$

Now, assume that $x_1, x_2, ..., x_{1000}$ *is a observed random sample from the double exponential distribution with density function of* $f_X(x) = e^{-(x-(T+t))}I(x > T+t)$ *with the size of* $n = 1000$*, it can be shown that the* ***FCR*** *is:*

$$\widetilde{C}_S(T,t)=\widetilde{R}_S(T+t)\oslash\widetilde{R}_S(t)=(\frac{R_S^L(T+t)}{R_S^U(t)},\frac{R_S(T+t)}{R_S(t)},\frac{R_S^U(T+t)}{R_S^L(t)})_T,$$

TABLE 6.10
Some specified values of $\widetilde{R}_S(t)$, $\widetilde{M}_S(t)$, and $\widetilde{C}_S(2t,t)$ in Example 6.6.

Time	$\widetilde{R}_S(t)$	$\widetilde{M}_S(t)$	$\widetilde{C}_S(2t,t)$
$t=1.5$	$(0.088, 0.187, 0.287,)_T$	$(0.268, 0.3029, 0.3204)_T$	$(0.3584, 0.379, 0.3931)_T$
$t=2$	$(0.027, 0.0618, 0.1003)_T$	$(0.327, 0.3557, 0.3732)_T$	$(0.3920, 0.4249, 0.4430)_T$
$t=2.5$	$(0.0123, 0.0273, 0.0456)_T$	$(0.3584, 0.3790, 0.393)_T$	$(0.4382, 0.4564, 0.4637)_T$
$t=3$	$(0.0066, 0.0144, 0.0244)_T$	$(0.371, 0.397, 0.4173)_T$	$(0.45110.4679, 0.4784)_T$
$t=3.5$	$(0.0039, 0.008, 0.0145)_T$	$(0.352, 0.389, 0.417293)_T$	$(0.4528, 0.4789, 0.4839)_T$
$t=4$	$(0.0025, 0.0054, 0.009)_T$	$(0.392, 0.4249, 0.443,)_T$	$(0.4630, 0.4842, 0.4983)_T$

where

1) $\frac{R_S^L(T+t)}{R_S^U(t)} \simeq$

$$\frac{\sum_{i=1}^{1000}(2(\frac{2}{x_i}\ln(\frac{(1+(3/2)x_i)}{(1+x_i)}))^3-(\frac{3}{x_i}\ln(\frac{1+x_i}{1+(2/3)x_i}))^4)e^{x_i-(T+t)}}{2(\frac{3}{t}\ln(\frac{(1+T+t)}{(1+(2/3)(T+t))}))^3-(\frac{2}{T+t}\ln(\frac{1+(3/2)(T+t)}{1+(T+t)}))^4},$$

2) $\frac{R_S(T+t)}{R_S(t)} \simeq \frac{(\frac{1}{1000})\sum_{i=1}^{1000}(2(\frac{1}{1+x_i})^3-(\frac{1}{1+x_i})^4)e^{x_i-(T+t)}}{2(\frac{1}{1+T+t})^3-(\frac{1}{1+T+t})^4}$, *and*

3) $\frac{R_S^U(T+t)}{R_S^L(t)} \simeq$

$$\frac{(\frac{1}{1000})\sum_{i=1}^{1000}(2(\frac{3}{t}\ln(\frac{(1+x_i)}{(1+(2/3)x_i)}))^3-(\frac{2}{x_i}\ln(\frac{1+(3/2)x_i}{1+x_i}))^4)e^{x_i-(T+t)}}{2(\frac{3}{T+t}\ln(\frac{(1+(3/2)(T+t))}{(1+T+t)}))^3-(\frac{3}{T+t}\ln(\frac{1+T+t}{1+(2/3)(T+t)}))^4}.$$

Some specified values of $\widetilde{R}_S(t)$, $\widetilde{M}_S(t)$, and $\widetilde{C}_S(2t,t)$ at $t \in \{1.5, 2, 2.5, 3, 3.5, 4\}$ are listed in Table 6.10.

6.3 Exercise

Exercise 6.1 *Consider a study monitoring the ph level of a basic solution in an industrial color production process based on fuzzy control charts (Table 6.11). Note that the ph shifts and deviations are small because of environmental chemical reactions. Therefore, process monitoring by* ***EWMA*** *chart in the fuzzy environment should detect the small changes in the ph of productions. Fuzzy ph contents are reported for 10* ***FRS****s (each size of 5) by fuzzy data as summarized in Table 6.11. Such quantities are reported by* ***STFN****s. It was assumed that $L = 2.7$, $\gamma = 0.3$, and $\widetilde{Z}_0 = \widetilde{\overline{\overline{x}}}$.*

1) Construct a ***FEWMA*** *chart.*

TABLE 6.11
Data set in Exercise 6.1

Sample number	$\widetilde{ph}$	Sample number	$\widetilde{ph}$
S_1-1	$(8.45, 9.45, 10.45)_T$	S_6-1	$(10.05, 11.05, 11.05)_T$
S_1-2	$(6.99, 7.99, 8.99)_T$	S_6-2	$(8.50, 10.50, 12.00)_T$
S_1-3	$(6.73, 9.29, 11.85)_T$	S_6-3	$(9.37, 11.62, 13.87)_T$
S_1-4	$(9.65, 11.66, 13.67)_T$	S_6-4	$(8.02, 11.31, 14.6)_T$
S_1-5	$(10.48, 12.16, 13.84)_T$	S_6-5	$(8.23, 10.52, 12.81)_T$
S_2-1	$(8.31, 10.18, 12.05)_T$	S_7-1	$(5.66, 7.69, 9.72)_T$
S_2-2	$(6.09, 8.04, 9.99)_T$	S_7-2	$(12.32, 15.23, 18.14)_T$
S_2-3	$(9.05, 11.46, 13.87)_T$	S_7-3	$(3.54, 4.47, 5.40)_T$
S_2-4	$(8.25, 9.20, 10.15)_T$	S_7-4	$(0.96, 2.84, 4.72)_T$
S_2-5	$(7.35, 10.34, 13.33)_T$	S_7-5	$(7.35, 10.34, 13.33)_T$
S_3-1	$(7.09, 9.03, 10.97)_T$	S_8-1	$(7.52, 10.5, 13.48)_T$
S_3-2	$(9.78, 11.47, 13.16)_T$	S_8-2	$(9.78, 11.47, 13.16)_T$
S_3-3	$(8.75, 10.51, 12.27)_T$	S_8-3	$(8.75, 10.51, 12.27)_T$
S_3-4	$(7.74, 9.4, 11.06)_T$	S_8-4	$(9.12, 12.35, 15.58)_T$
S_3-5	$(9.07, 10.8, 12.53)_T$	S_8-5	$(9.1, 10.54, 11.98)_T$
S_4-1	$(7.61, 9.37, 11.13)_T$	S_9-1	$(9.89, 12.3, 14.71)_T$
S_4-2	$(8.11, 10.62, 13.13)_T$	S_9-2	$(9.78, 11.75, 13.72)_T$
S_4-3	$(7.42, 10.31, 13.2)_T$	S_9-3	$(8.65, 12.32, 15.99)_T$
S_4-4	$(7.481, 8.52, 9.559)_T$	S_9-4	$(8.98, 10.76, 12.54)_T$
S_4-5	$(7.72, 10.84, 13.96)_T$	S_9-5	$(7.91, 10.36, 12.80)_T$
S_5-1	$(9.01, 10.9, 12.79)_T$	$S_{10}-1$	$(7.53, 9.38, 11.23)_T$
S_5-2	$(7.9, 9.33, 10.76)_T$	$S_{10}-2$	$(9.02, 10.42, 11.82)_T$
S_5-3	$(10.4, 12.29, 14.18)_T$	$S_{10}-3$	$(7.03, 8.67, 10.31)_T$
S_5-4	$(8.96, 11.50, 14.04)_T$	$S_{10}-4$	$(9.65, 11.04, 12.43)_T$
S_5-5	$(9.49, 10.90, 12.31)_T$	$S_{10}-5$	$(8.8, 9.92, 11.04)_T$

2) Compute the degrees of violence for all fifty fuzzy sample means to monitor the process.

Exercise 6.2 *The data set in Table 6.12 was applied to control the non-destructive measurements of bottle thickness. Such data were considered with the first fifteen samples each with size 5 (the total measurements is* $3\times 15 = 45$*) taken from the by different operators. Quality experts evaluated each value by a* ***FN*** *due to the variability of the measurement system including the operators and gauges.*

1) Construct a ***FEWMA*** *chart.*

2) Compute the degrees of violence for all fifty fuzzy sample means to monitor the process.

TABLE 6.12
Data set in Exercise 6.2

Sample No.	Subgroups reported by **TFNs**		
	$\widetilde{x}_{i1}$	$\widetilde{x}_{i2}$	$\widetilde{x}_{i3}$
1	$(5.71, 5.73, 5.75)_T$	$(5.50, 5.57, 5.64)_T$	$(5.43, 5.45, 5.46)_T$
2	$(5.41, 5.43, 5.44)_T$	$(5.52, 5.57, 5.58)_T$	$(5.25, 5.29, 5.30)_T$
3	$(5.25, 5.29, 5.32)_T$	$(5.51, 5.53, 5.54)_T$	$(5.00, 5.13, 5.17)_T$
4	$(5.42, 5.51, 5.62)_T$	$(5.26, 5.31, 5.38)_T$	$(5.42, 5.44, 5.46)_T$
5	$(5.19, 5.30, 5.32)_T$	$(5.18, 5.20, 5.31)_T$	$(5.25, 5.28, 5.32)_T$
6	$(5.36, 5.42, 5.63)_T$	$(5.31, 5.40, 5.55)_T$	$(5.18, 5.23, 5.31)_T$
7	$(5.26, 5.27, 5.31)_T$	$(5.53, 5.57, 5.62)_T$	$(5.41, 5.46, 5.49)_T$
8	$(5.43, 5.52, 5.63)_T$	$(5.28, 5.33, 5.40)_T$	$(5.44, 5.47, 5.50)_T$
9	$(5.69, 5.72, 5.75)_T$	$(5.45, 5.49, 5.56)_T$	$(5.32, 5.35, 5.41)_T$
10	$(5.31, 5.35, 5.42)_T$	$(5.26, 5.295.32)_T$	$(5.14, 5.21, 5.26)_T$
11	$(5.28, 5.32, 5.36)_T$	$(5.50, 5.56, 5.62)_T$	$(5.41, 5.46, 5.48)_T$
12	$(5.43, 5.46, 5.48)_T$	$(5.22, 5.26, 5.28)_T$	$(5.15, 5.18, 5.21)_T$
13	$(5.46, 5.52, 5.56)_T$	$(5.35, 5.39, 5.43)_T$	$(5.35, 5.42, 5.48)_T$
14	$(5.41, 5.44, 5.48)_T$	$(5.36, 5.40, 5.43)_T$	$(5.52, 5.56, 5.62)_T$
15	$(5.62, 5.64, 5.68)_T$	$(5.48, 5.55, 5.59)_T$	$(5.42, 5.46, 5.49)_T$

Exercise 6.3 *Consider a fuzzy statistical process control to control cards of GG25 cast iron production line using Carbon Equivalent values. Process control study was performed using the data obtained from fifteen samples with four groups. The obtained results are reported as fuzzy data in Table 6.13. The reason for using the fuzzy method is the possible deviations in Carbon Equivalent values of the final product due to the changes in the hearth entries (amount of scrap material, pig, and alloy elements).*

1) *Construct a fuzzy Schwartz $\widetilde{\overline{x}}$-chart.*

2) *Construct a s-chart.*

3) *Compute the degrees of violence for all fuzzy sample means to monitor the process for both (fuzzy Schwartz) $\widetilde{\overline{x}}$-chart and s-chart.*

Exercise 6.4 *To qualify for Carbon Equivalent values, assume that the lower and upper specific limits (standard values for calculations indexes) were obtained from the management of the plant as $\widetilde{USL} = (4.389, 4.394, 4.399)_T$ and $\widetilde{LSL} = (4.157, 4.162, 4.167)_T$. For such cases, it can be assumed that the process capability of $\widetilde{C}_p$ should exceed $\widetilde{C}_0 = (1.36, 1.40, 1.43)_T$. At a significance level of $\delta = 0.05$, verify the claim via a fuzzy hypothesis for $\widetilde{C}_p$.*

1) *Construct a fuzzy Schwartz $\widetilde{\overline{x}}$-chart.*

2) *Construct a s-chart.*

TABLE 6.13
Data set in Exercise 6.3

Sample No.	Subgroups reported by **TFNs**			
	$\widetilde{x}_{i1}$	$\widetilde{X}_{i2}$	$\widetilde{x}_{i3}$	$\widetilde{x}_{i4}$
1	$(4.149, 4.154, 4.159)_T$	$(4.112, 4.117, 4.122)_T$	$(4, 342, 4, 347, 4, 352)_T$	$(4.113, 4.118, 4.123)_T$
2	$(4.219, 4.224, 4.229)_T$	$(4.112, 4.117, 4.122)_T$	$(4.066, 4.071, 4.076)_T$	$(4.455, 4.46, 4.465)_T$
3	$(4.243, 4.248, 4.253)_T$	$(3.871, 3.876, 3.881)_T$	$(4.258, 4.263, 4.268)_T$	$(4.231, 4.236, 4.241)_T$
4	$(4.359, 4.364, 4.369)_T$	$(4.141, 4.146, 4.151)_T$	$(4.091, 4.096, 4.101)_T$	$(4.344, 4.349, 4.354)_T$
5	$3.957, 3.962, 3.967)_T$	$(4.130, 4.135, 4.140)_T$	$(4.198, 4.203, 4.208)_T$	$(4.019, 4.024, 4.029)_T$
6	$(3.999, 4.004, 4.009)_T$	$(4.173, 4.178, 4.183)_T$	$(4.103, 4.108, 4.113)_T$	$(4.205, 4.210, 4.215)_T$
7	$(3.948, 3.953, 3.958)_T$	$(4.269, 4.274, 4.279)_T$	$(4.332, 4.337, 4.342)_T$	$(4.327, 4.332, 4.337)_T$
8	$(3.990, 3.995, 4.000)_T$	$(4.183, 4.188, 4.193)_T$	$(4.055, 4.060, 4.065)_T$	$(4.147, 4.152, 4.157)_T$
9	$(3.886, 3.891, 3.896)_T$	$(4.231, 4.236, 4.241)_T$	$(4.222, 4.227, 4.232)_T$	$(4.300, 4.305, 4.310)_T$
10	$(4.001, 4.006, 4.011)_T$	$(4.177, 4.182, 4.187)_T$	$(4.151, 4.156, 4.161)_T$	$(4.044, 4.049, 4.054)_T$
11	$(3.713, 3.718, 3.723)_T$	$(4.146, 4.151, 4.156)_T$	$(4.283, 4.288, 4.293)_T$	$(4.292, 4.297, 4.302)_T$
12	$(3.759, 3.764, 3.769)_T$	$(4.215, 4.220, 4.225)_T$	$(4.352, 4.357, 4.362)_T$	$(4.183, 4.188, 4.193)_T$
13	$(3.817, 3.822, 3.827)_T$	$(4.194, 4.199, 4.204)_T$	$(4.042, 4.047, 4.052)_T$	$(4.336, 4.341, 4.346)_T$
14	$(4.081, 4.086, 4.091)_T$	$(4.237, 4.242, 4.247)_T$	$(4.071, 4.076, 4.081)_T$	$(4.079, 4.084, 4.089)_T$
15	$(4.348, 4.353, 4.358)_T$	$(4.344, 4.349, 4.354)_T$	$(4.169, 4.174, 4.179)_T$	$(4.325, 4.330, 4.335)_T$

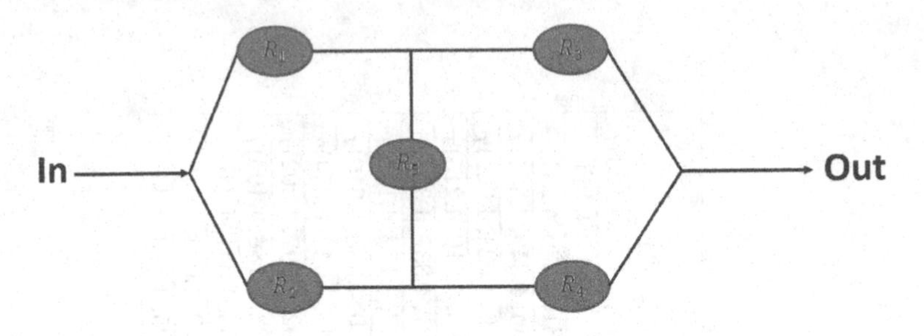

FIGURE 6.3
Diagram of five unit bridge network in Exercise 6.6.

3) Compute the degrees of violence for all fuzzy sample means to monitor the process for both fuzzy Schwartz $\widetilde{\overline{x}}$-chart and s-chart.

Exercise 6.5 *Suppose the minimal capability requirement for one fuzzy process is $\widetilde{C}_p \succ \widetilde{C}_0$. The specification limits are $\widetilde{LSL} = (14.5, 15, 16)_T$ and $\widetilde{USL} = (17, 18, 18.5)_T$. Fifty observations are taken to estimate the fuzzy process capability from a normal distribution. From the fuzzy sample data, the fuzzy sample mean and exact sample deviation are calculated as $\widetilde{\overline{\overline{x}}} = (17, 19, 20)_T$ and $s_{50} = 1$. At a significance level of $\delta = 0.05$, construct a fuzzy hypothesis to verify this claim.*

Exercise 6.6 *The bridge system is considered as a system of five components to find out the . The algebraic expression and pictorial form (Fig. 6.3) of the bridge are given. Assume that the lifetime of components $R_1 - R_5$ are distributed according to Weibull distribution as $R_i \sim W(\alpha_i, \beta_i)$, $i = 1, 2, \ldots, 5$ where*

1) $\alpha_1 = 1.1,\ \beta_1 = 1225,$

2) $\alpha_2 = 1.15,\ \beta_2 = 1227,$

3) $\alpha_3 = 1.20,\ \beta_3 = 1231,$

4) $\alpha_4 = 1.25,\ \beta_4 = 1233,$

5) $\alpha_5 = 1.30,\ \beta_5 = 1235.$

However, in real-life, the reliability of components will be more realistic if considered as fuzzy variables. The system reliability (1) of the bridge system can be expressed as fuzzy system reliability function as follows:

$$\widetilde{R}_S(t) = 2\bigotimes_{i=1}^{4} \widetilde{R}_i(t) \ominus \bigotimes_{i\in\{1,3,4,5\}} \widetilde{R}_i(t) \ominus \bigotimes_{i\in\{2,3,4,5\}} \widetilde{R}_i(t) \ominus \bigotimes_{i\in\{1,2,3,5\}} \widetilde{R}_i(t)$$

$$\ominus \bigotimes_{i\in\{1,2,4,5\}} \widetilde{R}_i(t) \oplus \bigotimes_{i\in\{1,4,5\}} \widetilde{R}_i(t) \oplus \bigotimes_{i\in\{2,3,5\}} \widetilde{R}_i(t) \oplus \bigotimes_{i\in\{1,3\}} \widetilde{R}_i(t) \oplus \bigotimes_{i\in\{2,5\}} \widetilde{R}_i(t).$$

According to the above assumption:

1) *Compute the fuzzy system reliability of* $\widetilde{R}_S(t)$ *for* $t \in \{1,2,3,4,5\}$.

2) *Compute the fuzzy components of* ***FRF***, ***IMTTF***, ***FMRLF***, *and* ***FCR***. *Then, interpret each fuzzy criterion.*

Exercise 6.7 *Consider a parallel system consisting of three independent and identical components. Assume that the lifetime of each electronic component can be modeled by* $\widetilde{X} = (X, UX)_T$ *where* $X \sim exp(1.5)$ *and* $U \sim U(2,3)$ *and* X *and* U *are independent random variables.*

1) *Compute the fuzzy system reliability of* $\widetilde{R}_S(t)$ *for* $t \in \{1,3,5\}$.

2) *Compute the fuzzy components of* ***FRF***, ***IMTTF***, ***FMRLF***, *and* ***FCR***. *Then, interpret each fuzzy criterion.*

Exercise 6.8 *Assume that an aircraft has three independent engines all of which must work normally for the aircraft to fly successfully. Assume that the lifetime of each engine can be modeled by* $\widetilde{X} = ((1-U)X, X, (1+U)X)_T$ *where* X *is distributed according to a modified Weibull extension distribution with a* ***CDF*** *of* $F_X(x) = 1 - \exp[-(\exp(x) - 1)]$, $U \sim U(1,2)$, *and* X *and* U *are independent random variables.*

1) *Compute the the fuzzy system reliability,* $\widetilde{R}_S(t)$ *for* $t = 2, 4,$ *and* 6.

2) *Compute the fuzzy components of* ***FRF***, ***IMTTF***, ***FMRLF***, *and* ***FCR***. *Then, interpret each fuzzy criterion.*

Exercise 6.9 *A broadcast station has three active independent and identical transmitters. At least one of these must function for the system's success. Assume that the life time of each transmitter is modeled as* $\widetilde{X} = (0.95X, X, 1.05X)_T$ *with* $F_X(x) = (4/\sqrt{\pi})x^2 \exp(-x^2)$.

1) *Compute the fuzzy system reliability* $\widetilde{R}_S(t)$ *for* $t \in \{0.5, 1.5, 2, 3\}$.

2) *Compute the fuzzy components of* ***FRF***, ***IMTTF***, ***FMRLF***, *and* ***FCR*** *and interpret each fuzzy criterion.*

Exercise 6.10 *300 samples of the measured thickness of tablets were collected from the record book of a company for the period of 30 months with the a sample size of 10 tablets per month. It is assumed that the thickness of tablets can be distributed according to a* ***FNRV*** *with a fuzzy mean of* $\widetilde{\mu}$ *and the exact variance of* $\sigma = 0.009$. *The specification limits for the thickness of the tablet are* $\widetilde{USL} = (3.96, 3.97, 3.98)_T$ *and* $\widetilde{LSL} = (3.565, 3.57, 3.575)_T$, *and the target value is placed as* $\widetilde{T} = (3.493, 3.5, 3.507)_T$. *According to the observed fuzzy sample assume that we get* $\widetilde{\overline{\overline{x}}} = (3.699, 3.799, 3.899)_T$.

1) Compute the fuzzy process capability $\widetilde{C}(u, \nu)$ *for all combinations of* $u, \nu \in \{0, 1\}$.

2) Interpret the fuzzy capability index $\widetilde{C}_p$ *according to the quality conditions of Table 6.6.*

Exercise 6.11 *A random sample of 45 flange (each of size 10) were collected from a factory to control their diameter. Assume that* $\widetilde{\overline{\overline{x}}} = (150.9, 151, 151.1)_T$ *and* $s_{10} = 1.75$. *The specification limits for the flange thickness are* $\widetilde{USL} = (151.72, 151.80, 151.88)_T$ *and* $\widetilde{LSL} = (151.33, 151.40, 151.47)_T$. *Compute the fuzzy process capability* C_p *and interpret it according to the quality conditions listed in Table 6.6.*

6.4 Glossary

Degree of violence: A preference degree to mark the point beyond which a fuzzy sample value is considered a special cause of variation.

Fuzzy control limits: Fuzzy quantities of the variation in the performance of the quality process.

Fuzzy control chart: A generalization of control charts to study how a process changes over time based on fuzzy random variables.

Fuzzy process capability: Generalization of a statistical measure of the inherent process variability based on fuzzy random variables.

Fuzzy reliability: An extension of e the reliability of a complex system from knowledge of the reliabilities of its components.

Fuzzy system component: Calculation the ability of a reliability system to function under stated conditions for a specified period of time based on fuzzy random variables.

7

Regression and Time Series Models for FRVs

This chapter is organized in two sections. First, some parametric and non-parametric regression models based on **FRV**s are introduced. Then, the proposed fuzzy regression models are utilized to construct some fuzzy parametric and non-parametric time series models.

7.1 Regression Analysis Based on FRVs

Regression analysis is a powerful tool for the prediction of the unknown values of the response variables from the known predictors. In real-life problems, such a relationship is usually unknown but can be estimated from a series of observations. Analysis of regression methods can be classified into parametric [172, 180] and non-parametric [51, 205] techniques. The fuzzy regression analysis may be a useful tool to express functional relationships between response variables and predictors when the available data are fuzzy quantities. The responses of the fuzzy regression are fuzzy quantities while the predictors are either exact or fuzzy quantities. In this chapter, we deal only with fuzzy predictors. Such cases are frequently occurred in real-life applications. For instance, consider a study to explain the relationship between specific supervisor characteristics and overall satisfaction with supervisors as perceived by the employees. The response is overall rating of job being done by supervisor, and predictors are handling employee complaints which do not allow special privileges provide opportunity to learn new things, is too critical of poor performances, and provide rate of advancing to better jobs. As seen, all such quantities can be represented by fuzzy quantities and thus a fuzzy parametric/non-parametric regression model should be employed to investigate the relationship between fuzzy responses and fuzzy predictors.

Fuzzy regression analysis has been addressed from different points of view using conventional statistical regression models [93]. In this regard, Chukhrova and Johannssen [28] presented a comprehensive systematic review on the latest fuzzy regression analyses and their applications, till 2019. For recent studies of fuzzy regression models, see [2, 3, 19, 59, 68, 80, 81, 82, 83, 92, 110, 104, 142].

DOI: 10.1201/9781003248644-7

The methods of fuzzy regression analysis have been suggested for linear and non-linear models. The methodologies of linear models can be classified as (1) approaches, (2) fuzzy least squares and fuzzy least absolute methods, and (3) evolutionary algorithms, support vector machines, and neural networks embedded in fuzzy regression analysis. The first class of the methodologies attempts to minimize a linear/non-linear programming model by minimizing the total spread of its fuzzy parameters to support the observations at some specific levels. The second class includes some of the most commonly fuzzy least squares and fuzzy least absolutes parametric/non-parametric methods, by which the diversity between predicted fuzzy values and available fuzzy data can be minimized relative to various distance measures between the two **FN**s. The other class relies on a combination of parametric and non-parametric fuzzy regression analysis called fuzzy semi-parametric regression models [69, 77].

In this section, some estimation procedures are suggested to evaluate the unknown components of some parametric and non-parametric regression models for **FRV**s. To simplicity in calculations and results, the proposed methods will be conducted based on **STFN**s.

7.1.1 A multivariate regression analysis based on FRVs

Multivariate regression analysis is the most basic and commonly used statistical technique to estimate the relationships between some independent predictor variables and a dependent response variable. Such techniques often rely on the least square errors and least absolute errors by which the diversity between predicted fuzzy responses and available fuzzy data can be minimized relative to various distance measures between the two **FN**s (see for example [24, 25, 33, 11, 107, 127]). In this section, a multivariate regression model with fuzzy predictors and fuzzy responses will be introduced.

Assume that the observed data on n **FRV**s are denoted by $(\widetilde{y}_i, (\widetilde{x}_{i1}, \widetilde{x}_{i2}, \ldots, \widetilde{x}_{ik}))$. Based on the aforementioned data set, the following fuzzy multiple linear regression model can be considered:

$$\widetilde{y}_i = \widetilde{\beta}_0 \bigoplus_{j=1}^{k} (\beta_j \otimes \widetilde{x}_{ij}) \oplus \widetilde{\epsilon}_i, \quad i = 1, 2, \ldots, n, \tag{7.1}$$

where

1. $\widetilde{y}_i = (y_i; l_{y_i})_T$, $P(l_{y_i} > 0) = 1$ denote the fuzzy responses,
2. $\widetilde{x}_i = (x_i; l_{x_i})_T$, $P(l_{x_i} > 0) = 1$ are fuzzy predictors,
3. $\widetilde{\beta}_0 = (\beta_0; l_{\beta_0})_T$, $l_{\beta_0} > 0$ denotes the fuzzy intercept,
4. $\beta_j \in \mathbb{R}$ present unknown non-fuzzy coefficients, and
5. $\widetilde{\epsilon}_i$ indicate **STFN**s with $\widetilde{E}(\widetilde{\epsilon}_i) = I\{0\}$.

The values of β_j's are unknown and must be estimated in the same sense that $\widetilde{\beta}_0 \in \mathbb{F}(\mathbb{R})$ must be estimated. A two-stage estimation procedure can be proposed as follows:

Step (1) Considering the fuzzy expected value of both sides of Eq. (7.1), we get:

$$\widetilde{\mu} = \widetilde{\beta}_0 \bigoplus_{j=1}^{k} (\beta_j \otimes \widetilde{\mu}_j), \tag{7.2}$$

where $\widetilde{\mu} = \widetilde{E}(\widetilde{y}_i)$ and $\widetilde{\mu}_j = \widetilde{E}(\widetilde{x}_{ij})$. By adding $\ominus_G(\bigoplus_{j=1}^{k}(\widehat{\beta}_j \otimes \widetilde{\mu}_j))$ to both sides of Eq. (7.2), we get:

$$\widetilde{\beta}_0 = \widetilde{\mu} \ominus_G (\bigoplus_{j=1}^{k} (\beta_j \otimes \widetilde{\mu}_j)). \tag{7.3}$$

Now, consider the fuzzy estimates of $\widetilde{\mu}$ and $\widetilde{\mu}_j$ corresponding their fuzzy unbiased estimators, i.e.,

$$\widetilde{\widehat{\mu}} = \widetilde{\overline{y}} = (\frac{1}{n}) \otimes (\bigoplus_{i=1}^{n} \widetilde{y}_i) = (\overline{y}; l_{\overline{y}})_T, \tag{7.4}$$

$$\widetilde{\widehat{\mu}}_j = \widetilde{\overline{x}}_j = (\frac{1}{n}) \otimes (\bigoplus_{i=1}^{n} \widetilde{x}_{ij}) = (\overline{x}_j; l_{\overline{x}_j})_T, \tag{7.5}$$

where $\overline{y} = (1/n)\sum_{i=1}^{n} y_i$, $l_{\overline{y}} = (1/n)\sum_{i=1}^{n} l_{y_i}$, $\overline{x}_j = (1/n)\sum_{i=1}^{n} x_{ij}$ and $l_{\overline{x}_j} = (1/n)\sum_{i=1}^{n} l_{x_{ij}}$. Let $\widehat{\beta}_1, \widehat{\beta}_2, ..., \widehat{\beta}_k$ be the estimated values of $\beta_1, \beta_2, ..., \beta_k$. Therefore, the unknown fuzzy intercept of $\widetilde{\beta}_0$ can be estimated as:

$$\widetilde{\widehat{\beta}}_0 = \widetilde{\overline{y}} \ominus_G (\bigoplus_{j=1}^{k} (\widehat{\beta}_j \otimes \widetilde{\overline{x}}_j)). \tag{7.6}$$

According to the arithmetic operations of **STFNs**, note that the fuzzy intercept of $\widetilde{\widehat{\beta}}_0$ can be shown by a **STFN**:

$$\widetilde{\widehat{\beta}}_0 = (\overline{y} - \sum_{j=1}^{k} \widehat{\beta}_j \overline{x}_j; |l_{\overline{y}} - \sum_{j=1}^{k} |\widehat{\beta}_j| l_{\overline{x}_j}|)_T. \tag{7.7}$$

Step (2) To continue the analysis, it is required to estimate $\beta_1, \beta_2, ..., \beta_k$. For this purpose, using the generalized difference ($\ominus_G$) between equations

(7.1) and (7.2), the following fuzzy regression model can be obtained:

$$\begin{aligned}\widetilde{y}_i \ominus_G \widetilde{\mu} &= (\widetilde{\beta}_0 \bigoplus_{j=1}^{k}(\beta_j \otimes \widetilde{x}_{ij}) \oplus \widetilde{\varepsilon}_i) \ominus_G (\widetilde{\beta}_0 \bigoplus_{j=1}^{k}(\beta_j \otimes \widetilde{\mu}_j)) \\ &= (\bigoplus_{j=1}^{k}(\beta_j \otimes \widetilde{x}_{ij}) \oplus \widetilde{\varepsilon}_i) \ominus_G (\bigoplus_{j=1}^{k}(\beta_j \otimes \widetilde{\mu}_j)) \\ &= \bigoplus_{j=1}^{k}(\beta_j \otimes (\widetilde{x}_{ij} \ominus_G \widetilde{\mu}_j) \oplus \widetilde{\varepsilon}_i.\end{aligned} \tag{7.8}$$

After substituting $\widetilde{\widehat{\mu}} = \widetilde{\overline{y}}$ and $\widetilde{\widehat{\mu}}_j = \widetilde{\overline{x}}_j$ in Eq. (7.35), and simplification, the following fuzzy multiple regression model can be written:

$$\begin{aligned}\widetilde{z}_i &= \widetilde{y}_i \ominus_G \widetilde{\overline{y}} \\ &= \bigoplus_{j=1}^{k}(\beta_j \otimes (\widetilde{x}_{ij} \ominus_G \widetilde{\overline{x}}_j)) \oplus \widetilde{\varepsilon}_i \\ &= (\sum_{j=1}^{k} \beta_j (x_{ij} - \overline{x}_j); \sum_{j=1}^{k} |\beta_j| |l_{x_{ij}} - l_{\overline{x}_j}|)_T \oplus \widetilde{\varepsilon}_i,\end{aligned} \tag{7.9}$$

where

$$\widetilde{z}_i = \widetilde{y}_i \ominus_G \widetilde{\overline{y}} = (y_i - \overline{y}; |l_{y_i} - l_{\overline{y}}|)_T. \tag{7.10}$$

To estimate the unknown regression coefficient $\boldsymbol{\beta} = (\beta_1, \beta_2, \ldots, \beta_k)$, the conventional mean square error in the fuzzy domain is extended as:

$$\begin{aligned}MSE(\boldsymbol{\beta}) =& \frac{1}{n}\sum_{i=1}^{n} d_2^2(\widetilde{z}_i, \bigoplus_{j=1}^{k}(\beta_j \otimes (\widetilde{x}_{ij} \ominus_G \widetilde{\overline{x}}_j))) \\ =& \frac{1}{n}\sum_{i=1}^{n}(|(y_i - \overline{y} - \sum_{j=1}^{k}\beta_j(x_{ij} - \overline{x}_j))|^2 \\ &+ 0.1|(|l_{y_i} - l_{\overline{y}}| - \sum_{j=1}^{k}|\beta_j||l_{x_{ij}} - l_{\overline{x}_j}|)|^2).\end{aligned} \tag{7.11}$$

Therefore, the unknown vector of regression coefficient $\boldsymbol{\beta}$ are chosen according to the following optimal problem:

$$\widehat{\boldsymbol{\beta}} = \arg\min_{\boldsymbol{\beta} \in \mathbb{R}^k} MSE(\boldsymbol{\beta}). \tag{7.12}$$

By the method described in the above steps, the predicted fuzzy response can be formulated as a **STFN**:

$$\widetilde{\widehat{y}}_i = \widetilde{\widehat{\beta}}_0 \bigoplus_{j=1}^{k} (\widehat{\beta}_j \otimes \widetilde{x}_{ij}) \tag{7.13}$$

$$= (\widehat{y}_i; l_{\widehat{y}_i})_T$$

$$= \Big(\overline{y} - \sum_{j=1}^{k} \widehat{\beta}_j \overline{x}_j; |l_{\overline{y}} - \sum_{j=1}^{k} |\widehat{\beta}_j| l_{\overline{x}_j}|\Big)_T \oplus \Big(\sum_{j=1}^{k} \widehat{\beta}_j x_{ij}; \sum_{j=1}^{k} |\widehat{\beta}_j| l_{x_{ij}}\Big)_T \tag{7.14}$$

$$= \Big(\overline{y} + \sum_{j=1}^{k} \widehat{\beta}_j (x_{ij} - \overline{x}_j); |l_{\overline{y}} - \sum_{j=1}^{k} |\widehat{\beta}_j| l_{\overline{x}_j}| + \sum_{j=1}^{k} |\widehat{\beta}_j| l_{x_{ij}}\Big)_T.$$

Remark 7.1 *To evaluate the performance of the fitted fuzzy linear regression model, some performance criteria are required to the adequacy. For this purpose, the following criteria are considered:*

1. *Root mean square error (RMSE):*

$$RMSE = \sqrt{\frac{\sum_{i=1}^{n} d_2^2(\widetilde{y}_i, \widetilde{\widehat{y}}_i)}{n}} = \sqrt{\frac{\sum_{i=1}^{n}(|y_i - \widehat{y}_i|^2 + 0.1|l_{y_i} - l_{\widehat{y}_i}|^2)}{n}}. \tag{7.15}$$

2. *Mean similarity measure (MSM):*

$$MSM = \frac{1}{n}\sum_{i=1}^{n}(1 - S(\widetilde{y}_i, \widetilde{\widehat{y}}_i)), \tag{7.16}$$

where

$$S(\widetilde{y}_i, \widetilde{\widehat{y}}_i) = \frac{|y_i - l_{y_i} - \widehat{y}_i + l_{\widehat{y}_i}| + |y_i - \widehat{y}_i| + |y_i + r_{y_i} - \widehat{y}_i - r_{\widehat{y}_i}|}{3(\max\{y_i + r_{y_i}, \widehat{y}_i + r_{\widehat{y}_i}\} - \min\{y_i - l_{y_i}, \widehat{y}_i - l_{\widehat{y}_i}\})}.$$

Example 7.1 *Based on the fuzzy data given in Table 7.1, let us fit a fuzzy linear regression model:*

$$\widetilde{y}_i = \widetilde{\beta}_0 \bigoplus_{j=1}^{2} (\beta_j \otimes \widetilde{x}_{ij}) \oplus \widetilde{\epsilon}_i, \quad i = 1, 2, \ldots, 10,$$

*relating the fuzzy response variable $\widetilde{y}$ and fuzzy predictor $\widetilde{x}$. According to **Step 2**, the unknown regression coefficients of β_1 and β_2 are estimated based upon the following fuzzy linear regression model:*

$$\widetilde{z}_i = \bigoplus_{j=1}^{2} (\beta_j \otimes \widetilde{y}_{ij}) \oplus \widetilde{\varepsilon}_i, \quad i = 1, 2, \ldots, 10.$$

TABLE 7.1
Data set in Example 7.1.

Observation	$\widetilde{y}$	$\widetilde{x}_1$	$\widetilde{x}_2$
1	$(54;3)_T$	$(5;2)_T$	$(5;1)_T$
2	$(57;2)_T$	$(10;3)_T$	$(10;3)_T$
3	$(60;4)_T$	$(12;3)_T$	$(15;2)_T$
4	$(62;3)_T$	$(18;4)_T$	$(20;4)_T$
5	$(63;3)_T$	$(25;4)_T$	$(25;3)_T$
6	$(65;2)_T$	$(30;5)_T$	$(25;5)_T$
7	$(68;2)_T$	$(36;4)_T$	$(30;4)_T$
8	$(70;2)_T$	$(40;6)_T$	$(30;4)_T$
9	$(69;3)_T$	$(45;7)_T$	$(25;5)_T$
10	$(72;3)_T$	$(48;8)_T$	$(30;7)_T$

where

1. $\widetilde{z}_i = \widetilde{y}_i \ominus_G \widetilde{\overline{y}} = (y_i - 64, |l_{y_i} - 2.7|)_T$,

2. $\widetilde{y}_{i1} = \widetilde{x}_{i1} \ominus_G \widetilde{\overline{x}_1} = (x_{i1} - 26.9; |l_{x_{i1}} - 4.6|)_T$,

3. $\widetilde{y}_{i2} = \widetilde{x}_{i2} \ominus_G \widetilde{\overline{x}_2} = (x_{i2} - 21.5; |l_{x_{i2}} - 3.8|)_T$.

Therefore, the unknown regression coefficients β_1 and β_2 can be estimated according to the principle of minimizing MSE, i.e.,

$$(\widehat{\beta_1}, \widehat{\beta_2}) = \arg \min_{(\beta_1,\beta_2)\in\mathbb{R}^2} \frac{1}{n} \sum_{i=1}^{10} d_2^2(\widetilde{z}_i, \bigoplus_{j=1}^{2}(\beta_j \otimes \widetilde{y}_{ij})) =$$

$$\arg \min_{\beta_1,\beta_2\in\mathbb{R}^2} \frac{1}{n} \sum_{i=1}^{10}(|(y_i - 64 - \sum_{j=1}^{2} \beta_j(x_{ij} - \overline{x}_j))|^2 + 0.1|(|l_{y_i} - 2.7| - \sum_{j=1}^{2} |\beta_j||l_{x_{ij}} - l_{\overline{x}_j}|)|^2).$$

*The reader can verify that $\widehat{\beta_1} = 0.25427$ and $\widehat{\beta_2} = 0.23386$. Therefore, the unknown fuzzy intercept can be estimated using **Step 1** as:*

$$\widetilde{\widehat{\beta}}_0 = \widetilde{\overline{y}} \ominus_G (\bigoplus_{j=1}^{2}(\widehat{\beta}_j \otimes \widetilde{\overline{x}}_j)) =$$

$$(64 - (0.25427 \times 26.9 + 0.23386 \times 21.5; |2.7 - (0.25427 \times 4.6 + 0.23386 \times 3.8)|))_T$$
$$= (35.4262; 0.64166)_T.$$

Thus, the fuzzy prediction model can be evaluated as:

$$\widetilde{\widehat{y}}_i = \widetilde{\widehat{\beta}}_0 \bigoplus_{j=1}^{2}(\widehat{\beta}_j \otimes \widetilde{x}_{ij}) = (\widehat{\beta}_0 + \sum_{j=1}^{2} \widehat{\beta}_j x_{ij}; l_{\widehat{\beta}_0} + \sum_{j=1}^{2} |\widehat{\beta}_j| l_{x_{ij}})_T =$$

$$(35.4262 + 0.25427x_{i1} + 0.23386x_{i2} - 12.1017; 0.64166 + 0.25427l_{x_{i1}} + 0.23386l_{x_{i2}})_T.$$

TABLE 7.2
Values of $d_2^2(\widetilde{y}_i, \widetilde{\widehat{y}}_i)$ and $S(\widetilde{y}_i, \widetilde{\widehat{y}}_i)$ in Example 7.1.

No.	$\widetilde{\widehat{y}}_i$	$d_2^2(\widetilde{y}_i, \widetilde{\widehat{y}}_i)$	$S(\widetilde{y}_i, \widetilde{\widehat{y}}_i)$
1	$(54.5727; 1.384)_T$	0.5890	0.2113
2	$(57.0134; 2.1060)_T$	0.0013	0.0178
3	$(58.6912; 1.8722)_T$	2.1656	0.2318
4	$(61.3862; 2.5942)_T$	0.3932	0.0988
5	$(64.3354; 2.3603)_T$	1.8242	0.1994
6	$(65.6068; 3.0823)_T$	0.4853	0.1498
7	$(68.3017; 2.5942)_T$	0.1263	0.0957
8	$(69.3188; 3.1027)_T$	0.5856	0.1550
9	$(69.4209; 3.5908)_T$	0.2120	0.0743
10	$(71.353; 4.3128)_T$	0.5909	0.1264

To check the performance of the fuzzy regression model, there is need to evaluate the goodness-of-fit RMSE, and MSM should be evaluated. For this purpose, the values of

$$d_2^2(\widetilde{y}_i, \widetilde{\widehat{y}}_i) = (|y_i - \widehat{y}_i|^2 + 0.1|l_{y_i} - l_{\widehat{y}_i}|^2),$$

and

$$S(\widetilde{y}_i, \widetilde{\widehat{y}}_i) = \frac{|y_i - l_{y_i} - \widehat{y}_i + l_{\widehat{y}_i}| + |y_i - \widehat{y}_i| + |y_i + r_{y_i} - \widehat{y}_i - r_{\widehat{y}_i}|}{3(\max\{y_i + r_{y_i}, \widehat{y}_i + r_{\widehat{y}_i}\} - \min\{y_i - l_{y_i}, \widehat{y}_i - l_{\widehat{y}_i}\})}$$

are given in Table 7.2. Accordingly, we get

$$RMSE = \sqrt{\frac{\sum_{i=1}^{10} d_2^2(\widetilde{y}_i, \widetilde{\widehat{y}}_i)}{10}} = 0.6973.$$

and

$$MSM = \frac{1}{n}\sum_{i=1}^{10}(1 - S(\widetilde{y}_i, \widetilde{\widehat{y}}_i)) = 0.8639.$$

As can be seen from the results, RMSE and MSM showed reasonable values within the corresponding range. Specifically, the MSM shows that there exists a very strong similarity between fuzzy responses and their estimated fuzzy values. This confirms that a fuzzy linear regression model is a proper choice for describing the relationship between fuzzy response and fuzzy predictors.

Example 7.2 *Consider the fuzzy data set listed in Table 7.3. This data set includes the students' grades and their family income. It is not unreasonable to regard both these variables as fuzzy to some extent, and the data were converted as* ***STFNs****. The model under consideration is:*

$$\widetilde{y}_i = \widetilde{\beta}_0 \oplus (\beta \otimes \widetilde{x}_i) \oplus \widetilde{\epsilon}_i, \quad i = 1, 2, \ldots, 10.$$

TABLE 7.3
Data set in Example 7.2.

No.	$\widetilde{y}_i$	$\widetilde{x}_i$
1	$(12; 1.8)_T$	$(18; 1.5)_T$
2	$(14; 2.2)_T$	$(17; 2)_T$
3	$(17; 2.6)_T$	$(14; 2.5)_T$
4	$(14; 2.6)_T$	$(17; 3)_T$
5	$(12; 2.4)_T$	$(20; 1.8)_T$
6	$(16; 2.3)_T$	$(16; 2)_T$
7	$(13; 2.2)_T$	$(18; 0.8)_T$
8	$(16; 4.8)_T$	$(14; 1.6)_T$
9	$(12; 1.9)_T$	$(19; 0.4)_T$
10	$(15; 2)_T$	$(16; 1.5)_T$

*According to **Step 2**, the unknown regression coefficient of β should be determined by minimizing MSE as follows:*

$$\widehat{\beta} = \arg\min_{\beta\in\mathbb{R}} \frac{1}{n}\sum_{i=1}^{10}(|(y_i-14.1-\beta(x_i-16.9))|^2+0.1|(|l_{y_i}-2.48|-|\beta||l_{x_i}-1.71|)|^2).$$

The calculations revealed that $\widehat{\beta} = -0.877988$. Therefore, the unknown fuzzy intercept can be expressed by Eq. (7.7) as:

$$\begin{aligned}\widetilde{\widehat{\beta}}_0 &= \overline{\widetilde{y}} \ominus_G (\widehat{\beta}\otimes\overline{\widetilde{x}})\\ &= (14.1-(-0.877988\times 16.9); |2.48-(-0.877988\times 1.71)|))_T\\ &= (28.938; 0.97864)_T.\end{aligned}$$

Finally, according to Eq. (7.13), the prediction model is:

$$\widehat{\widetilde{y}}_i = \widetilde{\widehat{\beta}}_0 \oplus (\widehat{\beta}\otimes\widetilde{x}_i) = (28.938-0.877988x_i; 0.97864+0.877988l_{x_i})_T.$$

The values of $d_2^2(\widetilde{y}_i, \widehat{\widetilde{y}}_i)$ and $S(\widetilde{y}_i, \widehat{\widetilde{y}}_i)$ are listed in Table 7.4. It can be concluded that

$$RMSE = \sqrt{\frac{\sum_{i=1}^{10} d_2^2(\widetilde{y}_i, \widehat{\widetilde{y}}_i)}{10}} = 0.44,$$

and

$$MSM = \frac{1}{10}\sum_{i=1}^{10}(1-S(\widetilde{y}_i, \widehat{\widetilde{y}}_i)) = 0.88.$$

The small value of $RMSE$ and large values of MSM identify that a fuzzy linear regression model performs well in predicting the fuzzy student grade based on fuzzy family income.

TABLE 7.4
Values of $d_2^2(\widetilde{y}_i, \widehat{\widetilde{y}}_i)$ and $S(\widetilde{y}_i, \widehat{\widetilde{y}}_i)$ in Example 7.2.

No.	$\widehat{\widetilde{y}}_i$	$d_2^2(\widetilde{y}_i, \widehat{\widetilde{y}}_i)$	$S(\widetilde{y}_i, \widehat{\widetilde{y}}_i)$
1	$(13.1342; 2.2956)_T$	1.311	0.2169
2	$(14.0122; 2.7346)_T$	0.0287	0.0659
3	$(16.6462; 3.1736)_T$	0.1581	0.0788
4	$(14.0122; 3.6126)_T$	0.1026	0.0939
5	$(11.3782; 2.5590)_T$	0.3891	0.1114
6	$(14.8902; 2.7346)_T$	1.2505	0.1806
7	$(13.1342; 1.6810)_T$	0.0449	0.0888
8	$(16.6462; 2.3834)_T$	1.0015	0.1902
9	$(12.2562; 1.3298)_T$	0.0981	0.1225
10	$(14.8902; 2.2956)_T$	0.0207	0.0509

7.1.2 Fuzzy non-parametric regression model

Non-parametric regression can be used in cases where the hypotheses about more classical linear regression methods cannot be verified. Several curve fitting methods have been proposed to estimate a function at a given point. These methods mainly include the k-nearest neighbor smoothing, spline and orthogonal series smoothing, and kernel smoothing (for more details, see [51, 205]). The topic of non-parametric curve fitting for fuzzy data has been investigated over the last decade by some authors from different points of view [21, 204, 98]. Since the kernel-based methods are the most popular ones in statistics, in this section, a non-parametric kernel-based regression model with fuzzy predictors and fuzzy responses is also introduced.

Based on the n **FRVs** denoted by $(\widetilde{y}_i, (\widetilde{x}_{i1}, \widetilde{x}_{i2}, \ldots, \widetilde{x}_{ik}))$, consider the following fuzzy non-parametric regression model:

$$\widetilde{y}_i = f(\widetilde{x}_{i1}, \widetilde{x}_{i2}, \ldots, \widetilde{x}_{ik}) \oplus \widetilde{\epsilon}_i, \tag{7.17}$$

where

1. $\widetilde{y}_i = (y_i; l_{y_i})_T$, $\widetilde{x}_{ij} = (x_{ij}; l_{x_{ij}})_T$,
2. $f(\widetilde{x}_{i1}, \widetilde{x}_{i2}, \ldots, \widetilde{x}_{ik}) = (f(x_{i1}, x_{i2}, ..., x_{ik}); g(l_{x_{i1}}, \ldots, l_{x_{ik}}))_T$, and
3. $\widetilde{\epsilon}_i = (\epsilon_i; l_{\epsilon_i})_T$'s are **STFNs** (say, fuzzy error terms).

Note that Eq. (7.26) can be rewritten as a **STFN** $(y_i; l_{y_i})_T = (f(x_{i1}, x_{i2}, ..., x_{ik}) + \epsilon_i; g(l_{x_{i1}}, \ldots, l_{x_{ik}}) + l_{\epsilon_i})_T$. This suggests estimating the unknown center and spread of $f(\widetilde{x}_{i1}, \widetilde{x}_{i2}, \ldots, \widetilde{x}_{ik})$ based on two ordinary nonlinear regression models $y_i = f(x_{i1}, x_{i2}, ..., x_{ik}) + \epsilon_i$, and $l_{y_i} = g(l_{x_{i1}}, \ldots, l_{x_{ik}}) + l_{\epsilon_i}$. Therefore, to estimate the unknown $\widetilde{f}$ at $\widetilde{\boldsymbol{x}} = (\widetilde{x}_1, \widetilde{x}_2, ..., \widetilde{x}_k)$ where $\boldsymbol{x} = (x_1, x_2, ..., x_k)$ and $l_{\boldsymbol{x}} = (l_{x_1}, l_{x_2}, ..., l_{x_k})$, the following two distinct stages are considered:

- **Step** (1) Consider the non-linear regression model based on spreads of fuzzy responses and fuzzy predictors as $l_{y_i} = g(l_{x_{i1}}, \ldots, l_{x_{ik}}) + l_{\epsilon_i}$. Based on the paired observations of $(l_{y_i}, l_{\boldsymbol{x}_i} = (l_{x_{i1}}, \ldots, l_{x_{ik}}))$. Similar to the conventional non-parametric regression analysis, we can employ the Nadarya-Watson estimator to evaluate g at $l_{\boldsymbol{x}} = (l_{x_1}, l_{x_2}, ..., l_{x_k})$ with the following form:

$$\widehat{g}(l_{\boldsymbol{x}}) = \sum_{i=1}^{n} w_i(l_{\boldsymbol{x}}, h_l) l_{y_i}, \tag{7.18}$$

where

$$w_i(l_{\boldsymbol{x}}, h_l) = \frac{\sum_{j=1}^{k} K\left(\frac{l_{x_{ij}} - l_{x_j}}{h_l}\right)}{\sum_{i=1}^{n} \sum_{j=1}^{k} K\left(\frac{l_{x_{ij}} - l_{x_j}}{h_l}\right)}. \tag{7.19}$$

In the above equation, $K(.)$ shows a kernel function and $h_l > 0$ is an unknown bandwidth to be estimated based on hole data. According to the generalized cross-validation criterion (**GCV**) [51], the optimal value of bandwidth of h_l can be estimated by minimizing **GCV** as follows:

$$\widehat{h}_l = \arg\min_{h_l>0} \mathbf{GCV}(h) = \arg\min_{h>0} \frac{1}{n} \sum_{i=1}^{n} \frac{\left(l_{y_i} - \sum_{j=1}^{n} w_j(l_{\boldsymbol{x}_i}; h_l) l_{y_j}\right)^2}{(1 - tr(W_h)/n)^2}, \tag{7.20}$$

where $tr(W_{h_l})$ represents the trace of the matrix of $W_h = [w_{ij}^{h}]$ in which $w_{ij}^{h_l} = w_i(l_{\boldsymbol{x}_j}, h_l)$, $i, j = 1, 2, \ldots, n$.

- **Step** (2) A similar argument can be explored for the centers of fuzzy responses and fuzzy predictors to evaluate f. Consider the non-linear regression model $y_i = f(x_{i1}, x_{i2}, ..., x_{ik}) + \epsilon_i$. Based on the paired observations of $(y_i, \boldsymbol{x}_i = (x_{i1}, x_{i2}, ..., x_{ik}))$, the extended Nadarya-Watson estimator of f at $\boldsymbol{x} = (x_1, x_2, ..., x_k)$ can be calculated as:

$$\widehat{f}(\boldsymbol{x}) = \sum_{j=1}^{n} w_i(\boldsymbol{x}, h) y_i, \tag{7.21}$$

where

$$w_i(\boldsymbol{x}, h) = \frac{\sum_{j=1}^{k} K\left(\frac{x_{ij} - x_j}{h_l}\right)}{\sum_{i=1}^{n} \sum_{j=1}^{k} K\left(\frac{x_{ij} - x_j}{h}\right)}, \tag{7.22}$$

and $h > 0$ is a bandwidth parameter. The optimization problem to estimate bandwidth of h can be formulated as:

$$\widehat{h} = \arg\min_{h>0} \mathbf{GCV}(h) \tag{7.23}$$

$$= \arg\min_{h>0} \frac{1}{n} \sum_{i=1}^{n} \frac{\left(y_i - \sum_{j=1}^{n} w_j(\boldsymbol{x}_i; h) y_j\right)^2}{(1 - tr(W_h)/n)^2},$$

TABLE 7.5
Data set in Example 7.3.

No.	$\widetilde{y}_i$	$\widetilde{x}_i$	No.	$\widetilde{y}_i$	$\widetilde{x}_i$
1	$(0.380; 0.14)_T$	$(0.20; 0.025)_T$	2	$(0.357; 0.16)_T$	$(0.15; 0.036)_T$
3	$(0.480; 0.12)_T$	$(0.28; 0.047)_T$	4	$(-0.442; 0.09)_T$	$(0.34; 0.015)_T$
5	$(0.173; 0.03)_T$	$(0.38; 0.010)_T$	6	$(0.457; 0.13)_T$	$(0.43; 0.034)_T$
7	$(-0.170; 0.07)_T$	$(0.49; 0.045)_T$	8	$(-0.489; 0.04)_T$	$(0.54; 0.041)_T$
9	$(-0.464; 0.16)_T$	$(0.57; 0.028)_T$	10	$(-0.132; 0.18)_T$	$(0.63; 0.018)_T$
11	$(0.062; 0.19)_T$	$(0.66; 0.013)_T$	12	$(0.307; 0.15)_T$	$(0.71; 0.014)_T$
13	$(0.423; 0.10)_T$	$(0.77; 0.022)_T$	14	$(0.370; 0.16)_T$	$(0.82; 0.028)_T$
15	$(0.235; 0.12)_T$	$(0.88; 0.024)_T$	16	$(0.134; 0.07)_T$	$(0.92; 0.030)_T$
17	$(0.480; 0.13)_T$	$(0.96; 0.025)_T$	18	$(0.371; 0.09)_T$	$(0.17; 0.017)_T$
19	$(0.285; 0.05)_T$	$(0.26; 0.019)_T$	20	$(0.180; 0.08)_T$	$(0.46; 0.028)_T$

in which $tr(W_h)$ shows the trace of $W_h = [w_{ij}^h]$ matrix with $w_{ij}^h = w_i(\boldsymbol{x}_j, h)$, $i, j = 1, 2, \ldots, n$.

Based on **Step** (1), one can observe that the spread of $\widetilde{f}$ is always a non-negative quantity. In this chapter, a popular kernel is utilized in our computational procedure (called the Epanechnikov kernel) [205] which is defined as:

$$K(\frac{x-y}{h}) = \begin{cases} \frac{3}{4}(1 - \frac{|x-y|^2}{h^2}), & \frac{|x-y|}{h} \leq 1, \\ 0, & \frac{|x-y|}{h} > 1. \end{cases} \tag{7.24}$$

To examine the performance of the proposed fuzzy non-parametric regression model, **RMSE** and **MSM** are employed.

Example 7.3 *Consider the data set given in Table 7.5. Let us fit a fuzzy non-linear regression model relating $\widetilde{y}$ and $\widetilde{x}$:*

$$\widetilde{y}_i = f(\widetilde{x}_i) \oplus \widetilde{\epsilon}_i, \tag{7.25}$$

where

1. $\widetilde{y}_i = (y_i; l_{y_i})_T$,
2. $f(\widetilde{x}_i) = (f(x_i); g(l_{x_i}))_T$,
3. $\widetilde{\epsilon}_i = (\epsilon_i; l_{\epsilon_i})_T$.

According to the estimation procedure, the unknown $\widetilde{f}$ should be estimated via the following steps:

- ***Step*** *(1) Consider the non-linear regression model of $l_{y_i} = g(l_{x_i}) + l_{\epsilon_i}$. The Nadarya-Watson estimator of $g(x)$ can be evaluated as:*

$$\widehat{g}(x) = \sum_{i=1}^{10} (\frac{K\left(\frac{l_{x_i} - x}{h_l}\right)}{\sum_{i=1}^{10} K\left(\frac{l_{x_i} - x}{h_l}\right)}) l_{y_i},$$

where the optimal value of h_l can be determined by:

$$\begin{aligned}\widehat{h}_l &= \arg\min_{h_l>0} \boldsymbol{GCV}(h_l)\\ &= \arg\min_{h_l>0} \frac{1}{10}\sum_{i=1}^{10} \frac{\left(l_{y_i} - \sum_{j=1}^{10} w_j(l_{x_i};h_l)l_{y_j}\right)^2}{(1-tr(W_h)/n)^2},\end{aligned}$$

where $tr(W_h)$ shows the trace of the matrix of $W_h = [w_{ij}^h]$ in which $w_{ij}^h = K(\frac{l_{x_i}-l_{x_j}}{h_l})/\sum_{l=1}^{n} K(\frac{l_{x_i}-l_{x_l}}{h_l})$, $i,j = 1,2,\ldots,10$.

- ***Step** (2) Considering the non-linear regression model $y_i = f(x_i)+\epsilon_i$, based on the paired observations of (y_i, x_i), the Nadarya-Watson estimator of $f(x)$ will be:*

$$\widehat{f}(x) = \sum_{i=1}^{10}\left(\frac{K\left(\frac{x_i - x}{\widehat{h}}\right)}{\sum_{i=1}^{10} K\left(\frac{x_i - x}{\widehat{h}}\right)}\right)y_i,$$

with

$$\begin{aligned}\widehat{h} &= \arg\min_{h>0} \boldsymbol{GCV}(h)\\ &= \arg\min_{h>0} \frac{1}{10}\sum_{i=1}^{10} \frac{\left(y_i - \sum_{j=1}^{10} w_j(x_i;h)y_j\right)^2}{(1-tr(W_h)/n)^2},\end{aligned}$$

where $tr(W_h)$ shows the trace of $W_h = [w_{ij}^h]$ matrix in which $w_{ij}^h = K(\frac{x_i-x_j}{h})/\sum_{j=1}^{10} K(\frac{x_i-x_j}{h})$, $i,j = 1,2,\ldots,10$.

The calculations show that $\widehat{h} = 0.03065$ and $\widehat{h}_l = 0.03977$. To calculate the performance measures, it is required to evaluate $d_2^2(\widetilde{y}_i, \widehat{\widetilde{y}}_i)$'s and $S(\widetilde{y}_i, \widehat{\widetilde{y}}_i)$'s as listed in Table 7.6. Therefore, the performance of the fitted fuzzy non-parametric regression model can be determined as:

$$RMSE = \sqrt{\frac{\sum_{i=1}^{20} d_2^2(\widetilde{y}_i, \widehat{\widetilde{y}}_i)}{20}} = 0.0205561,$$

and

$$MSM = \frac{1}{20}\sum_{i=1}^{20}(1 - S(\widetilde{y}_i, \widehat{\widetilde{y}}_i)) = 0.890678.$$

The goodness-of-fit measure of $RMSE$ shows a small error value in predicting fuzzy response values. Further, there is a strong similarity between fuzzy responses values and their predicted values. Now, instead of a fuzzy non-linear regression model, consider a fuzzy linear regression model for such fuzzy data:

$$\widetilde{y}_i = \widetilde{\beta}_0 \oplus (\beta \otimes \widetilde{x}_i) \oplus \widetilde{\epsilon}_i, \quad i = 1,2,\ldots,20.$$

TABLE 7.6
Values of $d_2^2(\widetilde{y}_i, \widehat{\widetilde{y}}_i)$ and $S(\widetilde{y}_i, \widehat{\widetilde{y}}_i)$ in Example 7.3.

No.	$\widehat{\widetilde{y}}_i = (\widehat{f}(x_i); \widehat{g}(l_{x_i}))_T$	$d_2^2(\widetilde{y}_i, \widehat{\widetilde{y}}_i)$	$S(\widetilde{y}_i, \widehat{\widetilde{y}}_i)$
1	$(0.374297; 0.121681)_T$	0.0000660851	0.0504064
2	$(0.358573; 0.11344)_T$	0.000219258	0.0986386
3	$(0.363129; 0.0852555)_T$	0.00213411	0.179393
4	$(-0.442; 0.111866)_T$	0.0000478133	0.0651559
5	$(0.173; 0.100778)_T$	0.000500954	0.234105
6	$(0.445497; 0.116422)_T$	0.00015076	0.0495638
7	$(-0.156108; 0.0918634)_T$	0.000240797	0.104538
8	$(-0.487987; 0.103101)_T$	0.000399201	0.205647
9	$(-0.465013; 0.121411)_T$	0.000149935	0.081448
10	$(-0.124142; 0.116541)_T$	0.000464443	0.124792
11	$(0.0541424; 0.107925)_T$	0.000735365	0.150883
12	$(0.307; 0.109978)_T$	0.000160174	0.0889372
13	$(0.423; 0.120467)_T$	0.0000418904	0.0566327
14	$(0.37; 0.121411)_T$	0.000148909	0.0803933
15	$(0.235; 0.121441)_T$	2.07683×10^{-7}	0.00395561
16	$(0.134; 0.120407)_T$	0.00025409	0.139547
17	$(0.48; 0.121681)_T$	6.92077×10^{-6}	0.0213311
18	$(0.37513; 0.115148)_T$	0.0000802991	0.0787766
19	$(0.329871; 0.11777)_T$	0.00247267	0.255315
20	$(0.177611; 0.121411)_T$	0.000177197	0.116974

Applying **Step** *(2) in the previous section, one can find that that* $\widehat{\beta} = 0.005714$. *Therefore, the prediction model will be:*

$$\widehat{\widetilde{y}}_i = \widehat{\widetilde{\beta}}_0 \oplus (\widehat{\beta} \otimes \widetilde{x}_i) = (0.143216 + 0.005714x_i; 0.112852 + 0.05714l_{x_i})_T,$$

with performance measures of $RMSE = 0.0959177$ *and* $MSE = 0.529741$. *Comparing the performance measures for both cases, it turns out that the fitted fuzzy non-parametric regression provides better results that fuzzy linear regression model for this data set.*

Example 7.4 *Consider a study evaluating the employees' engagement by assessing 25 officers working at an inland revenue board of a city in Iran. The main purpose of this study was to provide a non-linear regression model capable of enhancing the understanding of imprecision about the main individual factors of the employee's engagement in the final work outcomes. Four predicted variables were used for subjective evaluation of the employees' work quality* ($\widetilde{y}$)*:*

$\widetilde{x}_1$*: Ability to endure job stress,*

$\widetilde{x}_2$*: Frequency of delay,*

TABLE 7.7
Data set in Example 7.4.

Observation	$\widetilde{x}_1$	$\widetilde{x}_2$	$\widetilde{x}_3$	$\widetilde{x}_4$	
1	$(95;5)_T$	$(87;8)_T$	$(71;6)_T$	$(86;4)_T$	$(93;7)_T$
2	$(65;4)_T$	$(60;4)_T$	$(62;4)_T$	$(66;3)_T$	$(65;2)_T$
3	$(75;3)_T$	$(84;8)_T$	$(65;6)_T$	$(71;2)_T$	$(75;6)_T$
4	$(67;4)_T$	$(60;9)_T$	$(65;9)_T$	$(68;2)_T$	$(66;5)_T$
5	$(82;5)_T$	$(92;5)_T$	$(78;3)_T$	$(70;4)_T$	$(86;5)_T$
6	$(80;3)_T$	$(57;4)_T$	$(51;2)_T$	$(52;5)_T$	$(55;2)_T$
7	$(77;5)_T$	$(82;4)_T$	$(77;2)_T$	$(80;5)_T$	$(72;2)_T$
8	$(71;3)_T$	$(52;4)_T$	$(57;2)_T$	$(60;5)_T$	$(69;2)_T$
9	$85;6)_T$	$(85;3)_T$	$(77;6)_T$	$(71;5)_T$	$(68;7)_T$
10	$(68;4)_T$	$(66;3)_T$	$(60;2)_T$	$(56;3)_T$	$(73;4)_T$

$\widetilde{x}_3$: *Communication and coordination ability, and*

$\widetilde{x}_4$: *Performance.*

Each employee was asked to respond to a questionnaire with ***STFNs*** *of* $[0, 100]$. *This questionnaire contained 28 items and can be divided into 7 dimensions. Each dimension was relevant to the predictor* $\widetilde{x}_i$. *The fuzzy mean value of all dimensions of the questionnaire (with 4 items) was reported as a fuzzy observation corresponding to* $\widetilde{x}_i$. *The data set is listed in Table 7.7. The following fuzzy non-linear regression model was considered:*

$$(\widetilde{y}_i; l_{y_i})_T = (f(x_{i1}, x_{i2}, x_{i3}, x_{i4}); g(l_{x_{i1}}, l_{x_{i2}}, l_{x_{i3}}, l_{x_{i4}}))_T \oplus \widetilde{\epsilon}_i. \tag{7.26}$$

According to ***Step 1****, the unknown spread of* $\widetilde{f}$ *at* $l_{\boldsymbol{x}} = (l_{x_1}, l_{x_2}, l_{x_3}, l_{x_4})$ *can be estimated as:*

$$\widehat{g}(l_{\boldsymbol{x}}) = \sum_{i=1}^{10} \Big(\frac{\sum_{j=1}^{4} K\left(\frac{l_{x_{ij}} - l_{x_j}}{\widehat{h}_l}\right)}{\sum_{i=1}^{10}\sum_{j=1}^{4} K\left(\frac{l_{x_{ij}} - l_{x_j}}{\widehat{h}_l}\right)} \Big) l_{y_i},$$

where

$$\widehat{h}_l = \arg\min_{h_l>0} \frac{1}{10} \sum_{i=1}^{10} \frac{\left(l_{y_i} - \sum_{j=1}^{10} w_j(l_{\boldsymbol{x}_i}; h_l) l_{y_j}\right)^2}{(1 - tr(W_{h_l})/n)^2}.$$

The center of $\widetilde{f}$ *at* $\boldsymbol{x} = (x_1, x_2, x_3, x_4)$ *can be also evaluated, according to* ***Step 1****, as follows:*

$$\widehat{g}(\boldsymbol{x}) = \sum_{i=1}^{10} \Big(\frac{\sum_{j=1}^{4} K\left(\frac{x_{ij} - x_j}{\widehat{h}}\right)}{\sum_{i=1}^{10}\sum_{j=1}^{4} K\left(\frac{x_{ij} - x_j}{\widehat{h}}\right)} \Big) y_i,$$

TABLE 7.8
Values of $d_2^2(\widetilde{y}_i, \widehat{\widetilde{y}}_i)$ and $S(\widetilde{y}_i, \widehat{\widetilde{y}}_i)$ in Example 7.4.

No.	$\widehat{\widetilde{y}}_i = (\widehat{f}(x_{i1}, x_{i2}, x_{i3}, x_{i4}); \widehat{g}(l_{x_{i1}}, l_{x_{i2}}, l_{x_{i3}}, l_{x_{i4}}))_T$	$S(\widetilde{y}_i, \widehat{\widetilde{y}}_i)$	$d_2^2(\widetilde{y}_i, \widehat{\widetilde{y}}_i)$
1	$(95; 3.82488)_T$	0.07834	0.13809
2	$(63.126; 4.0736)_T$	0.18838	3.51238
3	$(75.5622; 3.53135)_T$	0.07925	0.34433
4	$(70.1216; 3.20727)_T$	0.30222	9.80735
5	$(81.0973; 4.51282)_T$	0.08666	0.83859
6	$(80; 3.63277)_T$	0.05806	0.04003
7	$(75.8843; 3.63277)_T$	0.12834	1.43177
8	$(70.8575; 3.63277)_T$	0.06459	0.06033
9	$(85.4142; 3.96208)_T$	0.12472	0.58688
10	$(67.9368; 0)_T$	0.33596	1.60399

where

$$\widehat{h} = \arg\min_{h>0} \frac{1}{10} \sum_{i=1}^{10} \frac{\left(y_i - \sum_{j=1}^{10} w_j(l_{\mathbf{x}_i}; h) l_{y_j}\right)^2}{(1 - tr(W_h)/n)^2}.$$

The numerical evaluations exhibit that $\widehat{h}_l = 3.9999$ and $\widehat{h} = 1.0194$. By computing $d_2^2(\widetilde{y}_i, \widehat{\widetilde{y}}_i)$ and $S(\widetilde{y}_i, \widehat{\widetilde{y}}_i)$ as given in Table 7.8, the performance measures of the fitted model are:

$$RMSE = \sqrt{\frac{\sum_{i=1}^{10} d_2^2(\widetilde{y}_i, \widehat{\widetilde{y}}_i)}{10}} = 1.35513, \text{ and}$$

$$MSM = \frac{1}{10} \sum_{i=1}^{10} (1 - S(\widetilde{y}_i, \widehat{\widetilde{y}}_i)) = 0.8553.$$

Both performance measures serve reasonable values in their domains. Specifically, $MSM = 0.8553$ shows a very strong similarity between fuzzy responses and their fuzzy predicted values. Therefore, the results strongly exhibit that a fuzzy non-linear regression model provides a good prediction procedure for the employees' work quality.

7.2 Time Series Models for FRVs

Time series analysis comprises methods intended to understand the underlying context of the data for prediction purposes. A time-series analysis-based prediction involves the use of a model to forecast future events based on known past events. The time series prediction method has a wide range of applications in business, finance, computer science, engineering, medicine, physics, chemistry, and many interdisciplinary fields [149, 183].

However, in the real world, many concepts cannot be expressed by precise values. A wide range of data might be vague and imprecise (fuzzy) either by their nature, e.g., environmental data or due to non-ideal measuring [97]. Many of these data and concepts (such as monthly energy consumption, weakly CO_2 emission constraints or annual global land-ocean temperature) can be expressed as sharp numbers or clear expressions of natural language. For such cases, fuzzy time series models based on imprecise information have gained considerable attention in the past decades. Many researchers have studied the conventional time series models for exact data to predict the future based on imprecise past relationships occurring in real-life applications. Soft computing techniques used in this context are mainly a combination of fuzzy sets, artificial neural networks, rough set, and evolutionary computation. Such approaches have been widely employed in real-life applications like enrollment, stock index prices, temperature, financial prediction, and electricity load (for a comprehensive review on these methods, see [1, 63, 23, 52, 64, 126, 146, 160, 166]). Concerning non-fuzzy time series data, these methods first provide some fuzzy time series prediction values followed by their defuzzification to exact prediction values as the final goal. Other methods relied on fuzzy quantity time series data and fuzzy prediction. In these methods, it was assumed that the available time series data can be reported by fuzzy data rather than exact values. The goal is to predict the future as fuzzy quantities. In this regard, Tseng and Tzeng [194] proposed a fuzzy seasonal ARIMA for non-fuzzy data in the cases where the future situations are predicted as fuzzy values. They suggested a fuzzy confidence region for future prediction. Moreover, Hesamian and Akbari [77] proposed a statistical time series model based on fuzzy data. They suggested a semi-parametric time series model with fuzzy data, non-fuzzy coefficients, and fuzzy smooth functions. Zarei et al. [220] applied a specific version of the Hesamian and Akbari model for triangular fuzzy data and the different distance measures for fuzzy data.

Noteworthy, as discussed above, the traditional time series models fail to address forecasting problems with fuzzy data. In this section, some procedures are introduced to construct a parametric/nonparametric time series model based on fuzzy data.

In this section, first a notion of fuzzy autocorrelation function (**ACF**) is introduced according to definition of fuzzy time series data inspired by Hesamian and Akbari [77]. Then, linear and non-linear fuzzy time series data are developed based on those techniques and will be used for regression analysis in the previous section.

7.2.1 Autocorrelation criterion based on a fuzzy time series

Autocorrelation of a random process refers to the correlation between members of a series of time-arranged numbers. Positive autocorrelation may be considered as a specific form of 'persistence' that is a tendency for a system to remain in the same state from one observation to the next one. Furthermore, negative autocorrelation indicates a tendency for positive departures to follow negative departures and vice versa. A common tool to assess the autocorrelation of a time series is the autocorrelation function (**ACF**) [15]. It is an important guide to the persistence in a time series, which is given by the series of quantities called the sample autocorrelation coefficients. These coefficients measure the correlation among observations at different times. The x-axis of the **ACF** plot indicates a lag at which the autocorrelation is computed and y-axis indicates the value of the correlation between -1 and 1. It is worth noting that a spike at a lag of k in an **ACF** plot implies a strong correlation between each value and its preceding one. Likewise, a spike at a lag of k indicates a strong correlation between each value and the one that occurred at two previous points, and so on.

Definition 7.1 *A fuzzy time series is a series of **FRVs** of* $\widetilde{\boldsymbol{X}}_{\boldsymbol{T}} = \{\widetilde{X}_1, \widetilde{X}_2, \ldots, \widetilde{X}_T\}$ *indexed in time order where* $\widetilde{X}_t = (X_t^L, X_t, X_t^U)_T$*. The observed values are shown by* $\widetilde{\boldsymbol{x}}_{\boldsymbol{T}} = \{\widetilde{x}_1, \widetilde{x}_2, \ldots, \widetilde{x}_T\}$*.*

In this section, the fuzzy expectation of $\widetilde{X}_t = (X_t^L, X_t, X_t^U)_T$ is considered as $\widetilde{E}(\widetilde{X}) = (E(X^L), E(X), E(X^U))_T$ with a variance of $\mathbf{var}(\widetilde{X}_t) = E(d_2^2(\widetilde{X}_t, \widetilde{E}(\widetilde{X}_t)))$ where $d_2^2(\widetilde{A}, \widetilde{B}) = ((a^L - b^L)^2 + (a - b)^2 + (a^U - b^U)^2)/3$. Therefore, the variance of $\widetilde{X}_t = (X_t^L, X_t, X_t^U)_T$ can be simplified by $\mathbf{var}(\widetilde{X}_t) = (var(X_t^L) + var(X_t) + var(X_t^U))/3$.

Definition 7.2 *Let* $\widetilde{\boldsymbol{X}}_{\boldsymbol{T}} = \{\widetilde{X}_1, \widetilde{X}_2, \ldots, \widetilde{X}_T\}$ *be a fuzzy time series with* $E(\widetilde{X}_t) = \widetilde{\mu}_t$*. The autocovariance of the generating process is defined as:*

$$\boldsymbol{cov}(\widetilde{X}_t, \widetilde{X}_{t-k}) = \frac{cov(X_t^L, X_{t-k}^L) + cov(X_t, X_{t-k}) + cov(X_t^U, X_{t-k}^U)}{3}, \quad (7.27)$$

where $cov(X, Y)$ *represents the conventional covariance between two random variables of* X *and* Y*.*

Lemma 7.1 *Let* $\widetilde{\boldsymbol{X}}_{\boldsymbol{T}} = \{\widetilde{X}_1, \widetilde{X}_2, \ldots, \widetilde{X}_T\}$ *be a fuzzy time series with* $\widetilde{E}(\widetilde{X}_t) = \widetilde{\mu}_t$*. Then,*

1. $\boldsymbol{cov}(\widetilde{X}_t, \widetilde{X}_{t-k}) = \boldsymbol{cov}(\widetilde{X}_{t-k}, \widetilde{X}_t)$,
2. $\boldsymbol{cov}(\widetilde{X}_t, \widetilde{X}_t) = \boldsymbol{var}(\widetilde{X}_t)$,
3. $\boldsymbol{cov}\big((a \otimes \widetilde{X}_t) \oplus \widetilde{b}, (c \otimes \widetilde{X}_{t-k}) \oplus \widetilde{d}\big) = ac\,\boldsymbol{cov}(\widetilde{X}_t, \widetilde{X}_{t-k})$ *provided that* $ac > 0$.

Proof *The proof is left for the reader.*

Definition 7.3 *Let* $\widetilde{\boldsymbol{X}}_{\boldsymbol{T}} = \{\widetilde{X}_1, \widetilde{X}_2, \ldots, \widetilde{X}_T\}$ *be a fuzzy time series with* $E(\widetilde{X}_t) = \widetilde{\mu}_t$*. Then, the autocorrelation of the generating fuzzy time series can be described by:*

$$\rho(\widetilde{X}_t, \widetilde{X}_{t-k}) = \frac{\boldsymbol{cov}(\widetilde{X}_t, \widetilde{X}_{t-k})}{\sqrt{\boldsymbol{var}(\widetilde{X}_t)\boldsymbol{var}(\widetilde{X}_{t-k})}}. \quad (7.28)$$

Lemma 7.2 *Let $\widetilde{\boldsymbol{X}_T} = \{\widetilde{X}_1, \widetilde{X}_2, \ldots, \widetilde{X}_T\}$ be a fuzzy time series with $E(\widetilde{X}_t) = \widetilde{\mu}_t$. Then,*

1) $|\boldsymbol{\rho}(\widetilde{X}_t, \widetilde{X}_{t-k})| \leq 1$,

2) $\boldsymbol{\rho}(\widetilde{X}_t, \widetilde{X}_{t-k}) = 1$ *if and only if there exists* $a_0 > 0$ *such that* $P(\widetilde{X}_{t-k} \oplus (a_0 \otimes \widetilde{\mu}_t) = \widetilde{\mu}_{t-k} \oplus (a_0 \otimes \widetilde{X}_t)) = 1$,

3) $\boldsymbol{\rho}(\widetilde{X}_t, \widetilde{X}_{t-k}) = -1$ *if and only if* $P(\widetilde{X}_{t-k} \oplus (a_0 \otimes \widetilde{X}_t) = \widetilde{\mu}_{t-k} \oplus (a_0 \otimes \widetilde{\mu}_t) = 1$.

Proof *The proof is left for the reader.*

Definition 7.4 *We say $\widetilde{\boldsymbol{X}_T} = \{\widetilde{X}_1, \widetilde{X}_2, \ldots, \widetilde{X}_T\}$ is a stationary fuzzy time series if*

1. $\widetilde{E}(\widetilde{X}_t) = \widetilde{\mu}$ *for any* t.
2. $c_k = \boldsymbol{cov}(\widetilde{X}_t, \widetilde{X}_{t-k})$ *for any* t *and* k.

Example 7.5 *Let $\widetilde{\boldsymbol{X}_T} = \{\widetilde{X}_1, \widetilde{X}_2, \ldots, \widetilde{X}_T\}$ be a independent fuzzy time series with $\widetilde{X}_t = (X_t - |\epsilon_t|, X_t, X_t + |\epsilon_t|)_T$ in which $X_t \sim N(0, \sigma_t^2)$ and $\epsilon_t \sim N(\mu_t, \tau^2)$. Therefore, $\boldsymbol{\rho}_k = 0$ for any $k \geq 1$ and $\widetilde{E}(\widetilde{X}_t) = (-\sqrt{\frac{2}{\pi}}\tau^2, 0, \sqrt{\frac{2}{\pi}}\tau^2)_T$. Therefore, $\widetilde{\boldsymbol{X}_T} = \{\widetilde{X}_1, \widetilde{X}_2, \ldots, \widetilde{X}_T\}$ is a stationary fuzzy time series.*

Definition 7.5 *Let $\widetilde{\boldsymbol{x}}_T = \{\widetilde{x}_1, \widetilde{x}_2, \ldots, \widetilde{x}_T\}$ be an observed value of a stationary fuzzy time series. Then, the sample autocovariance function of the generating fuzzy time series can be defined by:*

$$\widehat{\mathbf{c}}_k = \frac{\widehat{c}_k^L + \widehat{c}_k + \widehat{c}_k^U}{3}, \tag{7.29}$$

where

$$\widehat{c}_k^L = \frac{1}{T} \sum_{t=k+1}^{T} (x_t^L - \overline{x}^L)(x_{t-k}^L - \overline{x}^L),$$

$$\widehat{c}_k = \frac{1}{T} \sum_{t=k+1}^{T} (x_t - \overline{x})(x_{t-k} - \overline{x}),$$

$$\widehat{c}_k^U = \frac{1}{T} \sum_{t=k+1}^{T} (x_t^U - \overline{x}^U)(x_{t-k}^U - \overline{x}^U).$$

in which $\overline{x}^L = (1/T)\sum_{t=1}^T x_t^L$, $\overline{x} = (1/T)\sum_{t=1}^T x_t$, *and* $\overline{x}^U = (1/T)\sum_{t=1}^T x_t^U$.

TABLE 7.9
Data set in Example 7.6.

t	$\widetilde{x}_t$	t	$\widetilde{x}_t$
1989	$(21.9; 0.25)_T$	1990	$(20.5; 0.260)_T$
1991	$(21.6; 0.380)_T$	1992	$(22.5; .240)_T$
1993	$(23.25; 0.270)_T$	1994	$(22.7; .200)_T$
1995	$(23.42; 0.290)_T$	1996	$(24.7; .180)_T$
1997	$(23.3; 0.210)_T$	1998	$(25.2; 0.190)_T$
1999	$(26.1; 0.240)_T$	2000	$(27.5; 0.260)_T$
2001	$(26.5; 0.190)_T$	2002	$(28.2; 0.220)_T$
2003	$(27.4; 0.190)_T$	2004	$(29.8; 0.250)_T$
2005	$(30.4; 0.200)_T$	2006	$(31.3; 0.230)_T$
2007	$(32.3; 0.270)_T$	2008	$(33.2; 0.270)_T$
2009	$(31.4; 0.30)_T$	2010	$(32.5; 0.30)_T$
2011	$(33.1; 0.40)_T$	2012	$(35.45; 0.50)_T$
2013	$(36.2; 0.23)_T$	2014	$(35.6; 0.35)_T$
2015	$(35.3; 0.38)_T$	2016	$(35.7; 0.29)_T$
2017	$(35.4; 0.23)_T$	2018	$(35.6; 0.35)_T$

Definition 7.6 *Let $\widetilde{\boldsymbol{X}}_{\boldsymbol{T}} = \{\widetilde{x}_1, \widetilde{x}_2, \ldots, \widetilde{x}_T\}$ be an observed value of a stationary fuzzy time series. Then, the sample autocorrelation function (**ACF**) of the generating process is defined by:*

$$\widehat{\boldsymbol{\rho}}_k = \frac{\widehat{\mathbf{c}}_k}{\widehat{\mathbf{c}}_0}, \tag{7.30}$$

where $\widehat{\mathbf{c}}_k$ is given in Definition 7.5.

Example 7.6 *The data set given in Table 7.9 denotes the rent prices in a commercial complex in the center of a town for 30 years. The rents are reported by fuzzy data. The values of $\widehat{\boldsymbol{\rho}}_k$ for $k = 1, 2, ..., 30$ are listed in Table 7.11. For instance, for $k = 1$, Table 7.10 shows that*

$$\widehat{c}_1^L = \frac{1}{30}\sum_{t=2}^{30}(x_t^L - 28.6633)(x_{t-1}^L - 28.6633) \qquad = 23.6889,$$

$$\widehat{c}_1 = \frac{1}{30}\sum_{t=2}^{30}(x_t - 28.934)(x_{t-1} - 28.934) \qquad = 23.974, \text{ and}$$

$$\widehat{c}_1^U = \frac{1}{30}\sum_{t=2}^{30}(x_t^U - 29.2047)(x_{t-1}^U - 29.2047) \qquad = 24.2624.$$

Therefore, $\widehat{\mathbf{c}}_1 = (\widehat{c}_1^L + \widehat{c}_1 + \widehat{c}_1^U)/3 = (23.6889 + 23.974 + 24.2624)/3 = 23.9751$ *and*

$$\widehat{\mathbf{c}}_0 = \frac{\widehat{c}_0^L + \widehat{c}_0 + \widehat{c}_0^U}{3} = \frac{25.8975 + 26.21 + 26.5332}{3} = 26.2136.$$

TABLE 7.10
Calculation the elements of $\widehat{c}_1$ in Example 7.6.

No.	$(X_t^L - 28.6633)(x_{t-1}^L - 28.6633)$	$(x_t - 28.934)(x_{t-1} - 28.934)$	$(x_t^U - 29.2047)(x_{t-1}^U - 29.2047)$
2	59.0756	59.3248	59.5743
3	62.6977	61.855	61.0099
4	47.6621	47.187	46.7051
5	36.3923	36.5709	36.7495
6	35.0283	35.4341	35.8399
7	34.1038	34.3743	34.642
8	22.9264	23.3463	23.7626
9	23.0922	23.8544	24.6275
10	20.3612	21.0374	21.7233
11	10.2415	10.5822	10.9277
12	3.99008	4.06396	4.13849
13	3.34958	3.49036	3.63286
14	1.60811	1.78656	1.97318
15	0.993111	1.12596	1.26698
16	-1.28862	-1.32844	-1.36493
17	1.36251	1.26956	1.17952
18	3.69824	3.46856	3.24462
19	8.10244	7.96396	7.82552
20	14.3644	14.3594	14.3543
21	10.3964	10.52	10.6434
22	8.61768	8.79376	8.97156
23	14.2763	14.856	15.4432
24	25.3772	27.1457	28.9735
25	45.9346	47.3453	48.7373
26	48.1266	48.4352	48.7373
27	41.2106	42.4358	43.6783
28	42.2116	43.0724	43.9373
29	43.8983	43.749	43.598
30	42.8572	43.1024	43.341

TABLE 7.11
Values of $\widehat{\rho}_k$ for some k in Example 7.6.

k	1	2	3	4	5	6	7	8	9	10
$\widehat{\rho}_k$	0.92	0.82	0.73	0.64	0.55	0.45	0.33	0.25	0.14	0.06
k	11	12	13	14	15	16	17	18	19	20
$\widehat{\rho}_k$	−0.03	−0.10	−0.18	−0.25	−0.337	−0.36	−0.39	−0.41	−0.42	−0.41
k	21	22	23	24	25	26	27	28	29	30
$\widehat{\rho}_k$	−0.42	−0.41	−0.39	−0.35	−0.29	−0.24	−0.19	−0.13	−0.06	0

This concludes that $\widehat{\rho}_1 = 23.9751/26.2136 \simeq 0.91$. *By plotting* $\widehat{\rho}_k$*'s versus* k*, an **ACF** plot can be constructed as shown in Fig. 7.1.*

7.2.2 Multivariate time series model based on fuzzy time series data

Let $\widetilde{\boldsymbol{x}}_{\boldsymbol{T}} = \{\widetilde{x}_1, \widetilde{x}_2, \ldots, \widetilde{x}_T\}$ be a fuzzy time series. The fuzzy linear time series model based on a fuzzy time series $\widetilde{\boldsymbol{x}}_{\boldsymbol{T}}$ can be defined by:

$$\widetilde{x}_t = \widetilde{\beta}_0 \bigoplus_{j=1}^{p} (\beta_j \otimes \widetilde{x}_{t-j}) \oplus \widetilde{\epsilon}_t, \quad t = 1, 2, \ldots, T, \tag{7.31}$$

where

1. $\widetilde{E}(\widetilde{x}_t) = \widetilde{\mu}_t$,

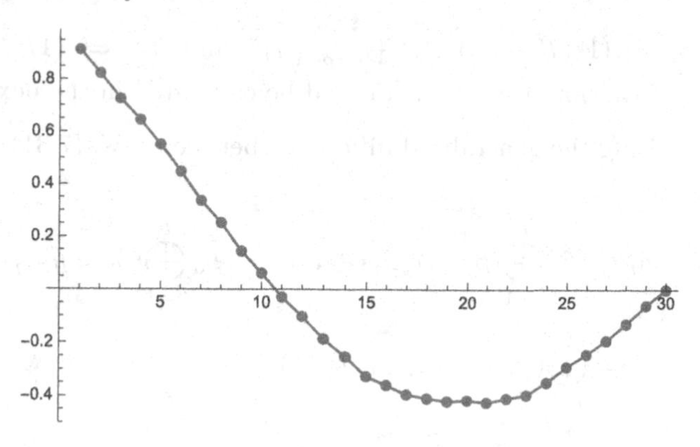

FIGURE 7.1
ACF plot in Example 7.6.

2. $\widetilde{\beta}_0 = (\beta_0; l_{\beta_0})_T$, $l_{\beta_0} > 0$ denotes the fuzzy intercept,
3. $\beta_j \in \mathbb{R}$ present unknown non-fuzzy coefficients, and
4. $\widetilde{\epsilon}_i$ indicate **TFRVs** with $\widetilde{E}(\widetilde{\epsilon}_i) = I\{0\}$.

To estimate the unknown coefficients of $\widetilde{\beta}_0$ and $(\beta_1, \ldots, \beta_k) \in \mathbb{R}^k$, a two-step estimation procedure is suggested based on a within-sample forecast of the size of $T^{'} < T$. It should be pointed out that a within-sample forecast utilizes a subset of the available data to forecast values outside of the estimation period and compare them with their corresponding known or actual outcomes. This is done to assess the ability of the model to forecast known values.

- **Step** (1) Considering the expected value of both sides of Eq. (7.31), we get:

$$\widetilde{\mu}_t = \widetilde{\beta}_0 \bigoplus_{j=1}^{p} (\beta_j \otimes \widetilde{\mu}_{t-j}), \tag{7.32}$$

where $\widetilde{\mu}_t = \widetilde{E}(\widetilde{x}_t)$ and $\widetilde{\mu}_{t-j} = \widetilde{E}(\widetilde{x}_{t-j})$. The fuzzy intercept can be obtained by $\ominus_G(\bigoplus_{j=1}^{p}(\widehat{\beta}_j \otimes \widetilde{\mu}_{t-j}))$ on both sides of Eq. (7.32) as:

$$\widetilde{\beta}_0 = \widetilde{\mu}_t \ominus_G (\bigoplus_{j=1}^{p} (\beta_j \otimes \widetilde{\mu}_{t-j})). \tag{7.33}$$

By substituting the unbiased values of $\widetilde{\mu}_t$ and $\widetilde{\mu}_{t-j}$ in the above equation, the estimate of $\widetilde{\beta}_0$ can be obtained as:

$$\widehat{\widetilde{\beta}}_0 = \overline{\widetilde{x}} \ominus_G (\bigoplus_{j=1}^{p} (\widehat{\beta}_j \otimes \overline{\widetilde{x}}_j)), \tag{7.34}$$

where $\widetilde{\overline{x}} = (1/(T-p)) \otimes (\bigoplus_{t=p+1}^{T} \widetilde{x}_t)$ and $\widetilde{\overline{x}}_j = (1/(T^{'}-j)) \otimes (\bigoplus_{t=p+j+1}^{T} \widetilde{x}_{t-j})$ and $\beta_1, \beta_2, ..., \beta_k$ will be estimated in the next step.

- **Step** (2) Using the generalized difference between Eqs. (7.31) and (7.32), one get:

$$\widetilde{x}_t \ominus_G \widetilde{\mu}_t = (\widetilde{\beta}_0 \bigoplus_{j=1}^{p} (\beta_j \otimes \widetilde{x}_{t-j}) \oplus \widetilde{\epsilon}_t) \ominus_G (\widetilde{\beta}_0 \bigoplus_{j=1}^{p} (\beta_j \otimes \widetilde{\mu}_{t-j})) \tag{7.35}$$
$$= \bigoplus_{j=1}^{p} (\beta_j \otimes (\widetilde{x}_{t-j} \ominus_G \widetilde{\mu}_{t-j}) \oplus \widetilde{\epsilon}_t.$$

By substituting $\widetilde{\widehat{\mu}}_t = \widetilde{\overline{x}}$ and $\widetilde{\widehat{\mu}}_{t-j} = \widetilde{\overline{x}}_j$ in Eq. (7.35), the following fuzzy multiple time series model can be obtained:

$$\widetilde{z}_t = \bigoplus_{j=1}^{p} (\beta_j \otimes \widetilde{z}^*_{t-j}) \oplus \widetilde{\epsilon}_t, \tag{7.36}$$

where $\widetilde{z}_t = \widetilde{x}_t \ominus_G \widetilde{\overline{x}}$ and $\widetilde{z}^*_{t-j} = \widetilde{x}_{t-j} \ominus_G \widetilde{\overline{x}}_j$.

In the above steps, the unknown vector of the regression coefficient of $\boldsymbol{\beta}_p$ and the autoregressive parameter of p should be simultaneously evaluated. For this purpose, one can follow the below steps to estimate such parameters:

- **Step** (1) Let $p = 1$.
- **Step** (2) Compute

$$\widehat{\boldsymbol{\beta}}_p = \arg \min_{\boldsymbol{\beta} \in \mathbb{R}^k} MSE(\boldsymbol{\beta}_p). \tag{7.37}$$

where

$$MSE(\boldsymbol{\beta}_p) = \frac{1}{T-p} \sum_{t=p+1}^{T} d_2^2(\widetilde{z}_t, \widetilde{z}^*_t) \tag{7.38}$$
$$= \sum_{t=p+1}^{T} (\frac{|z_t - z^*_t|^2 + 0.1|l_{z_t} - l_{z^*_t}|^2}{T-p}).$$

- **Step** (3) Let $p = p + 1$ and proceed to **Step 2** until $|MSE(\boldsymbol{\beta}_{p+1}) - MSE(\boldsymbol{\beta}_p)| < \epsilon$ where ϵ is a small number.
- **Step** (4) Let $\widehat{p} = \min_{p \geq 1} MSE(\boldsymbol{\beta}_p)$.

Let $\widehat{\boldsymbol{\beta}}$ and autoregressive parameter of $\widehat{p}$ be the estimates of $\boldsymbol{\beta}_p$ and autoregressive parameter of p, respectively. Therefore, future prediction can be evaluated as a **STFN**:

$$\widetilde{\widehat{x}}_t = \widetilde{\widehat{\beta}}_0 \bigoplus_{j=1}^{p} (\widehat{\beta}_j \otimes \widetilde{\widehat{x}}_{t-j}) = (\widehat{\beta}_0 + \sum_{j=1}^{p} \widehat{\beta}_j x_{t-j}; l_{\widehat{\beta}_0} + \sum_{j=1}^{p} |\widehat{\beta}_j| l_{x_{t-j}})_T. \tag{7.39}$$

TABLE 7.12
The center values of fuzzy data in Example 7.7.

No.	x_t	No.	x_t	No.	x_t	No.	x_t	No.	x_t
1	327.5	6	332.3	11	338.8	16	341.2	21	339.2
2	331.3	7	334.2	12	335.6	17	339.2	22	341.4
3	333.4	8	329.7	13	334.5	18	337.8	23	343.7
4	328.6	9	336.2	14	336.8	19	340.5	24	342.6
5	330.6	10	335.7	15	340.7	20	342.4	25	344.2

The performance of the time series model is examined for the out-sample forecast data set using the $RMSE$ and MSM criteria as:

1. Root mean square error (RMSE):

$$RMSE = \sqrt{\frac{\sum_{t=T'+1}^{T} d_2^2(\widetilde{x}_t, \widehat{\widetilde{x}}_t)}{T-T'}} = \sqrt{\frac{\sum_{t=T'+1}^{T}(|x_t - \widehat{x}_t|^2 + 0.1|l_{x_t} - l_{\widehat{x}_t}|^2)}{(T-T')}}. \tag{7.40}$$

2. Mean similarity measure (MSM)

$$MSM = \frac{1}{T-T'} \sum_{t=T'+1}^{T'} (1 - S(\widetilde{x}_t, \widehat{\widetilde{x}}_t)), \tag{7.41}$$

where

$$S(\widetilde{x}_t, \widehat{\widetilde{x}}_t) = \frac{|x_t - l_{x_t} - \widehat{x}_t + l_{\widehat{x}_t}| + |x_t - \widehat{x}_t| + |x_t + r_{x_t} - \widehat{x}_t - r_{\widehat{x}_t}|}{3(\max\{x_t + r_{x_t}, \widehat{x}_t + r_{\widehat{x}_t}\} - \min\{x_t - l_{x_t}, \widehat{x}_t - l_{\widehat{x}_t}\})}$$

Example 7.7 *Consider a study to predict CO_2 concentrations in a town. For this purpose, the CO_2 concentration values for past 25 months were reported by **STFNs** with the form of $\widetilde{x}_t = (x_t; 0.1x_t)_T$ where the center values are given in Table 7.12. These data are partitioned into two sets with 76% of the data assigned to the within-sample forecast and 24% assigned to the out-sample forecast. Consider the following fuzzy linear time series model:*

$$\widetilde{x}_t = \widetilde{\beta}_0 \bigoplus_{j=1}^{p} (\beta_j \otimes \widetilde{x}_{t-j}) \oplus \widetilde{\epsilon}_t.$$

Thus the fuzzy predicted values for out-sample forecast can be evaluated as:

$$\widehat{\widetilde{x}}_t^p = \widehat{\widetilde{\beta}}_0 \bigoplus_{j=1}^{p} (\widehat{\beta}_j \otimes \widetilde{x}_{t-j}) = (\widehat{\beta}_0 + \sum_{j=1}^{p} \widehat{\beta}_j x_{t-j}; l_{\widehat{\beta}_0} + \sum_{j=1}^{p} |\widehat{\beta}_j| l_{x_{t-j}})_T,$$
$$t = 20, 21, \ldots, 25,$$

TABLE 7.13
Preference measures and estimated values of coefficients in Example 7.7.

p	Coefficients	$RMSE_p$	MSM_p
1	$\widehat{\widetilde{\beta}}_0 = (325.23, 32.523)_T$, $\widehat{\beta}_1 = 0.0319$	4.2957	0.9476
2	$\widehat{\widetilde{\beta}}_0 = (333.061, 31.2771)_T$, $\widehat{\beta}_1 = 0.0426$, $\widehat{\beta}_2 = -0.0343$	4.3904	0.9459
3	$\widehat{\widetilde{\beta}}_0 = (333.713, 30.843)_T$, $\widehat{\beta}_1 = 0.0525$, $\widehat{\beta}_2 = -0.02510$, $\widehat{\beta}_3 = -0.0209$	4.3582	0.9467
4	$\widehat{\widetilde{\beta}}_0 = (333.806, 28.5344)_T$, $\widehat{\beta}_1 = 0.0755$, $\widehat{\beta}_2 = -3.8114 \times 10^{-9}$, $\widehat{\beta}_3 = -0.09629$, $\widehat{\beta}_4 = 0.02818$	4.1658	0.9495
5	$\widehat{\widetilde{\beta}}_0 = (337.216, 31.2071)_T$, $\widehat{\beta}_1 = 0.000031$, $\widehat{\beta}_2 = 0.0110$, $\widehat{\beta}_3 = -0.0544$, $\widehat{\beta}_4 = 0.0419$, $\widehat{\beta}_5 = -2.4573 \times 10^{-6}$	4.0998	0.9502
6	$\widehat{\widetilde{\beta}}_0 = (342.539, 31.5831)_T$, $\widehat{\beta}_1 = -1.92971 \times 10^{-9}$, $\widehat{\beta}_2 = -5.17969 \times 10^{-9}$, $\widehat{\beta}_3 = -1.26974 \times 10^{-8}$, $\widehat{\beta}_4 = 0.0387502$, $\widehat{\beta}_5 = 0.0359076$, $\widehat{\beta}_6 = -0.000016$	7.3747	0.915951

where the estimates of $\beta_1, \beta_2, ..., \beta_p$ can be evaluated via steps (1) – (4). The performance of the proposed fuzzy time series model for the out-sample forecast can be then examined by

$$RMSE_p = \sqrt{\frac{\sum_{t=20}^{25} d_2^2(\widetilde{x}_t, \widehat{\widetilde{x}}_t^p)}{6}},$$

and

$$MSM_p = \frac{1}{6}\sum_{t=20}^{25}(1 - S(\widetilde{x}_t, \widehat{\widetilde{x}}_t^p)).$$

Table 7.13 summarizes the estimates of the model components for $p = 1, 2, 3, 4, 5$ based on the within-sample forecast. The second column exhibits the estimated regression coefficients. The third and fourth columns show the performance measures relevant to each p. As can be seen, the best results are relevant to $p = 5$. For this case, the values of $d_2^2(\widetilde{x}_t, \widehat{\widetilde{x}}_t^5)$ and $S(\widetilde{x}_t, \widehat{\widetilde{x}}_t^5)$ for the out-sample forecast are listed in Table 7.14. The performance results indicate that a fuzzy linear time series model performs well for such fuzzy time series data.

Example 7.8 *Consider the data set in Table 7.15. These data are partitioned into two sets with 71.43% of the data assigned to the within-sample forecast*

TABLE 7.14
Values of $d_2^2(\widetilde{x}_t, \widetilde{\widehat{x}}_t^5)$ and $S(\widetilde{x}_t, \widehat{\widetilde{x}}_t^5)$ in Example 7.7.

No.	$d_2^2(\widetilde{x}_t, \widetilde{\widehat{x}}_t^p)$	$S(\widetilde{x}_t, \widehat{\widetilde{x}}_t^p)$
20	39.6273	0.0848
21	9.1971	0.0429
22	28.4737	0.0729
23	29.5892	0.0753
24	0.6486	0.0118
25	3.1841	0.0259

TABLE 7.15
Fuzzy time series data in Example 7.8.

t	$\widetilde{x}_t$	t	$\widetilde{x}_t$
1	$(2.1; 0.8)_T$	15	$(2.9; 1.1)_T$
2	$(2.5; 1)_T$	16	$(2.6; 0.7)_T$
3	$(2.3; 0.7)_T$	17	$(3; 0.4)_T$
4	$(4.5; 0.8)_T$	18	$(6.8; 0.9)_T$
5	$(1.4; 1.2)_T$	19	$(4.9; 0.8)_T$
6	$(4.2; 0.9)_T$	20	$(2.5; 0.6)_T$
7	$(2.8; 1.1)_T$	21	$(2.1; 1.1)_T$
8	$(4.7; 1.2)_T$	22	$(2.5; 0.9)_T$
9	$(4.1; 1.1)_T$	23	$(2.6; 0.8)_T$
10	$(4.4; 1)_T$	24	$(2.7; 0.4)_T$
11	$(1.3; 1)_T$	25	$(2.6; 1.2)_T$
12	$(3.2; 0.7)_T$	26	$(2.8; 1.4)_T$
13	$(3.1; 0.5)_T$	27	$(3.1; 1)_T$
14	$(3.5; 0.9)_T$	28	$(3.2; 0.7)_T$

and 28.57% assigned to the out sample forecast. Consider the following fuzzy time series model:

$$\widetilde{x}_t = \widetilde{\beta}_0 \bigoplus_{j=1}^{p} (\beta_j \otimes \widetilde{x}_{t-j}) \oplus \widetilde{\epsilon}_t, \quad t = 1, 2, \ldots, 20.$$

The performance of the proposed fuzzy time series model for out-sample forecast can be examined by

$$RMSE_p = \sqrt{\frac{\sum_{t=21}^{28} d_2^2(\widetilde{x}_t, \widetilde{\widehat{x}}_t^p)}{8}},$$

and

$$MSM_p = \frac{1}{8} \sum_{t=21}^{28} (1 - S(\widetilde{x}_t, \widetilde{\widehat{x}}_t^p)).$$

TABLE 7.16
Performance measures and estimated values of coefficients in Example 7.7.

p	Coefficient	$RMSE_p$	MSM_p
1	$\widetilde{\widehat{\beta}}_0 = (3.51554, 0.845316)_T$, $\widehat{\beta}_1 = -0.03368$	0.8045,	0.7237
2	$\widetilde{\widehat{\beta}}_0 = (3.71381, 0.801514)_T$, $\widehat{\beta}_1 = -0.07144$, $\widehat{\beta}_2 = -0.0110927$	0.8677	0.7041
3	$\widetilde{\widehat{\beta}}_0 = (5.00064, 0.47469)_T$, $\widehat{\beta}_1 = -0.01098$ $\widehat{\beta}_2 = -0.0848134$, $\widehat{\beta}_3 = -0.445111$	0.7829	0.7421
4	$\widetilde{\widehat{\beta}}_0 = (3.66449, 0.209358)_T$, $\widehat{\beta}_1 = 0.0783594$ $\widehat{\beta}_2 = -0.128269$, $\widehat{\beta}_3 = -0.387354$, $\widehat{\beta}_4 = 0.360484$	0.9939	0.7133
5	$\widetilde{\widehat{\beta}}_0 = (3.95193, 0.114506)_T$, $\widehat{\beta}_1 = 0.187673$, $\widehat{\beta}_2 = -0.192414$, $\widehat{\beta}_3 = -0.457993$ $\widehat{\beta}_4 = 0.323296$, $\widehat{\beta}_5 = -0.0148356$	1.0094	0.6956

TABLE 7.17
The values of $d_2^2(\widetilde{x}_t, \widetilde{\widehat{x}}_t^3)$ and $S(\widetilde{x}_t, \widetilde{\widehat{x}}_t^3)$ for $p = 3$ in Example 7.7.

No.	$d_2^2(\widetilde{x}_t, \widetilde{\widehat{x}}_t^3)$	$S(\widetilde{x}_t, \widetilde{\widehat{x}}_t^3)$
21	0.799746	0.303306
22	0.00415307	0.0346799
23	1.11249	0.390976
24	1.02412	0.405755
25	1.08116	0.385866
26	1.42947	0.342432
27	0.208777	0.206563
28	0.169282	0.200981

Table 7.16 shows the estimates of regression coefficients along with their performance measures for $p = 1, 2, 3, 4, 5$ based on the within-sample forecast. The performance results reveal that a fuzzy multivariate time series model with $p = 3$ exhibited the lowest RMSE and highest MSM. The values of $d_2^2(\widetilde{x}_t, \widetilde{\widehat{x}}_t^3)$ and $S(\widetilde{x}_t, \widetilde{\widehat{x}}_t^3)$ for the within-sample forecast corresponding to $p = 3$ are listed in Table 7.17.

7.2.3 Non-parametric time series model based on fuzzy time series data

Let $\widetilde{\boldsymbol{x}}_{\boldsymbol{T}} = \{\widetilde{x}_1, \widetilde{x}_2, \ldots, \widetilde{x}_T\}$ be a fuzzy time series. It is assumed that there is a non-linear relationship between fuzzy time series data as follows:

$$\widetilde{x}_t = f(\widetilde{x}_t, \widetilde{x}_{t-1}, \ldots, \widetilde{x}_{t-p}) \oplus \widetilde{\epsilon}_t, \; t = 1, 2, ..., T, \tag{7.42}$$

where

1. $\widetilde{x}_t = (x_t; l_{x_t})_T$,
2. $f(\widetilde{x}_t, \widetilde{x}_{t-1}, \ldots, \widetilde{x}_{t-p}) = (f(x_{t-1}, x_{t-2}, ..., x_{t-p}); g(l_{x_{t-1}}, \ldots, l_{x_{t-p}}))_T$, and
3. $\widetilde{\epsilon}_t = (\epsilon_t; l_{\epsilon_t})_T$'s are **TF** error terms.

To continue the analysis, the data set, the fuzzy predicted values are estimated based on a within-sample forecast $\widetilde{x}_1, \widetilde{x}_2, ..., \widetilde{x}_{T'}$ with $T' < T$. From Eq. (7.42), one can obtain two ordinary non-parametric time series model as 1) $f(x_{t-1}, x_{t-2}, ..., x_{t-p}) + \epsilon_t$, and 2) $l_{x_t} = g(l_{x_{t-1}}, \ldots, l_{x_{t-p}}) + l_{\epsilon_t}$ for $t = 1, 2, ..., T'$. Therefore, to estimate the unknown fuzzy smooth function at $\widetilde{\boldsymbol{x}} = (\boldsymbol{x}; l_{\boldsymbol{x}})_T$ with $\boldsymbol{x} = (x_1, x_2, ..., x_p)$ and $l_{\boldsymbol{x}} = (l_{x_1}, l_{x_2}, ..., l_{x_p})$, we have two distinct steps:

- **Step** (1) Consider the non-linear time series model $f(x_{t-1}, x_{t-2}, ..., x_{t-p}) + \epsilon_t$. Based on the within-sample forecast fuzzy time series of $l_{\boldsymbol{x}_t} = (l_{x_{t-1}}, \ldots, l_{x_{t-p}})$, the fuzzy weighted Nadarya-Watson estimator of g at $l_{\boldsymbol{x}} = (l_{x_1}, \ldots, l_{x_p}))$ can be evaluated as:

$$\widehat{g}(l_{\boldsymbol{x}}) = \sum_{s=p+1}^{T} w_s(l_{\boldsymbol{x}}, h_l) l_{x_s}, \tag{7.43}$$

where

$$w_s(l_{\boldsymbol{x}}, h_l) = \frac{\sum_{i=1}^{p} K(\frac{l_{x_i} - l_{x_{s-i}}}{h_l})}{\sum_{s=p+1}^{T} \sum_{i=1}^{p} K(\frac{l_{x_i} - l_{x_{s-i}}}{h_l})}, \tag{7.44}$$

$K(.)$ shows a kernel function and $h_l > 0$ is a bandwidth parameter. The optimal value of bandwidths h_l can be estimated using the generalized cross-validation criterion as follows:

$$\begin{aligned} \widehat{h}_l &= \arg\min_{h_l > 0} \mathbf{GCV}(h) \qquad (7.45) \\ &= \arg\min_{h_l > 0} \frac{1}{T' - p} \sum_{t=p+1}^{T'} \frac{\left(l_{x_t} - \sum_{s=p+1}^{T'} w^{h_l}(s,t) l_{x_s}\right)^2}{(1 - tr(W_{h_l})/(T' - p))^2}, \end{aligned}$$

where $tr(W_{h_l})$ shows the trace of the matrix $W_{h_l} = [w^{h_l}(s,t)]$ with

$$w^h(t,s) = \frac{\sum_{i=1}^{p} K(\frac{l_{x_{t-i}} - l_{x_{s-i}}}{h_l})}{\sum_{j=p+1}^{T'} \sum_{i=1}^{p} K(\frac{l_{x_{j-i}} - l_{x_{s-i}}}{h_l})}. \tag{7.46}$$

- **Step** (2) Consider the time series model of $x_T = f(x_{t-1}, x_{t-2}, ..., x_{t-p}) + \epsilon_t$. Similar to the previous step, the weighted Nadarya-Watson estimator of f, according to the within-sample forecast data $x_1, x_2, ..., x_{T'}$ is:

$$\widehat{f}(\boldsymbol{x}) = \sum_{s=p+1}^{T} w_s(\boldsymbol{x}, h) x_s, \tag{7.47}$$

where

$$w_s(\boldsymbol{x}, h) = \frac{\sum_{i=1}^{p} K(\frac{x_i - x_{s-i}}{h})}{\sum_{s=p+1}^{T} \sum_{i=1}^{p} K(\frac{x_i - x_{s-i}}{h})} \tag{7.48}$$

$K(.)$ shows a kernel function and $h > 0$ is a bandwidth parameter. The optimal value of h can be also estimated by minimizing the **GCV** criterion as:

$$\begin{aligned} \widehat{h} &= \arg\min_{h>0} \mathbf{GCV}(h) \qquad (7.49) \\ &= \arg\min_{h>0} \frac{1}{T' - p} \sum_{t=p+1}^{T'} \frac{\left(x_t - \sum_{s=p+1}^{T'} w^h(s,t) x_s\right)^2}{(1 - tr(W_h)/(T' - p))^2}, \end{aligned}$$

where $tr(W_h)$ shows the trace of the matrix $W_h = [w^h(s,t)]$ in which

$$w^h(t, s) = \frac{\sum_{i=1}^{p} K(\frac{x_{t-i} - x_{s-i}}{h})}{\sum_{j=p+1}^{T'} \sum_{i=1}^{p} K(\frac{x_{j-i} - x_{s-i}}{h})}. \tag{7.50}$$

Taking into account the above steps, the one step ahead fuzzy prediction of $\widetilde{x}_{T'+1}$ can be evaluated by a **STFN** $\widehat{\widetilde{x}}_{T'+1} = (\widehat{x}_{T'+1}; l_{\widehat{x}_{T'+1}})_T$ with the center of

$$\widehat{x}_{T'+1} = \sum_{s=p+1}^{T'} \frac{\sum_{i=1}^{p} K(\frac{x_{T'-i+1} - x_{s-i}}{\widehat{h}})}{\sum_{s=p+1}^{T} \prod_{i=1}^{p} K(\frac{x_{T'-i+1} - x_{s-i}}{\widehat{h}})} x_s, \tag{7.51}$$

and spread of

$$l_{\widehat{x}_{T'+1}} = \sum_{s=p+1}^{T'} \frac{\sum_{i=1}^{p} K(\frac{l_{x_{T'-i+1}} - l_{x_{s-i}}}{\widehat{h}})}{\sum_{s=p+1}^{T} \prod_{i=1}^{p} K(\frac{l_{x_{T'-i+1}} - l_{x_{s-i}}}{\widehat{h}})} l_{x_s}. \tag{7.52}$$

Therefore, the performance of the above fuzzy non-linear time series model can be examined for fuzzy the out-sample forecast of $\widetilde{x}_{T'+1}, \widetilde{x}_2, ..., \widetilde{x}_T$ as:

1. Root mean square error ($RMSE$):

$$\begin{aligned} RMSE_p &= \sqrt{\frac{\sum_{t=T'+1}^{T} d_2^2(\widetilde{x}_t, \widehat{\widetilde{x}}_t^p)}{T - T'}} \qquad (7.53) \\ &= \sqrt{\frac{\sum_{t=T'+1}^{T} (|x_t - \widehat{x}_t^p|^2 + 0.1|l_{x_t} - l_{\widehat{x}_t^p}|^2)}{T - T'}}. \end{aligned}$$

TABLE 7.18
Performance measures and estimated values of bandwidths in Example 7.9.

Autoregressive order	$\widehat{h}^p$	$\widehat{h}^p_l$	$RMSE_p$	MSM_p
$p=1$	3.4,	0.34,	4.5158,	0.9513
$p=2$	3.4,	0.11,	4.3710,	0.9527
$p=3$	3.9,	0.39,	4.4024,	0.9515
$p=4$	3.9,	0.39,	4.4245,	0.9521
$p=5$	1.1,	0.11,	5.13661,	0.9449

2. Mean similarity measure (MSM):

$$MSM_p = \frac{1}{T-T'} \sum_{t=T'+1}^{T} (1 - S(\widetilde{x}_t, \widetilde{\widehat{x}}^p_t)), \tag{7.54}$$

where

$$S(\widetilde{x}_t, \widetilde{\widehat{x}}^p_t) = \frac{|x_t - l_{x_t} - \widehat{x}^p_t + l_{\widehat{x}_t}| + |x_t - \widehat{x}^p_t| + |x_t + r_{x_t} - \widehat{x}^p_t - r_{\widehat{x}^p_t}|}{3(\max\{x_t + r_{x_t}, \widehat{x}^p_t + r_{\widehat{x}^p_t}\} - \min\{x_t - l_{x_t}, \widehat{x}^p_t - l_{\widehat{x}^p_t}\})}.$$

Example 7.9 *Recall the data set in Example 7.7 given in Table 7.12. Consider the following fuzzy non-parametric time series model for the within-sample forecast:*

$$\widetilde{x}_t = f(\widetilde{x}_t, \widetilde{x}_{t-1}, \ldots, \widetilde{x}_{t-p}) \oplus \widetilde{\epsilon}_t, \quad t = 1, 2, \ldots, 19.$$

The performance of the proposed fuzzy time series model can be examined by

$$RMSE_p = \sqrt{\frac{\sum_{t=20}^{25} d_2^2(\widetilde{x}_t, \widetilde{\widehat{x}}^p_t)}{6}},$$

and

$$MSM_p = \frac{1}{6} \sum_{t=19}^{25} (1 - S(\widetilde{x}_t, \widetilde{\widehat{x}}^p_t)),$$

for the out-sample forecast. Table 7.18 shows the estimates of bandwidths of $\widehat{h}^p$ and $\widehat{h}^p_l$ and their corresponding performance measures for some specific values of $p = 1, 2, 3, 4, 5$. Accordingly, the best results are relevant to $p = 2$ ($RMSE_2 = 4.37101$ and $MSM_2 = 0.952761$). For this case, the values of $d_2^2(\widetilde{x}_t, \widetilde{\widehat{x}}^2_t)$ and $S(\widetilde{x}_t, \widehat{\widehat{x}}^2_t)$ for the within-sample forecast are listed in Table 7.19.

Example 7.10 *Recall the data set in Example 7.6. The following non-linear fuzzy time series model is considered based on within-forecast data:*

$$\widetilde{x}_t = f(\widetilde{x}_t, \widetilde{x}_{t-1}, \ldots, \widetilde{x}_{t-p}) \oplus \widetilde{\epsilon}_t, \quad t = 1, 2, \ldots, 25.$$

TABLE 7.19
Values of $d_2^2(\widetilde{x}_t, \widehat{\widetilde{x}}_t^2)$ and $S(\widetilde{x}_t, \widehat{\widetilde{x}}_t^2)$ in Example 7.9.

No.	$d_2^2(\widetilde{x}_t, \widehat{\widetilde{x}}_t^2)$	$S(\widetilde{x}_t, \widehat{\widetilde{x}}_t^2)$
20	15.5951	0.05469
21	0.0046	0.00208
22	4.93445	0.03151
23	74.5316	0.11378
24	2.8505	0.02427
25	16.7182	0.05708

TABLE 7.20
Performance measures and estimated values of bandwidths in Example 7.10.

p	$\widehat{h}^p$	$\widehat{h}_l^p$	$RMSE_p$	MSM_p
$p = 1$	2.7,	0.1051,	0.39901,	0.7021
$p = 2$	3.3,	0.01083,	0.4264,	0.6687
$p = 3$	3.59186,	0.01240,	1.03589,	0.3793
$p = 4$	3.25,	0.03,	1.05983,	0.3595
$p = 5$	2.7,	0.06,	1.07082,	0.3631

Table 7.20 shows the estimated bandwidths and performance measures based on the within-sample forecast $\widetilde{x}_1, \widetilde{x}_2, \ldots, \widetilde{x}_{T'}$ for the first five autoregressive parameters ($p = 1, 2, 3, 4, 5$). The performance of the proposed fuzzy time series model can be examined by

$$RMSE_p = \sqrt{\frac{\sum_{t=26}^{30} d_2^2(\widetilde{x}_t, \widehat{\widetilde{x}}_t^p)}{5}},$$

and

$$MSM_p = \frac{1}{5} \sum_{t=26}^{30} (1 - S(\widetilde{x}_t, \widehat{\widetilde{x}}_t^p)),$$

for the out-sample forecast. Accordingly, the best results are relevant to $p = 1$. The values of $d_2^2(\widetilde{x}_t, \widehat{\widetilde{x}}_t^1)$ and $S(\widetilde{x}_t, \widehat{\widetilde{x}}_t^1)$ for this case are listed in Table 7.21. Now, instead of a fuzzy non-linear time series model, consider the following fuzzy linear time series model:

$$\widetilde{x}_t = \widetilde{\beta}_0 \bigoplus_{j=1}^{p} (\beta_j \otimes \widetilde{x}_{t-j}) \oplus \widetilde{\epsilon}_t, \quad t = 1, 2, \ldots, 25.$$

Table 7.22 shows the estimates of regression coefficients along with performance measures for $p = 1, 2, 3, 4, 5$ based on the within-sample forecast. Compared to the fuzzy linear time series model, it can be concluded that a fuzzy

TABLE 7.21
Values of $d_2^2(\widetilde{x}_t, \widehat{\widetilde{x}}_t^1)$ and $S(\widetilde{x}_t, \widehat{\widetilde{x}}_t^1)$ in Example 7.10.

t	$d_2^2(\widetilde{x}_t, \widehat{\widetilde{x}}_t^1)$	$S(\widetilde{x}_t, \widehat{\widetilde{x}}_t^1)$
26	0.36025	0.52192
27	0.34733	0.48186
28	0.000159	0.04305
29	0.08708	0.37780
30	0.00121	0.06437

TABLE 7.22
Performance measures and estimated values of coefficients for some p based on the fuzzy linear time series model in Example 7.10.

p	Coefficient(s)	$RMSE_p$	MSM_p
1	$\widehat{\widetilde{\beta}}_0 = (2.30217, 0.0171662)_T$, $\widehat{\beta}_1 = 0.9698$	22.374	0.2017
2	$\widehat{\widetilde{\beta}}_0 = (6.81031, 0.0625303)_T$, $\widehat{\beta}_1 = 0.4188$, $\widehat{\beta}_2 = 0.426425$,	22.3758	0.1997
3	$\widehat{\widetilde{\beta}}_0 = (12.7552, 0.117322)_T$, $\widehat{\beta}_1 = 0.0670$, $\widehat{\beta}_2 = 0.429029$, $\widehat{\beta}_3 = 0.156971$	22.3634	0.3083
4	$\widehat{\widetilde{\beta}}_0 = (18.1708, 0.11106)_T$, $\widehat{\beta}_1 = 0.0894$, $\widehat{\beta}_2 = -0.136118$, $\widehat{\beta}_3 = 0.164775$, $\widehat{\beta}_4 = 0.354716$	22.3847	0.238388
5	$\widehat{\widetilde{\beta}}_0 = (22.1597, 0.0896721)_T$, $\widehat{\beta}_1 = 0.4462$, $\widehat{\beta}_2 = -0.301848$, $\widehat{\beta}_3 = -0.398653$ $\widehat{\beta}_4 = 0.641076$, $\widehat{\beta}_5 = -0.0719253$	22.445	0.1589

non-linear time series performed well compared to a fuzzy linear time series model.

7.3 Exercise

Exercise 7.1 *Consider the data set as shown in Table 7.23.*

1) *Construct a fuzzy univariate regression model and evaluate its goodness-of-fit measures (RMSE and MSM).*

TABLE 7.23
Data set in Exercise 7.1.

No.	$\widetilde{x}$	$\widetilde{y}$
1	$(6.0; 0.5)_T$	$(1.5; 0.5)_T$
2	$(9.0; 1.0)_T$	$(2.0; 0.5)_T$
3	$(7.0; 0.5)_T$	$(2.2; 0.5)_T$
4	$(10.0; 1.0)_T$	$(2.3; 0.5)_T$
5	$(8.0; 0.5)_T$	$(2.5; 0.5)_T$
6	$(12.0; 1.0)_T$	$(2.5; 0.5)_T$
7	$(13.0; 1.0)_T$	$(2.8; 0.5)_T$
8	$(12.0; 1.0)_T$	$(3.0; 1.0)_T$
9	$(15.0; 1.5)_T$	$(3.0; 1.0)_T$
10	$(15.0; 1.5)_T$	$(3.5; 1.0)_T$
11	$(17.0; 1.5)_T$	$(3.5; 1.0)_T$
12	$(18.0; 1.5)_T$	$(3.5; 1.0)_T$
13	$(21.0; 1.5)_T$	$(4.0; 1.0)_T$
14	$(20.0; 1.5)_T$	$(4.0; 1.0)_T$
15	$(23.0; 1.5)_T$	$(4.8; 1.0)_T$

2) *Construct a fuzzy non-parametric regression model and compare its goodness-of-fit measures to that of fuzzy univariate regression model in part 1.*

Exercise 7.2 *The data set in Table 7.24 consists of two fuzzy predictors decision on cooking ($\widetilde{x}_1$) and decision on the environment ($\widetilde{x}_2$) and a fuzzy response variable decision on the cellar ($\widetilde{y}$). These data are the performances of the 20 good-quality tourist restaurant s in a towns.*

1) *Construct a fuzzy multivariate regression model and its relevant goodness-of-fit measures.*

2) *Construct a fuzzy non-parametric regression model and compare its goodness-of-fit measures to that of part 1.*

Exercise 7.3 *Table 7.25 shows a data set on recently-produced cement of a factory. Such a data set comes from an experimental investigation of the heat evolved during the setting and hardening of cement with various compositions and its dependence on the percentage of four compounds in the clinkers from which the cement was produced. The four predictors are $\widetilde{x}_1$: tricalcium aluminate, $\widetilde{x}_2$: tricalcium silicate, $\widetilde{x}_3$: tetracalcium aluminoferrite and $\widetilde{x}_4$: α^1-dicalcium silicate. The heat evolved after 120 days of curing, $\widetilde{y}$, was measured in calories per gram of cement. Construct a fuzzy multivariate regression model and compute its relevant goodness-of-fit measures.*

TABLE 7.24
Data set in Exercise 7.2.

No.	$\widetilde{x}_1$	$\widetilde{x}_2$	$\widetilde{y}$
1	$(7;0.50)_T$	$(8;0.75)_T$	$(8;1)_T$
2	$(7;1.25)_T$	$(7;1)_T$	$(6;0.25)_T$
3	$(6;0.65)_T$	$(7;1.20)_T$	$(6;0.35)_T$
4	$(8;1.00)_T$	$(9;1)_T$	$(9;0.45)_T$
5	$(8;0.55)_T$	$(8;0.15)_T$	$(8;0.85)_T$
6	$(6;0.55)_T$	$(7;0.5)_T$	$(5;0.45)_T$
7	$(7;0.85)_T$	$(8;0.25)_T$	$(7;0.75)_T$
8	$(7;0.65)_T$	$(7;0.5)_T$	$(5;0.3)_T$
9	$(7;0.60)_T$	$(8;0.25)_T$	$(7;0.5)_T$
10	$(6;0.25)_T$	$(7;1)_T$	$(6;0.40)_T$
11	$(7;0.50)_T$	$(8;1.00)_T$	$(8;0.75)_T$
12	$(7;0.60)_T$	$(6;0.50)_T$	$(6;0.50)_T$
13	$(7;0.60)_T$	$(8;0.3)_T$	$(9;0.75)_T$
14	$(7;0.8)_T$	$(8;1)_T$	$(8;0.35)_T$
15	$(7;0.25)_T$	$(7;1)_T$	$(7;0.70)_T$
16	$(7;0.40)_T$	$(7;0.65)_T$	$(7;0.45)_T$
17	$(6;0.20)_T$	$(7;0.30)_T$	$(6;0.45)_T$
18	$(7;0.50)_T$	$(8;0.65)_T$	$(7;0.55)_T$
19	$(7;1)_T$	$(7;0.55)_T$	$(8;0.80)_T$
20	$(7;0.15)_T$	$(9;1.25)_T$	$(7;1)_T$

Exercise 7.4 *Recall data set in Exercise 7.3.*

1) *Perform a fuzzy nonparametric regression to predict* $\widetilde{y}$ *based on* $\widetilde{x}_1$,$\widetilde{x}_2$,$\widetilde{x}_3$ *and* $\widetilde{x}_4$ *and compute its performance measures.*

2) *Beside the Epanechnikov kernel, compare the performance measures of* $RMSE$ *and* MSM *for some common kernels listed in Table 7.26.*

3) *For each Epanechnikov, triweight, and gaussian kernel (Table 7.26) compare the performance measures with the fuzzy multivariate regression model in Part 1.*

Exercise 7.5 *Recall data set in Example 7.1.*

1) *Perform a fuzzy multivariate time series model and compute its performance measures.*

2) *Perform a fuzzy non-parametric time series model and compare its performance measures with the previous case.*

Exercise 7.6 *Recall data set in Example 7.7.*

1) *Compute* $\widehat{\rho}(\widetilde{x}_i, \widetilde{x}_{i-k})$ *for* $k = 2, 5, 8$.

TABLE 7.25
Data set in Exercise 7.3.

No.	$\widetilde{x}_1$	$\widetilde{x}_2$	$\widetilde{x}_3$	$\widetilde{y}$
1	$(0.7; 0.1)_T$	$(1.1; 0.2)_T$	$(1.0; 0.2)_T$	$(-1.4; 0.3)_T$
2	$(0.8; 0.2)_T$	$(0.5; 0.1)_T$	$(0.6; 0.1)_T$	$(-1.0; 0.2)_T$
3	$(-0.6; 0.1)_T$	$(-0.8; 0.2)_T$	$(-0.8; 0.1)_T$	$(0.9; 0.2)_T$
4	$(0.4; 0.1)_T$	$(0.2; 0.1)_T$	$(0.17; 0.05)_T$	$(-0.7; 0.1)_T$
5	$(-0.3; 0.1)_T$	$(-0.21; 0.05)_T$	$(-0.2; 0.1)_T$	$(0.9; 0.2)_T$
6	$(0.6, 0.1, 0.1)_T$	$(0.7, 0.1, 0.1)_T$	$(0.7, 0.1, 0.2)_T$	$(-2.1, 0.3, 0.4)_T$
7	$(1.3; 0.2)_T$	$(1.0; 0.2)_T$	$(1.1; 0.2)_T$	$(-2.3; 0.5)_T$
8	$(0.5; 0.1)_T$	$(0.9; 0.2)_T$	$(0.9; 0.1)_T$	$(-1.2; 0.2)_T$
9	$(-0.9; 0.2)_T$	$(-1.2; 0.2)_T$	$(-1.3; 0.3)_T$	$(3.6; 0.8)_T$
10	$(0.0; 0.1)_T$	$(0.2; 0.1)_T$	$(0.1; 0.1)_T$	$(0.2; 0.1)_T$
11	$(-0.8; 0.2)_T$	$(-0.6; 0.1)_T$	$(-0.8; 0.2)_T$	$(1.3; 0.3)_T$
12	$(-0.2; 0.2)_T$	$(-0.10.05)_T$	$(0.1; 0.2)_T$	$(-0.3; 0.1)_T$
13	$(0.0; 0.1)_T$	$(-0.3; 0.1)_T$	$(-0.2; 0.05)_T$	$(1.9; 0.4)_T$
14	$(-0.7; 0.2)_T$	$(-1.0; 0.3)_T$	$(-0.9; 0.1)_T$	$(0.8; 0.2)_T$
15	$(-0.6; 0.1)_T$	$(-1.0; 0.2)_T$	$(-1.0; 0.1)_T$	$(5.3; 1.1)_T$
16	$(0.8; 0.3)_T$	$(0.9; 0.3)_T$	$(1.1; 0.2)_T$	$(-1.8; 0.4)_T$
17	$(-0.4; 0.1)_T$	$(-0.7; 0.2)_T$	$(-0.7; 0.2)_T$	$(1.8; 0.4)_T$
18	$(-0.6; 0.1)_T$	$(-0.9; 0.2)_T$	$(-0.8; 0.2)_T$	$(0.5; 0.1)_T$
19	$(1.2; 0.2)_T$	$(1.2; 0.2)_T$	$(1.3; 0.3)_T$	$(-0.9; 0.2)_T$
20	$(1.0; 0.2)_T$	$(1.2; 0.3)_T$	$(1.5; 0.2)_T$	$(-2.7; 0.6)_T$
21	$(0.1; 0.2)_T$	$(-0.2; 0.3)_T$	$(-0.1; 0.1)_T$	$(1.5, ; 0.3)_T$
22	$(2.8; 0.6)_T$	$(2.5; 0.5)_T$	$(3.2; 0.7)_T$	$(-4.1; 0.9)_T$
23	$(-0.7; 0.1)_T$	$(-0.7; 0.2)_T$	$(-0.8; 0.2)_T$	$(1.5; 0.3)_T$
24	$(-0.5; 0.2)_T$	$(-0.7; 0.2)_T$	$(-0.6; 0.1)_T$	$(0.6; 0.1)_T$
25	$(-1.4; 0.4)_T$	$(-1.4; 0.2)_T$	$(-1.5; 0.3)_T$	$(1.7; 0.4)_T$

TABLE 7.26
Kernel functions in Exercise 7.4.

Triweight	$K(y) = \begin{cases} \frac{35}{32}(1-y^2)^3, & \lvert y\rvert \leq 1, \\ 0, & \lvert y\rvert > 1. \end{cases}$
Gaussian	$K(y) = \frac{1}{\sqrt{2\pi}} e^{-y^2/2}, \ y \in \mathbb{R}.$

2) Construct an ***AFC*** *plot and interpret it.*

Exercise 7.7 *Recall data set in Example 7.8. Compare the performance measures of $RMSE$ and MSM in cases where the triweight and Gaussian kernels, beside the Epanechnikov kernel, were employed.*

Exercise 7.8 *Recall data set in Example 7.8.*

1) Compute $\widehat{\boldsymbol{\rho}}_k$ for $k = 1, 4, 7$.

TABLE 7.27
Data set in Exercise 7.9.

No.	$\widetilde{x}_1$	$\widetilde{x}_2$	$\widetilde{y}$
1	$(2948; 15)_T$	$(42.7; 2.2)_T$	$(88.8; 4.6)_T$
2	$(3738; 16)_T$	$(43.8; 3.1)_T$	$(100.3; 3.5)_T$
3	$(3077; 12)_T$	$(41.7; 4.2)_T$	$(97.5; 2.8)_T$
4	$(4020; 14)_T$	$(35.8; 2.8)_T$	$(102.1; 3.7)_T$
5	$(3571; 18)_T$	$(44.7; 3.4)_T$	$(94.2; 2.8)_T$
6	$(3377; 15)_T$	$(40.4; 4.7)_T$	$(95.8; 3.4)_T$
7	$(4007; 14)_T$	$(42.2; 4.5)_T$	$(93.6; 3.6)_T$
8	$(3677; 11)_T$	$(38.9; 2.7)_T$	$(96.3; 2.7)_T$
9	$(3223; 19)_T$	$(36.6; 2.6)_T$	$(90.4; 2.6)_T$
10	$(3478; 13)_T$	$(40.3; 3)_T$	$(98.7; 3.1)_T$

2) Construct an ***AFC*** *plot and interpret it.*

Exercise 7.9 *Consider the data set in Table 7.27. Perform fuzzy linear and non-linear regression models (adopted with non-parametric evaluations) and compute its performance measures.*

Exercise 7.10 *Prove Lemma 1.8.*

Exercise 7.11 *Prove Lemma 1.10.*

7.4 Glossary

Fuzzy autocorrelation function: A way to measure the (linear) relationship between an fuzzy observation at time t and the fuzzy observations at previous times.

Fuzzy regression: An extension of regression modeling based on fuzzy-valued responses and predictors.

Fuzzy time series data: Fuzzy-values of a variable recorded over a long period of time.

Fuzzy time series model: An extension of time series forecasting to predict future values based on previously observed values as fuzzy quantities.

Defuzzification: A process of obtaining a single number from the output of the aggregated fuzzy set.

Bibliography

[1] S.S.G. Abhishekh and S.R. Singh. A score function-based method of forecasting using intuitionistic fuzzy time series. *New Mathematics and Natural Computation*, 14:91–111, 2018.

[2] M.G. Akbari and G. Hesamian. A partial-robust-ridge-based regression model with fuzzy predictors-responses. *Journal of Computational and Applied Mathematics*, 351:290–301, 2019.

[3] M.G. Akbari and G. Hesamian. Elastic net oriented to fuzzy semiparametric regression model with fuzzy explanatory variables and fuzzy responses. 1:1–11, 2019.

[4] Abbasi G. Al-Refaie, A. and D. Ghanim. Proposed α-cut CUSUM and EWMA Control Charts for Fuzzy Response Observations. *International Journal of Reliability, Quality and Safety Engineering*, 28:21–28, 2021.

[5] I.M. Aliev and Z. Kara. Fuzzy system reliability analysis using time dependent fuzzy set. *Control and Cybernetics*, 33:653–662, 2004.

[6] B.F. Arnold. Testing fuzzy hypothesis with crisp data. *Fuzzy Sets and Systems*, 9:323–333, 1998.

[7] Z. Artstein and J.C. Hansen. Convexification in limit laws of random sets in Banach spaces. *Annals of Probability*, 13:307–309, 1985.

[8] S Ashraf and T Rashid. *Fuzzy similarity measures*. LAP LAMBERT Academic Publishing, 2010.

[9] Bantan R.A.R. Aslam, M. and N. Khan. Design of S^2-NNEWMA control chart for monitoring process having indeterminate production data. *Processes*, 7:36–44, 2019.

[10] Lawry J. Baldwin, J.F. and T.P. Martin. A mass assignment theory of the probability of fuzzy events. *Fuzzy Sets and Systems*, 83:353–367, 1996.

[11] Pedrycz W. Bargiela, A. and T. Nakashima. Multiple regression with fuzzy data. *Fuzzy sets and systems*, 158:2169–2188, 2007.

[12] B. Bede. *Mathematics of Fuzzy Sets and Fuzzy Logic*. Springer Heidelberg, Berlin, 2013.

[13] P Billingsley. *Probability and measure.* John Wiley and Sons, New York, 2008.

[14] G. Bortolan and R.A. Degani. A review of some methods for ranking fuzzy subsets. *Fuzzy Sets and Systems*, 15:1–19, 1985.

[15] G.E.P. Box and G.M. Jenkins. *Time series analysis: forecasting and control.* Holden-Day, San Francisco, 1976.

[16] Barrenechea E. Bustince, H. and M. Pagola. Relationship between restricted dissimilarity functions, restricted equivalence functions and normal EN-functions: Image thresholding invariant. *Pattern Recognition Letters*, 29:525–536, 2008.

[17] C. Chakraborty and D. Chakraborty. A theoretical development on a fuzzy distance measure for fuzzy numbers. *Mathematical and Computer Modeling*, 43:254–261, 2006.

[18] K.S. Chen and T.C. Chang. Construction and fuzzy hypothesis testing of Taguchi Six Sigma quality index. *International Journal of Production Research*, 58:3110–3125, 2020.

[19] L.H. Chen and S.H. Nien. A new approach to formulate fuzzy regression models. *Applied Soft Computing*, 86:23–36, 2020.

[20] S.J. Chen and S.M. Chen. Fuzzy risk analysis based on similarity measures of generalized fuzzy numbers. *IEEE Transactions on Fuzzy Systems*, 11:45–56, 2003.

[21] C.B. Cheng and E.S. Lee. Non-parametric fuzzy regression k-NN and kernel smoothing techniques. *Computers and Mathematics with Applications*, 38:239–251, 1999.

[22] C.H. Cheng. A new approach for ranking fuzzy numbers by distance method. *Fuzzy Sets and Systems*, 95:307–317, 1998.

[23] C.H. Cheng and C.H. Chen. Fuzzy time series model based on weighted association rule for financial market forecasting. *Expert Systems*, 35:23–30, 2018.

[24] S.H. Choi and J.J. Buckley. Fuzzy regression using least absolute deviation estimators. *Soft Computing*, 12:257–263, 2008.

[25] S.H. Choi and J.H. Yoon. General fuzzy regression using least squares method. *International Journal of Systems Science*, 41:477–485, 2010.

[26] C.c. Chou. A new similarity measure of fuzzy numbers. *Journal of Intelligent and Fuzzy Systems*, 26:287–294, 2014.

[27] T.C. Chu and C.T. Tsao. Ranking fuzzy numbers with an area between the centroid point and original point. *Computers and Mathematic Application*, 43:111–117, 2002.

[28] N. Chukhrova and A. Johannssen. Fuzzy regression analysis: Systemtic review and bibliography. *Applied Soft Computing*, 84:1–20, 2019.

[29] N. Chukhrova and A. Johannssen. Fuzzy hypothesis testing for a population proportion based on set-valued information. *Fuzzy Sets and Systems*, 387:127–157, 2020.

[30] N. Chukhrova and A. Johannssen. Fuzzy hypothesis testing: Systematic review and bibliography. *Applied Soft Computing*, page https://doi.org/10.1016/j.asoc.2021.107331, 2021.

[31] KL Chung. *A Course in Probability Theory (3rd Ed.)*. Academic Press, San Diego, 2001.

[32] Dominguez-Menchero J.S. Lopez-Diaz M. Colubi, A. and D.A. Ralescu. On the formalization of fuzzy random variables. *Information Science*, 133:3–6, 2001.

[33] D'Urso-P. Giordani P. Coppi, R. and A. Santoro. Least squares estimation of a linear regression model with LR-fuzzy response. *Computational Statistics and Data Analysis*, 51:267–286, 2006.

[34] Bustince-H. Fernndez J. Couso, I. and L. Snchez. The Null Space of Fuzzy Inclusion Measures. *IEEE Transactions on Fuzzy Systems*, 29:641–648, 2021.

[35] W.H. Woodall Champ-C.W. Cynthia, A. Lowry and E.R. Steven. A multivariate exponentially weighted moving average control chart. *Technometrics*, 34:46–53, 1992.

[36] J. de Andrés-Sánchez. Claim reserving with fuzzy regression and the two ways of ANOVA. *Applied Soft Computing*, 12, 2012.

[37] G Debara. *Measure theory and integration.* Ellis Horwood Limited, 1981.

[38] V.G. Degaribay. Behaviour of fuzzy ANOVA. *Kybernetes*, 16:107–112, 1987.

[39] Masson M.H. Denoeux, T. and P.H. Herbert. Non-parametric rank-based statistics and significance tests for fuzzy data. *Fuzzy Sets and Systems*, 153:1–28, 2005.

[40] P. Diamond and P. Kloeden. *Metric Spaces of Fuzzy Sets.* World Scientific, Singapore, 1994.

[41] D. Dubois and H. Prade. *Fuzzy Sets and Systems: Theory and Applications.* Academic, New York, 1980.

[42] D. Dubois and H. Prade. Ranking fuzzy numbers in the setting of possibility theory. *Information Sciences*, 30:183–224, 1983.

[43] D Dubois and H Prade. *Possibility Theory.* New York: Plenum, 1988.

[44] N Erginel and S Senturk. *Fuzzy EWMA and fuzzy CUSUM control charts. In: Kahraman C., Kabak O. (eds) Fuzzy Statistical Decision-Making. Studies in Fuzziness and Soft Computing, vol. 343., pp. 281-295.* Springer, Cham, 2016.

[45] A. Faraz and A.F. Shapiro. An application of fuzzy random variables to control charts. *Fuzzy Sets and Systems*, 161:2684–2694, 2010.

[46] TS Ferguson. *A Course in Large Sample Theory.* London: Chapman & Hall, 1996.

[47] T. Flaminio and L. Godo. A logic for reasoning about the probability of fuzzy events. *Fuzzy Sets and Systems*, 158:625–638, 2007.

[48] S. Fruhwirth-Schnatter. On statistical inference for fuzzy data with applications to descriptive statistics. *Fuzzy Sets and Systems*, 50:143–165, 1992.

[49] M. Ganbari and R. Nuraei. Revision of a fuzzy distance measure. *International Journal of Industrial Mathematics*, 5:143–147, 2013.

[50] A.H. Ganie and S. Singh. An innovative picture fuzzy distance measure and novel multi-attribute decision-making method. *Complex and Intelligent Systems*, 7:781–805, 2021.

[51] G.D. Garson. *Curve fitting and non-linear regression.* Asheboro, NC: Statistical Associates Publishers, 2012.

[52] S.S. Gautam and S. Singh. A refined method of forecasting based on high-order intuitionistic fuzzy time series data. *Progress in Artificial Intelligence*, 7:339–350, 2018.

[53] JD Gibbons and S Chakraborti. *Non-parametric Statistical Inference (4th ed.).* Marcel Dekker, New York, 2003.

[54] JD Gibbons and S Chakraborti. *Non-parametric Statistical Inference (5th ed.).* Marcel Dekker, New York, 2003.

[55] Lpez-Daz M. Gil, M.A. and D.A. Ralescu. Overview on the development of fuzzy random variables. *Fuzzy Sets and Systems*, 157:2546–2557, 2006.

[56] Montenegro-M. Rodríguez G.-Colubi A. Gil, M.A. and M.R. Casals. Bootstrap approach to the multi-sample test of means with imprecise data. *Computer Statistics and Data Analysis*, 51:148–162, 2006.

[57] Hahn-M.G. Gine, E. and J. Zinn. Limit theorems for random sets, an application of probability in Banach space results, Probab. *ACM International Conference Proceeding Series*, pages 112–135, 1983.

[58] Belyayev-YK Gnedenko, BV and AD Solovyev. *Mathematical methods of reliability theory*. Academic Press, 2014.

[59] Xiang-L. Gong, Y. and G. Liu. Fuzzy regression model based on geometric centroid and incentre points and application to performance evaluation. *International Journal of Uncertainty, Fuzziness and Knowledge-Based Systems*, 28:269–288, 2020.

[60] P Grzegorzewski. *Distribution-free tests for vague data, In: (Lopez-Diaz M, et al. (Eds)) Soft Methodology and Random Information Systems*. Springer, Heidelberg, 2004.

[61] P. Grzegorzewski. K-sample median test for vague data. *International Journal of Intelligent Systems*, 24:529–539, 2009.

[62] P. Grzegorzewski. Two-sample dispersion problem for fuzzy data. *Information Processing and Management of Uncertainty in Knowledge-Based Systems*, 1239:82–96, 2020.

[63] Dai-Z. Zhao A. Guan, H. and J. He. A novel stock forecasting model based on High-order-fuzzy-fluctuation trends and back propagation neural network. *PLoS ONE*, 13:40–51, 2018.

[64] Jain-G. Tayal D.K. Gupta, C. and O. Castillo. ClusFuDE: Forecasting low dimensional numerical data using an improved method based on automatic clustering, fuzzy relationships and differential evolution. *Engineering Applications of Artificial Intelligence*, 71:175–189, 2018.

[65] E. Haktanir and C. Kahraman. Z-fuzzy hypothesis testing in statistical decision making. *Journal of Intelligent and Fuzzy Systems*, 37:6545–6555, 2019.

[66] Yabing-Z.H.A. Zhang R.-Sun Q. He, Q. and T. Liu. Reliability analysis for multi-state system based on triangular fuzzy variety subset bayesian networks. *Eksploatacja Niezawodnosc-Maintenance and Reliability*, 19:152–165, 2017.

[67] S. Heilpern. Fuzzy subsets of the space of probability measures and expected value of fuzzy variable. *Fuzzy Sets and Systems*, 54:301–309, 1993.

[68] Akbari-M.G. Hesamian, G. and M. Asadollahi. Fuzzy semi-parametric partially linear model with fuzzy inputs and fuzzy outputs. *Expert Systems with Applications*, 71:230–239, 2017.

[69] Akbari-M.G. Hesamian, G. and M. Asadollahi. Fuzzy semi-parametric partially linear model with fuzzy inputs and fuzzy outputs. *Expert Systems With Applications*, 71:230–239, 2017.

[70] Akbari-M.G. Hesamian, G. and E. Ranjbar. Exponentially weighted moving average control chart based on normal fuzzy random variables. *International Journal of Fuzzy Systems*, 21:1187–1195, 2019.

[71] Akbari-M.G. Hesamian, G. and V. Ranjbar. Some inequalities and limit theorems for fuzzy random variables adopted with a-values of fuzzy numbers. *Soft Computing*, 24:3797–3807, 2020.

[72] Akbari-M.G. Hesamian, G. and R. Yaghoobpoor. Quality control process based on fuzzy random variables. *IEEE Transactions on Fuzzy Systems*, 27:671–685, 2019.

[73] Akbari-M.G. Hesamian, G. and R. Yaghoobpoor. Quality Control Process based on Fuzzy Random Variables. *IEEE Transactions on Fuzzy Systems*, 27:671–685, 2019.

[74] Akbari-M.G. Hesamian, G. and J. Zendehdel. Location and scale fuzzy random variables. *International Journal of Systems Science*, 51:229–241, 2020.

[75] Akbari-M.G. Hesamian, G. and J. Zendehdel. Location and scale fuzzy random variables. *International Journal of Systems Science*, 51:229–241, 2020.

[76] G. Hesamian. Fuzzy similarity measure based on fuzzy sets. *Control and Cybernetics*, 46:23–40, 2017.

[77] G. Hesamian and M.G. Akbari. A semi-parametric model for time series based on fuzzy data. *IEEE Transactions on Fuzzy Systems*, 26:2953–2966, 2018.

[78] G. Hesamian and M.G. Akbari. Fuzzy absolute error distance measure based on a generalised difference operation. *International Journal of Systems Science*, 49:2454–2462, 2018.

[79] G. Hesamian and M.G. Akbari. Fuzzy absolute error distance measure based on a generalized difference operation. *International Journal of Systems Science*, 49:2454–2462, 2018.

[80] G. Hesamian and M.G. Akbari. Fuzzy Lasso regression model with exact explanatory variables and fuzzy responses. *International Journal of Approximate Reasoning*, 115:290–300, 2019.

[81] G. Hesamian and M.G. Akbari. Fuzzy Quantile Linear Regression model adopted with a semi-parametric technique based on fuzzy predictors and fuzzy responses. *Expert systems with applications*, 118:585–597, 2019.

[82] G. Hesamian and M.G. Akbari. A robust varying coefficient approach to fuzzy multiple regression model. *Journal of Computational and Applied Mathematics*, 371:23–34, 2020.

[83] G. Hesamian and M.G. Akbari. Fuzzy spline univariate regression with exact predictors and fuzzy responses. *Journal of Computational and Applied Mathematics*, 375:31–44, 2020.

[84] G. Hesamian and M.G. Akbari. Testing hypotheses for multivariate normal distribution with fuzzy random variables. *International Journal of Systems Science*, page DOI: 10.1080/00207721.2021.1936274, 2021.

[85] G. Hesamian and J. Chachi. Two-sample kolmogorov-smirnov fuzzy test for fuzzy random variables. *Statistical Papers*, 56:61–82, 2015.

[86] G. Hesamian and J. Chachi. Two-sample Kolmogorov-Smirnov fuzzy test for fuzzy random variables. *Statistical Papers*, 56:61–82, 2015.

[87] G. Hesamian and Akbari. M.G. A preference index for ranking closed intervals and fuzzy numbers. *International Journal of Uncertainty, Fuzziness and Knowledge-Based Systems*, 25:741–757, 2017.

[88] G. Hesamian and M. Shams. A Note on Fuzzy Probability of a Fuzzy Event. *International Journal of Intelligent*, 32:676–685, 2017.

[89] G. Hesamian and S.M. Taheri. Fuzzy empirical distribution: properties and applications. *Kybernetika*, 49:962–982, 2013.

[90] D.H. Hong and H.J. Kim. Marcinkiewicz-type law of large numbers for fuzzy random variables. *Fuzzy Sets Systems*, 64, 1994.

[91] D.H. Hong and H.J. Kim. Weak law of large numbers for i.i.d. fuzzy random variables. *Kybernetika*, 43:87–96, 2007.

[92] E. Hosseinzadeh and H. Hassanpour. Estimating the parameters of fuzzy linear regression model with crisp inputs and Gaussian fuzzy outputs: A goal programming approach. *Soft Computing*, 25:2719–2728, 2021.

[93] V. Hrissanthou and M. Spiliotis. *Conventional and Fuzzy Regression: Theory and Engineering Applications*. Nova Science Pub Inc, New York, 2018.

[94] O. Hryniewicz. Goodman-Kruskal measure of dependence for fuzzy ordered categorical data. *Computational Statistics and Data Analysis*, 51:323–334, 2006.

[95] O. Hryniewicz. Possibilistic decisions and fuzzy statistical tests. *Fuzzy Sets and Systems*, 157:2665–2673, 2006.

[96] Zuo-M.J. Huang, H.Z. and Z.Q. Sun. Bayesian reliability analysis for fuzzy lifetime data. *Fuzzy Sets and Systems*, 157:1674–1686, 2006.

[97] M. Hudec and D. Praenka. Collecting and managing fuzzy data in statistical relational databases. *Statistical Journal of the IAOS*, 32:245–255, 2016.

[98] M.A. Hussian and E.A. Amin. Fuzzy reliability estimation for exponential distribution using ranked set sampling. *International Journal of Contemporary Mathematical Sciences*, 12:31–42, 2017.

[99] H. Inoue. A strong law of large numbers for fuzzy random sets. *Fuzzy Sets Systems*, 41:285–291, 1991.

[100] C. Jershan and J.S. Yao. Fuzzy probability over fuzzy σ-field with fuzzy topological spaces. *Fuzzy Sets and Systems*, 116:201–223, 2000.

[101] Parchami-A. Jiryaei, A. and M. Mashinchi. One-Way ANOVA and least squares method based on fuzzy random variables. *Turkish Journal of Fuzzy Systems*, 4:18–33, 2013.

[102] S.Y. Joo. Week laws of large numbers for fuzzy random variables. *Fuzzy Sets Systems*, 147:453–464, 2004.

[103] Mc Clelland GH Judd, CM and SR Carey. *Data Analysis: A Model Comparison Approach To Regression, ANOVA, and Beyond (3rd ed.)*. Routledge, New York, 2017.

[104] Lee W.J. Jung H.Y. and Choi S.H. Hybrid fuzzy regression analysis using the F-Transform. *Applied Sciences*, 10:14–24, 2020.

[105] Bozdag C.F. Kahraman, C. and D. Ruan. Fuzzy sets approaches to statistical parametric and non-parametric tests. *International Journal of Intelligent Systems*, 19:1069–1078, 2004.

[106] D. Kalpanapriya and P. Pandian. Fuzzy hypothesis testing Of ANOVA model with fuzzy data. *International Journal of Modern Engineering Research*, 2:2951–2956, 2007.

[107] C. Kao and C.L. Chyu. Least-squares estimates in fuzzy regression analysis. *European Journal of Operational Research*, 148:426–435, 2003.

[108] Khan M.F. Aslam M.-Niaki S.T.A. Khan, M.Z. and A.R. Mughal. A fuzzy EWMA attribute control chart to monitor process mean. *Information*, 9:24–33, 2018.

[109] Z. Khan, Hashim R. Yaqoob N. Gulistan, M., and W. Chammam. Design of S-control chart for neutrosophic data: An application to manufacturing industry. *Journal of Intelligent and Fuzzy Systems*, 38:4743–4751, 2020.

[110] H. Kim and H.Y. Jung. Ridge fuzzy regression modelling for solving multicollinearity. *Mathematics*, 8:63–77, 2020.

[111] Y.K. Kim. A strong law of large numbers for fuzzy random variables. *Fuzzy Sets Systems*, 111:319–323, 2000.

[112] E.P. Klement. Some remarks on a paper of R.R. Yager. *Information Sciences*, 27:211–220, 1982.

[113] Puri M.L. Klement, E.P. and D.A. Ralescu. Limit theorems for fuzzy random variables. *Proceedings of the Royal Society A.*, 407:171–182, 1986.

[114] GJ Klir and B Yuan. *Fuzzy sets and fuzzy logic: theory and applications.* Prentice Hall, 1995.

[115] JE Kolassa. *An Introduction to Nonparametric Statistics.* Chapman and Hall/CRC, 2020.

[116] Okuda T. Konishi, M. and K. Asai. Analysis of variance based on fuzzy interval data using moment correction method. *International Journal of Innovative Computing*, 2:83–99, 2006.

[117] V. Kratschmer. A unified approach to fuzzy random variables. *Fuzzy Sets Systems*, 123:1–9, 2001.

[118] R Kruse and KD Meyer. *Statistics with vague data.* Amsterdam, The Netherlands: Reidel, 1987.

[119] A. Kumar and P. Dhiman. Reliability estimation of a network structure using generalized trapezoidal fuzzy numbers. *Journal of Kobin*, 51:225–241, 2021.

[120] Ram M. Kumar, A. and O.P. Yadav. *Advancements in Fuzzy Reliability Theory.* Hershey, PA : IGI Global, 2021.

[121] H. Kwakernaak. Fuzzy random variables. Part I: definitions and theorems. *Information Science*, 19:1–15, 1987.

[122] H.T. Lee. C_{pk} index estimation using fuzzy numbers. *European Journal of Operational Research*, 129:683–688, 2001.

[123] KH Lee. *First course on fuzzy theory and applications.* Springer-Verlag: Berlin, 2005.

[124] Pedrycz W. Lee, S.H. and G. Sohn. Design of similarity and dissimilarity measures for fuzzy sets on the basis of distance measure. *International Journal of Fuzzy System*, 11:67–72, 2009.

[125] EL Lehmann and G Casella. *Theory of Point Estimation (2nd ed.)*. Springer, 1998.

[126] R. Li. Water quality forecasting of Haihe River based on improved fuzzy time series model. *Desalination and Water Treatment*, 108:285291, 2018.

[127] Zeng W. Xie J. Li, J. and Q. Yin. A new fuzzy regression model based on least absolute deviation. *Engineering Applications of Artificial Intelligence*, 52:54–64, 2016.

[128] Wu B. Lin, P. and J. Watada. Kolmogorov-Smirnov two sample test with continuous fuzzy data. *Advances in Intelligent and Soft Computing*, 68:175–186, 2010.

[129] B Liu. *Uncertainty Theory*. Springer-Verlag: Berlin, 2004.

[130] Y.K. Liu and J. Gao. Convergence criteria and convergence relations for sequences of fuzzy random variables. *Fuzzy Knowledge Based Networks*, 3613:321–331, 2005.

[131] Y.K. Liu and B. Liu. Fuzzy random variables: A scalar expected value operator. *Fuzzy Optimization and Decision Making*, 2:143–160, 2003.

[132] M. Loeve. *Probability Theory I*. Springer-Verlag. New York, 1977.

[133] Magott J. Lower, M. and J. Skorupski. Analysis of air traffic incidents using event trees with fuzzy probabilities. *Fuzzy Sets and Systems*, 293:5079, 2016.

[134] M.A. Lubiano and W. Trutschnig. ANOVA for fuzzy random variables using the R-package SAFD; Combining soft Computing and statistical methods in data. *Analysis Advances in Intelligent and Soft Computing*, 77:449–456, 2010.

[135] A.S. Mendes, Rizol P. Machado, M.A.G., and M.S. Rocha. Fuzzy control chart for monitoring mean and range of univariate processes. *Pesquisa Operacional*, 39:339–357, 2019.

[136] Jain D. Mishra, A.R. and D.S. Hooda. On fuzzy distance and induced fuzzy information measures. *Journal of Information and Optimization Sciences*, 37:193–211, 2016.

[137] I. Molchanov. On strong laws of large numbers for random upper semicontinuous. *Journal of Mathematical Analysis and Applications*, 235:349–355, 1999.

[138] Rodriguez G Gil MA-Colubi A Montenegro, M and MR Casals. *Introduction to ANOVA with fuzzy random variables, In Soft Methodology and Random Information Systems, eds. M. Lopez-Diaz, M.A. Gil, P. Grzegorzewski, O. Hryniewicz, and J. Lawry.* Berlin, Springer, pp. 487–494, 2004.

[139] D.C. Montgomery. *Introduction to statistical quality control.* John Wiley and Sons, New York, 2007.

[140] J Munkres. *Topology (2nd ed.).* Prentice-Hall, Upper Saddle River, NJ, USA, 2000.

[141] N. Mylonas and B. Papadopoulos. Unbiased fuzzy estimators in fuzzy hypothesis testing. *Algorithms*, 14:185–192, 2021.

[142] Behzad M.H. Razzaghnia T. Naderkhani, R. and R. Farnoosh. Fuzzy regression analysis based on fuzzy neural networks using trapezoidal data. *International Journal of Fuzzy Systems*, In press https://doi.org/10.1007/s40815-020-01033-2, 2021.

[143] K. Nakamura. Preference relations on a set of fuzzy utilities as a basis for decision making. *Fuzzy Sets and Systems*, 20:147–162, 1986.

[144] Riaz M. Nasir, A. and R.J.M.M. Does. An EWMA-type control chart for monitoring the process mean using auxiliary information. *Communications in Statistics; Theory and Methods*, 43:3485–3498, 2014.

[145] Mashinchi M. Nourbakhsh, N. and A. Parchamic. Analysis of variance based on fuzzy observations. *International Journal of Systems Science*, 44:714–726, 2013.

[146] V. Novak. Detection of structural breaks in time series using fuzzy techniques. *International Journal of Fuzzy Logic and Intelligent Systems*, 18:1–12, 2018.

[147] TW O'Gorman. *Applied Adaptive Statistical Methods; Tests of Significance and Confidence Intervals.* SIAM, 2004.

[148] Parham G.A. Pak, A. and M. Saraj. Reliability estimation in Rayleigh distribution based on fuzzy lifetime data. *International Journal of System Assurance Engineering and Management*, 5, 2014.

[149] W Palma. *Time Series Analysis.* John Wiley and Sons, New York, 2016.

[150] D.C. Panda. A new method for evaluation of fuzzy reliability of multistage interconnection. *International Journal of Research in Applied Science*, 3:14–20, 2016.

[151] P. Pandian and D. Kalpanapriya. Two-factor ANOVA technique for fuzzy data having membership grades. *Research Journal of Pharmacy and Technology*, 9:2394–2402, 2016.

[152] A. Parchami. Fuzzy decision in testing hypotheses by fuzzy data: Two case studies. *Iranian Journal of Fuzzy Systems*, 17:127–136, 2020.

[153] A. Parchami and M. Mashinchi. Fuzzy estimation for process capability indices. *Information Sciences*, 177:1452–1462, 2007.

[154] Mashinchi M. Yavari A.R. Parchami, A. and H.R. Maleki. Fuzzy confidence intervals for fuzzy process capability index. *Journal of Intelligent and Fuzzy Systems*, 17:287–295, 2006.

[155] K. Patra. Fuzzy risk analysis using a new technique of ranking of generalized trapezoidal fuzzy numbers. *Granular Computing*, In press https://doi.org/10.1007/s41066-021-00255-5, 2021.

[156] W.L. Pearn and S. Kotz. *Encyclopedia and Handbook of Process Capability Indices. Series on Quality. Reliability and Engineering Statistics (vol. 12).* World Scientific publishing Co, Pte. Ltd, 2006.

[157] W Pedrycz. *An Introduction to Computing with Fuzzy Sets: Analysis, Design, and Applications.* Springer, 2021.

[158] N. Pekin Alakoc. Fuzzy X-bar and S-control charts based on confidence intervals. *Journal of Advanced Research in Natural and Applied Sciences*, 7:114–131, 2021.

[159] N. Pekin Alakoc and A. Apaydin. A fuzzy control chart approach for attributes and variables. *Engineering, Technology and Applied Science Research*, 8:3360–3365, 2018.

[160] Bigand A. Phan, T.T.H. and E.P. Caillault. A new fuzzy logic-based similarity measure applied to large gap imputation for uncorrelated multivariate time series. *Applied Computational Intelligence and Soft Computing*, 2018:1–15, 2018.

[161] U.M. Pirzada and D.C. Vakaskar. Existence of Hukuhara differentiability of fuzzy-valued functions. *Journal of the Indian Mathematical Society*, 87, 2017.

[162] K. Plasecki. Probability of fuzzy events defined as denumerable additivity measure. *Fuzzy Sets and Systems*, 17:271–287, 1985.

[163] O.I. Provotar and O.O. Provotar. Fuzzy probabilities of fuzzy events. *Cybernetics and Systems Analysis*, 56:171180, 2020.

[164] M.L. Puri and D. Ralescu. Strong law of large numbers for Banach space valued random variables. *Annals of Probability*, 11:222–224, 1983.

[165] M.L. Puri and D.A. Ralescu. Differentials of fuzzy functions. *Journal of Mathematical Analysis and Applications*, 91:552–558, 1983.

[166] Othman M. Sokkalingam R. Rahim, N.F. and E.A. Kadir. Forecasting crude palm oil prices using fuzzy rule-based time series method. *IEEE Access*, 2:32216–32224, 2018.

[167] S.W. Roberts. Control chart tests based on geometric moving averages. *Technometrics*, 1:239–250, 1959.

[168] Colubi A. Rodríguez, G. and M.A. Gil. Fuzzy data treated as functional data: a one-way ANOVA test approach. *Computational Statistics and Data Analysis*, 56:943–955, 2011.

[169] Montenegro M. Colubi A. Rodríguez, G. and M.A. Gil. Bootstrap techniques and fuzzy random variables: Synergy in hypothesis testing with fuzzy data. *Fuzzy Sets and Systems*, 157:2608–2613, 2006.

[170] R Rodriguez-Lopez. *On boundary value problems for fuzzy differential equations, Soft Methods for Handling Variability and Imprecision.* Advances in Soft Computing, V. 48, Springer, Berlin, Heidelberg, pp. 218225, 2008.

[171] Tahir M. Ubaid U.R.-Zeeshan A. Ronnason, C. and I. Aiyared. Some novel cosine similarity measures based on complex hesitant fuzzy sets and their applications. *Journal of Mathematics*, 4:1–9, 2021.

[172] TP Ryan. *Modern regression methods.* John Wiley and Sons, New York, 2008.

[173] H. Sabahno, S.M. Mousavi, and A. Amiri. A new development of an adaptive X-R control chart under a fuzzy environment. *International Journal of Data Mining, Modelling and Management*, 11:19–44, 2019.

[174] H. Scheffe. *The analysis of variance.* John Wiley and Sons, New York, 1999.

[175] H Scott E, Maxwell and D Delaney. *Designing Experiments and Analyzing Data: A Model Comparison Perspective (2nd ed.).* Routledge, New York, Erlbaum.

[176] Patra K. Sen, S. and S.K. Mondal. Similarity Measure of Gaussian Fuzzy Numbers and Its Application. *International Journal of Applied and Computational Mathematics*, 96:https://doi.org/10.1007/s40819–021–01040–3, 2021.

[177] M. Shafi and R. Viertl. Empirical reliability functions based on fuzzy life time data. *Journal of Intelligent and Fuzzy Systems*, 28:707–711.

[178] Heydari A. Kazemi-M. Shahsavari-Pour, N. and M. Karami. A novel method for ranking fuzzy numbers based on the different areas fuzzy number. *International Journal of Mathematics in Operational Research*, 11:544–566, 2017.

[179] F.A. Shapiro. Fuzzy random variables. *Insurance: Journal of Mathematical Economics*, 44:307–314, 2009.

[180] AK Sharma. *Text Book of Correlations and Regression.* Discovery Publishing House, New Delhi, 2005.

[181] NZ Shi and J Tao. *Statistical Hypothesis Testing: Theory and Methods.* World Scientific Publishing Company, Singapore, 2008.

[182] Nguyen T.L. Shu, H.M and B.M. Hsu. Fuzzy MaxGWMA chart for identifying abnormal variations of on-line manufacturing processes with imprecise information. *Expert System With Application*, 41:1342–1356, 2013.

[183] H Shumway and DS Stoffer. *Time series analysis and its applications.* Springer-Verlag Berlin, Heidelberg, 2011.

[184] P. Smets. Probability of a fuzzy event: An axiomatic approach. *Fuzzy Sets and Systems*, 7:153–164, 1982.

[185] M. Smithson and J. Verkuilen. *Fuzzy set theory: Applications in the social sciences.* Sage, USA, 2006.

[186] L. Stefanini and B. Bede. Generalized Hukuhara differentiability of interval-valued functions and interval differential equations. *Nonlinear Analysis: Theory Methods and Applications*, 71:1311–1328, 2009.

[187] A. Syropoulos and T. Grammenos. *A Modern Introduction to Fuzzy Mathematics.* John Wiley and Sons, New York, 2020.

[188] A Taff. *Hypothesis Testing: The Ultimate Beginner's Guide to Statistical Significance.* CreateSpace Independent Publishing Platform, 2018.

[189] J. Talašová and O. Pavlačka. Fuzzy probability spaces and their applications in decision making. *Austrian Journal of Statistics*, 35, 2006.

[190] Seymour L. Taylor, R.L. and Y. Chen. Week laws of large numbers for fuzzy random sets. *Nonlinear Analysis*, 47:1245–1256, 2001.

[191] H. Toth. Probabilities and fuzzy events: An operational approach. *Fuzzy Sets and Systems*, 48:113–127, 1992.

[192] L. Tran and L. Duckstein. Comparison of fuzzy numbers using a fuzzy distance measure. *Fuzzy Sets and Systems*, 130:331–341, 2002.

[193] Ye J. Unver M. Trkarslan, E. and M. Olgun. Consistency fuzzy sets and a cosine similarity measure in fuzzy multiset setting and application to medical diagnosis. *Mathematical Problems in Engineering*, pages 1–12, 2021.

[194] F.M. Tseng and G.H. Tzeng. A fuzzy seasonal ARIMA model for forecasting. *Fuzzy Sets and Systems*, 126:367–376, 2002.

[195] T. Uemura. A law of large numbers for random sets. *Fuzzy Sets and Systems*, 59:181–188, 1993.

[196] K. Vännman. A unified approach to the capability indices. *Statistica Sinica*, 5:805–820, 1995.

[197] R. Viertl. Univariate statistical analysis with fuzzy data. *Computational Statisticsand Data Analysis*, 51:133–147, 2006.

[198] R. Viertl. *Statistical methods for fuzzy data.* John Wiley and Sons, New York, 2011.

[199] W. Voxman. Some remarks on distances between fuzzy numbers. *Fuzzy Sets and Systems*, 100:353–365, 1998.

[200] X. Wang and E.E. Kerre. Reasonable properties for the ordering of fuzzy quantities (I). *Fuzzy Sets and Systems*, 118:375–385, 2001.

[201] Xu Y. Wang, G. and S. Qin. Basic fuzzy event space and probability distribution of probability fuzzy space. *Mathematics*, 7:542–553, 2019.

[202] Y. Wang, G. Xu and S. Qin. Basic fuzzy event space and probability distribution of probability fuzzy space. *Mathematics*, 2019:1–15, 2019.

[203] Y.M. Wang and Y. Luo. Area ranking of fuzzy numbers based on positive and negative ideal points. *Computers and Mathematic Application*, 58:1769–1779, 2009.

[204] Zhang W.X. Wang, N. and C.L. Mei. Fuzzy non-parametric regression based on local linear smoothing technique. *Information Sciences*, 177:3882–3900, 2007.

[205] L Wasserman. *All of non-parametric statistics.* Springer, New York, 2007.

[206] C.W. Wu. Decision-making in testing process performance with fuzzy data. *European Journal of Operational Research*, 193:499–509, 2009.

[207] H.C. Wu. The central limit theorems for fuzzy random variables. *Information Science*, 120:239–256, 1999.

[208] H.C. Wu. The laws of large numbers for fuzzy random variables. *Fuzzy Sets and Systems*, 116:245–262, 2000.

[209] H.C. Wu. Statistical hypotheses testing for fuzzy data. *Fuzzy Sets and Systems*, 175:30–56, 2005.

[210] H.C. Wu. Analysis of variance for fuzzy data. *International Journal of Systems Science*, 38:235–246, 2007.

[211] R.R. Yager. A note on probabilities of fuzzy events. *Information Sciences*, 18:113–122, 1979.

[212] R.R. Yager. A representation of the probability of a fuzzy subset. *Fuzzy Sets and Systems*, 13:273–283, 1984.

[213] L. Yang and B. Liu. On inequalities and critical values of fuzzy random variables. *International Journal of Uncertainty, Fuzziness and Knowledge-Based Systems*, 13:163–175, 2005.

[214] J.S. Yao and K. Wu. Ranking fuzzy numbers based on decomposition principle and signed distance. *Fuzzy Sets and Systems*, 116:275–288, 2000.

[215] J. Yen and R. Langari. *Fuzzy Logic: Intelligence, Control, and Information.* Prentice Hall, 1999.

[216] Y. Yuan. Criteria for evaluating fuzzy ranking methods. *Fuzzy Sets and Systems*, 43:139–157, 1991.

[217] S Zacks. *Introduction to reliability analysis: probability models and statistical methods.* Springer Science and Business Media, 2012.

[218] L.A. Zadeh. Probability measures of fuzzy events. *Journal of Mathematical Analysis and Applications*, 23:421–427, 1968.

[219] L.A. Zadeh. Fuzzy sets as a basis for a theory of possibility. *Fuzzy Sets and Systems*, 1:3–28, 1978.

[220] M.Gh. Zarei, R. Akbari and J. Chachi. Modeling autoregressive fuzzy time series data based on semi-parametricmethods. *Soft Computing*, 24:7295–7304, 2020.

[221] Mirzazadeha A. Salehiana-S. Zavvar Sabegha, M.H. and G.W. Weber. A literature review on the fuzzy control chart; classifications and analysis. *International Journal of Supply and Operations Management*, 1:167–189, 2014.

[222] X.R. Zhao and B.Q. Hu. Fuzzy and interval-valued fuzzy decision-theoretic rough set approaches based on fuzzy probability measure. *Information Science*, 298:534554, 2015.

[223] H.J. Zimmermann. *Fuzzy set theory and its applications.* (4th ed.) Kluwer, 2001.

Index

Printed in the United States
by Baker & Taylor Publisher Services